Innovation, Technology and Hypercompetition

In an increasingly technologically led century, the striking pattern that emerges in firms' innovative activities is that companies compete for a technological leadership position in situations best described as races. In high-technology industries, where customers are willing to pay a premium for advanced technology, leadership translates into increasing returns in the market through positive network externalities.

Innovation, Technology and Hypercompetition synthesizes and unifies the various methodological approaches for the industry-specific analysis of fast-changing competitive positions driven by relentless innovation (hypercompetition).

Game-theoretic and agent-based tools are applied to competitive industries in various market settings and in a global context. Rivalry of this sort is seen to extend to the catching up and forging ahead of regions and nations.

The book provides the behavioural foundations for what is driving globalization. It will be of interest to a broad range of people, including academic economists, legal experts, management consultants and practitioners.

Hans-Werner Gottinger is at Kwansei Gakuin University, Sanda-Kobe, Japan.

Routledge Studies in Global Competition

Edited by John Cantwell, The University of Reading, UK and David Mowery, University of California, Berkeley, USA

Japanese Firms in Europe
Edited by Frédérique Sachwald

Technological Innovation, Multinational Corporations and New International Competitiveness
The case of intermediate countries
Edited by José Molero

Global Competition and the Labour Market
Nigel Driffield

The Source of Capital Goods Innovation
The role of user firms in Japan and Korea
Kong-Rae Lee

Climates of Global Competition
Maria Bengtsson

Multinational Enterprises and Technological Spillovers
Tommaso Perez

Governance of International Strategic Alliances
Technology and transaction costs
Joanne E. Oxley

Strategy in Emerging Markets
Telecommunications establishments in Europe
Anders Pehrsson

Going Multinational
The Korean experience of direct investment
Edited by Frédérique Sachwald

Multinational Firms and Impacts on Employment, Trade and Technology
New perspectives for a new century
Edited by Robert E. Lipsey and Jean-Louis Mucchielli

Multinational Firms
The global-local dilemma
Edited by John H. Dunning and Jean-Louis Mucchielli

MIT and the Rise of Entrepreneurial Science
Henry Etzkowitz

Technological Resources and the Logic of Corporate Diversification
Brian Silverman

The Economics of Innovation, New Technologies and Structural Change
Cristiano Antonelli

European Union Direct Investment in China
Characteristics, challenges and perspectives
Daniel Van Den Bulcke, Haiyan Zhang and Maria do Céu Esteves

Biotechnology in Comparative Perspective
Edited by Gerhard Fuchs

Technological Change and Economic Performance
Albert N. Link and Donald S. Siegel

Multinational Corporations and European Regional Systems of Innovation
John Cantwell and Simona Iammarino

Knowledge and Innovation in Regional Industry
An entrepreneurial coalition
Roel Rutten

Local Industrial Clusters
Existence, emergence and evolution
Thomas Brenner

The Emerging Industrial Structure of the Wider Europe
Edited by Francis McGowen, Slavo Radosevic and Nick Von Tunzelmann

Entrepreneurship
A new perspective
Thomas Grebel

Evaluating Public Research Institutions
The U.S. Advanced Technology Program's intramural research initiative
Albert N. Link and John T. Scott

Location and Competition
Edited by Steven Brakman and Harry Garretsen

Entrepreneurship and Dynamics in the Knowledge Economy
Edited by Charlie Karlsson, Börje Johansson and Roger R. Stough

Evolution and Design of Institutions
Edited by Christian Schubert and Georg von Wangenheim

The Changing Economic Geography of Globalization
Reinventing space
Edited by Giovanna Vertova

Economics of the Firm
Analysis, evolution and history
Edited by Michael Dietrich

Innovation, Technology and Hypercompetition
Review and synthesis
Hans-Werner Gottinger

Innovation, Technology and Hypercompetition

Review and synthesis

Hans-Werner Gottinger

Routledge
Taylor & Francis Group
LONDON AND NEW YORK

First published 2006
by Routledge
2 Park Square, Milton Park, Abingdon, Oxon OX14 4RN

Simultaneously published in the USA and Canada
by Routledge
270 Madison Ave, New York, NY 10016

Routledge is an imprint of the Taylor & Francis Group, an informa business

Transferred to Digital Printing 2009

Typeset in Times New Roman by
Keyword Group Ltd

British Library Cataloguing in Publication Data
A catalogue record for this book is available from the British Library

Library of Congress Cataloging in Publication Data
Gottinger, Hans-Werner.
Innovation, technology, and hypercompetition/Hans Gottinger.
p. cm.
Includes bibliographical references and index.
1. Technological innovations–Economic aspects. 2. Evolutionary economics. 3. Competition, International. I. Title.

HD45.G596 2006
338′.064–dc22 2005035563

ISBN10: 0–415–40002–3 (hbk)
ISBN10: 0–415–49408–7 (pbk)

ISBN13: 978–0–415–40002–2 (hbk)
ISBN13: 978–0–415–49408–3 (pbk)

Contents

Foreword

This is an important and ambitious book. And, for those of us in the profession trying to get an overall structure to understand the emerging economic environment, it is also a welcome book.

Economists tend to feel uncomfortable with 'patches' or 'quick-fixes' that seemingly resolve an outstanding issue but which do not nicely fit into an overall 'framework' – *ad hoc* explanations bother us. Yet between the aesthetically pleasing yet lacking *micro* economics and the cobbled together and mostly pragmatic *macro* economics lies a whole area of what can be called *meso* economics, where most of the world lives and on which we had little of coherence to say. Why are industries structured the way they are? What determines the size distribution of firms? What do we have to say about the growth of the firm? What is the role of history? What is the role of geography? What is the role of expectation formation? What is the role of innovative activity? How do we incorporate and account for 'knowledge'? What really is the impact of 'increasing' returns? Come to think of it, what actually is this unit of analysis called *the* firm?

Of course, over the years we have developed partial explanations, some of them ingenious, which have shed light on some aspect or other of the issues raised. Mostly, though, attempts at extending our traditional framework, in addition to being incomplete, were crippled with little or no generalizability. And this smacked of having constructed a 'theory of the special case' which, although quite common in the field of strategy, tends to leave the economist with a bitter aftertaste.

Micro theory, for all its faults and shortcomings, was a theory of price formation under well-defined conditions. Its modelling of one of the agents involved in that process, the 'consumer', is, despite critiques of 'rationality', fairly robust in that the construct of the 'utility function' does allow for considerable latitude before becoming vacuous.

The modelling of the other agent, the 'firm', however, has been problematic nearly from the start. The 'firm', as a unit of analysis in micro theory was (let's call this an E-firm), and is, nothing but a well-behaved, exogenous, production function. It is not the legal entity called a 'firm' which is the actor on markets (let us call this an L-firm); and the difference is not just due to the normal process of 'stylizing' facts while modelling something. An L-firm generally consists of many E-firms. To take a simple case, the pharmaceuticals 'firm' of everyday life economically consists of a 'molecule generating' firm and a 'bottling' firm – each with very different production functions. The use of the same word for two very different concepts continues to lead to needless confusion to this day.

But the apparatus of micro analysis gave no indication as to where this production function came from, how or why it got modified, or, really, what its arguments were,[1] nor did it provide any clue as to why they would accumulate in creating the legal entity called 'the firm'.

A strand of literature tried to answer this last question by reference to contracts. Beginning with Coase's original insight which shifted emphasis from production constraints to contractual ones, and elaborated by other researchers, the main hypothesis was that a 'firm' is a locus of contracts, implicit and explicit, structured in such a way as to minimize transaction costs. Thus an L-firm might incorporate a multitude of E-firms (production functions with their concomitant inputs) if it were cheaper to organize the transactions between them internally as opposed to between independent entities.

In answering the question of how many such E-firms would be incorporated into an L-firm, the issue of firm size, traditional micro theory reverted to the technological consideration of scale economies and its mirror image, the long-run cost function. This took several forms, from contestable markets, to adjustment costs, to coordination–communication trade-offs.

But it quickly became obvious that the main source of transaction costs is information; or more precisely, the asymmetry thereof. Treating the issue of information spawned a vast literature but turned out to be more problematic than originally imagined. Once information was explicitly thrown into the picture, not only did the traditional micro edifice begin to crumble; well-established results on the E-firm began to break down, market clearing and price determination began to pose problems, and the welfare implications began to look shaky, but the requisite coordination of what people knew, especially about themselves, also began to pose problems for the contractual theories of the firm. The obvious fact that 'human capital' was an inalienable asset, the ownership of which could not be transferred, created a particular set of problems for contracting theories.

Although the contractual view of the firm has led to considerable increase in our understanding and a restructuring of the E-firm, the 'agency' considerations generated by the observation in the previous paragraph

[1] Joan Robinson's critique of half a century ago of 'capital' as a factor of production have never been answered; in publications, this tended to be referred to in an initial footnote as an issue and then conveniently left aside as if it never existed. Similar issues exist with labour as an input also.

touch upon a large number of issues from financing structure, to motivation and incentives of management, to the critical issue of the 'objective' of the 'firm' and firm behaviour. Indeed, the logical chain from contracts to information asymmetries (or, equivalently the resultant probability distribution of inputs and outcomes), to incompleteness of labour markets, to the resulting lack of unanimity among 'owners', means that we effectively have no reliable 'objective' for the firm. This is disturbing in that it leaves the conceptual corpus of contracts without a head.

It is at this juncture that the conceptual apparatus of this book will prove to be critical. Indeed, in contrast to the increasingly 'internal' focus of contract-based approaches, with almost no reference to the industrial/competitive context a firm finds itself in, Professor Gottinger resolutely focuses on the impact of these factors. The idea of a 'technological race' is enticing; it takes into account 'patterns of interaction' but does not presuppose any particular market structure, it effectively endogenizes the production function, does not need an objective function that would lead to agency-type issues, and critically, leads to empirically observable indicators of 'success' along various dimensions.

This by itself would have been sufficient to consider this book as an important contribution to our understanding of firms and industries, but I suspect there is a fertile area of research in combining the contractual approaches to the 'racing' approach treated in this book to begin a reconstruction of the 'coherence' we are currently lacking.

Ahmet Aykac
Professor of Economics, Theseus-EDHEC, France

Preface

> ... monopolistic structure is most conducive to innovation in fields with a slow pace of scientific advance or limited opportunities for product differentiation whereas the effect of monopoly power is weak or even negative in high opportunity fields.
>
> F.M. Scherer, *Innovation and Growth: Schumpeterian Perspectives* (1984)

This book is about how technological racing, rivalry and competition instigates a process of innovation, industrial and market evolution, and how it extends to larger entities than firms and industries to regions and national economies or economy networks. It shows what drives globalization, which industries are most significantly affected and how technological racing results in value generation in increasing returns and network industries. Furthermore, we explore how the emergence of selective managerial strategies are most likely to carry success in the pursuit of corporate and industrial policies.

The striking pattern that emerges in firms' innovative activities is that firms' rivalries for a technological leadership position in situations are best described as 'races'. A 'race' is an interactive pattern characterized by firms constantly trying to get ahead of their rivals, or trying not to fall too far behind. In high-technology industries, where customers are willing to pay a premium for advanced technology, leadership translates into increasing returns in the market through positive network externalities.

Racing behaviour is also a dynamic story of how technology unfolds in an industry. In contrast to any existing way of looking at the evolution of technology, racing behaviour recognizes the fundamental importance of strategic interactions between competing firms. Thus firms take their rivals' actions into account when formulating their own decisions. The importance of this characterization is at least twofold. At one level, racing behaviour has implications for understanding technology strategy at the level of the individual firm; at the other level, for understanding the impact of policies that aim to spur technological innovation in an industry or country. In this high-intensity, dynamic environment, under hypercompetition we understand a set of technological races in different market settings of high-technology industries inducing (economic) welfare-enhancing innovation processes. What then are the strategic choices that cause firms to become entrenched in relentless hypercompetition? Following this line of reasoning, if technology racing leads to hypercompetition with a tendency of monopolizing market structures, then what about the converse – that exceedingly competitive rivalry encourages a too-rapid rate of technology advance and also the undertaking of excessively risky research and development (R&D) prospects or too much correlation (parallelism) in similar projects?

Although one may envisage conditions where this may be true, in general, welfare-enhancing technology racing as a constituent element of the capitalist process reinforced by globalization provides social benefits far exceeding the costs. Even more important, any alternative path, other than the competitive, would likely be inferior given the costs, in that it would generate a less-valued and less-welfare-producing technology portfolio. That is, even if the competitive process is wasteful (for example, in parallel or correlated technology development), its unique, high-value innovation outcome far exceeds the benefits of any alternative path. There is historical, observational and analytical evidence given in this book.

An interesting problem of the new economic literature is to understand how technology races can be induced endogenously, e.g. by changes in economic variables (such as costs, prices and productivity).

On a national scale, simple catchup hypotheses have put emphasis on the great potential of adopting unexploited technology in the early stage and the increase of self-limiting power in the later stage. However, the actual growth path of the technological trajectory of a specific economy may be overwhelmingly constrained by social capability. And the capability endogenously changes as states of the economy and technology evolve. The success of economic growth due to diffusion of advanced technology or the possibility of leapfrogging is mainly attributable to how the social capability evolves, i.e. which effects become more influential: growing responsiveness to competition or growing obstacles to it on account of vested interests and established positions.

Some observations on industrial patterns in Europe, the United States or Asia point to which type of racing behaviour is prevalent in global high-technology industries, as exemplified by ICT (information and communications technology) industries. The pattern evolving from such racing behaviour could be benchmarked against the frontier racing type of the global technological leaders.

Another observation relates to policy inferences on market structure, entrepreneurship, innovation activity, industrial policy and regulatory frameworks in promoting and hindering industry frontier races in a global industrial context. Does lagging behind one's closest technological rivals cause a firm to increase its innovative effort? The term 'race' suggests that no single firm would want to fall too far behind, and that every firm would like to get ahead. If a firm tries to innovate more when it is behind than when it is ahead, then 'catch-up' behaviour will be the dominant effect. Once a firm gets far enough ahead of its rivals, then rivals will step up their efforts to catch up. The leading firm will slow down its innovative efforts until its rivals have drawn uncomfortably close or have surpassed it. This process repeats itself every time a firm gets far enough ahead of its rivals. An alternative behaviour pattern would correspond to a firm increasing its innovative effort if it gets far enough ahead, thus making catch-up by the lagging firms increasingly difficult. For any of these forms there appears to be a clear link to market and industry structure, as termed 'intensity of rivalry'. We investigate two different kinds of races: one that is a frontier race among leaders and 'would-be' leaders, and another that is a catch-up race among laggards and imitators.

Furthermore, it is interesting to distinguish between two kinds of catch-up behaviour. A lagging firm might simply try to close the gap between itself and the technological leader at any point in time ('frontier-sticking' behaviour), or it might try to actually usurp the position of the leader by 'leapfrogging' it. When there are disproportionately large payoffs to being in the technical lead (relative to the payoffs that a firm can realize if it is simply close enough to the technical frontier), then one would expect that leapfrogging behaviour would occur more frequently than frontier-sticking behaviour. Alternatively, racing toward the frontier creates the 'reputation' of being an innovation leader hoping to maintain and increase market share in the future. All attempts to leapfrog the current technogical leader might not be successful since many lagging firms might be attempting to leapfrog the leader simultaneously and the leader might be trying to get further ahead simultaneously. Correspondingly, one could distinguish between attempted leapfroggings and realized leapfroggings.

Among the key issues to be addressed is the apparent inability of technology-oriented corporations to maintain leadership in fields that they pioneered. There is a presumption that firms fail to remain competitive because of agency problems or other suboptimal managerial behaviour within these organizations. An alternative explanation is that technologically trailing firms, in symmetric competitive situations, will devote greater effort to innovation, so that a failure of technological leaders to maintain their position is an appropriate response to the competitive environment. In asymmetric situations, with entrants challenging incumbents, research does demonstrate that startup firms show a stronger endeavour to close up to or leapfrog the competitors. Such issues highlight the dynamics of the race within the given market structure in any of the areas concerned. We observe two different kinds of market asymmetries with bearing on racing behaviour: (a) risk-driven and (b) resource-based asymmetries.

When the incumbents' profits are large enough and do not vary much with the product characteristics, the entrant is likely to choose the faster option in each stage as long as he has not fallen behind in the race.

In view of resource-based asymmetries, we observe, as a firm's stage resource endowment increases, it could use the additional resources to either choose more aggressive targets or to attempt to finish the stage sooner, or both.

Previous work has suggested that a firm that surges ahead of its rival increases its investment in R&D and speeds up, while a lagging firm reduces its investment in R&D and slows down. Consequently, previous work suggests that the lead continues to increase. However, based on related work for the US and Japanese telecommunications industry when duopolistic and monopolistic competition and product system complexity for new products are accounted for, the speeding up of a leading firm occurs only under rare circumstances. For example, a firm getting far enough ahead such that the (temporary) monopoly term dominates its payoff expression will always choose the fast strategy, while a firm that gets far enough behind will always choose the aggressive approach. Then the lead is likely to continue to increase. If, on the other hand, both monopoly and duopoly profits increase substantially with increased aggressiveness then even large leads can vanish with significant probability.

Overall, this characterization highlights two forces that influence a firm's choices in the various stages: proximity to the finish line and distance between the firms. This probability of reaping monopoly profits is higher the farther ahead a firm is of its rival, and even more so the closer the firm is to the finish line. If the lead firm is far from the finish line, even a sizeable lead may not translate into the dominance of the monopoly profit term, since there is plenty of time for the lead situation to be reversed, and failure to finish first remains a probable outcome. In contrast, the probability that the lagging firm will get to be a monopolist becomes smaller as it falls behind the lead firm. This raises the following question: what kind of actions cause a firm to get ahead? Intuitively, one would expect that a firm that is ahead of its rival at any time *t*, in the sense of having completed more stages by time *t*, is likely to have chosen the faster strategy more often. We will construct numerical estimates of the probability that a leading firm is more likely to have chosen a strategy faster to verify this intuition.

Moving away from the firm-led race patterns revolving in a particular industry to a clustering of racing on an industry level is putting industry in different geoeconomic zones against each other and becoming dominant in strategic product/process technologies. Here racing patterns among industries in a relatively free-trade environment could lead to competitive advantages and more wealth creating and accumulating dominance in key product/process technologies in one region at the expense of others. There appears to be a link that individual races on the firm level induce similar races on the industry level and will be a contributing factor to the globalization of network industries.

Thus, similar catch-up processes are taking place between leaders and followers within a group of industrialized countries in pursuit of higher levels of productivity. Supposing that the level of labour productivity were governed entirely by the level of technology embodied in capital stock, one may consider that the differentials in productivities among countries are caused by the 'technological age' of the stock used by a country relative to its 'chronological age'. The technological age of capital is the age of technology at the time of investment plus years elapsing from that time. Since a leading country may be supposed to be furnished with the capital stock embodying, in each vintage, technology which was 'at the very frontier' at the time of investment, the technological age of the stock is, so to speak, the same as its chronological age. While a leader is restricted in increasing its productivity by the advance of new technology, trailing countries have the potential to make a larger leap as they are provided with the privilege of exploiting the backlog in addition of the newly developed technology. Hence, followers being behind with a larger gap in technology will have a stronger potential for growth in productivity. The potential, however, will be reduced as the catch-up process goes on because the unexploited stock of technology becomes smaller and smaller. However, as new technologies arise and are rapidly adopted in a Schumpeterian process of 'creative destruction', their network effects induce rapid accelerating and cumulative growth potentials are catalysed through firm and industry racing.

In the absence of such a process we can explain the tendency to convergence of productivity levels of follower countries. Historically, it fails to answer alleged puzzles as to why a country, the United States, has preserved the standing of the technological leader for a long time since taking over leadership from Britain in around the end of the nineteenth century and why the shifts have taken place in the ranks of follower countries in their relative levels of productivitiy, i.e. technological gaps between them and the leader. The American economist Abramovitz (1986) poses some extensions and qualifications on this simple catch-up hypothesis in the attempt to explain these facts. Among other factors than technological backwardness, he lays stress on a country's 'social capability', i.e. years of education as a proxy of technical competence and its political, commercial, industrial, and financial institutions. The social capability of a country may become stronger or weaker as technological gaps close and thus, he states, the actual catch-up process 'does not lend itself to simple formulation'. This view has a common understanding to what another economist, Olson (1996), expresses to be 'public policies and institutions' as his explanation of the great differences in per capita income across countries, stating that 'any poorer countries that adopt relatively good economic policies and institutions enjoy rapid catch-up growth'.

The suggestion should be taken seriously when we wish to understand the technological catching-up to American leadership by Japan, in particular during the post-war period, and explore the possibility of a shift in standing between these two countries. This consideration will directly bear on the future trend of the state of the art which exerts a crucial influence on the development of the world economy.

These explanations notwithstanding, I venture as a major factor for divergent growth processes the level of intensity of the racing process within the most prevalent value-added industries with cross-sectional spillovers. These are the communications and information industries which have been shaped and led by leading American firms and where the rewards benefited their industries and country. Although European and Japanese companies were part of the race they were left behind in core markets reaping lesser benefits. (Since ICT investment relative to GDP is only less than half in countries such as Japan, Germany and France campared to the US, 2% vs more than 4% in 1999, this does not bode well for a rapid catch-up in those countries.)

Steering or guiding the process of racing through the pursuit of industrial policies aiming to increase competitive advantage of respective industries, as having been practiced in Japan, would stimulate catch-up races but appears to be less effective in promoting frontier racing. Another profound reason lies in the phenomenon of network externalities affecting ICT industries. That is, racing ahead of rivals in respective industries may create external economies to the effect that such economies within dominant industries tend to improve their international market position and therefore pull ahead in competitiveness *vis-à-vis* their (trading) partners.

As P. Krugman (1997) observed: ‘It is probably true that external economies are a more important determinant of international trade in high technology sectors than elsewhere’. The point is that racing behaviour in leading high-growth network industries by generating frontier positions create critical cluster and network externalities pipelining through other sectors of the economy and creating competitive advantages elsewhere, as supported by the ‘increasing returns’ debate (Arthur, 1996). In this sense we can speak of positive externalities endogenizing growth of these economies and contributing to competitive advantage. All these characteristics lay the foundations of the ‘Network Economy’. The Network Economy is formed through an ever-emerging and interacting set of increasing returns industries, it is about high-intensity, technology driven-racing, dynamic entrepreneurship, focused risk-taking through (free) venture capital markets endogenized by societal and institutional support. With the exception of pockets of activity in some parts of Europe (the UK and Scandinavia), and in specific areas such as mobile communications, these ingredients for the Network Economy are only in the early stage of emerging in Continental Europe, and the political mindset in support of the Network Economy is anything but prevalent. As long as we do not see a significant shift toward movements in this direction, Europe will not see the full benefits of the Network Economy within a Global Economy.

Racing behaviour on technological positions among firms in high-technology industries, as exemplified by the globally operating telecommunications and computer industries, produce spillover benefits in terms of increasing returns and widespread productivity gains. Because of relentless competition among technological leaders the network effects lead to significant advantages in the value added to this industry contributing to faster growth of GDP, and through a flexible labour market, also to employment growth. This constitutes a new paradigm in economic thinking through network economies and is a major gauge to compare the wealth-creating power of the US economy against the European and advanced Asian economies over the past decade.

It is interesting to speculate on the implications of the way the firms in major high-technology markets, such as telecommunications, split clearly into the two major technology races, with one set of firms clearly lagging the other set technologically. The trajectories of technological evolution certainly seem to suggest that firms from one frontier cannot simply jump to another trajectory. Witness, in this regard, the gradual process necessary for the firm in the catch-up race to approach those in the frontier race. There appears to be a frontier 'lock-in', in that once a firm is part of a race, the group of rivals within that same race are the ones whose actions influence the firm's strategy the most. Advancing technological capability is a cumulative process. The ability to advance to a given level of technical capability appears to be a function of existing technical capability. Given this path dependence, the question remains: why do some firms apparently choose a path of technological evolution that is less rapid than others? Two sets of possible explanations could be derived from our case analysis, which need not be mutually exclusive. The first explanation lingers primarily on the expensive nature of R&D in industries like telecommunications and computers which rely on novel discovery for their advancement. Firms choosing the catch-up race will gain access to a particular technical level later than those choosing the frontier, but will do so at a lower cost.

Summary of chapter contents

The first chapter highlights scope and scale of the underlying technological racing situation. Based on industry cases it shows that racing could develop along different dimensions and intensity as in frontier and catch-up racing or leapfrogging. This gives rise to a finely tuned typology across knowledge-based industries for multi-stage racing in several unique 'environmental' situations.

Chapter 2 first provides the basic notions and tools for a conceptual understanding of technological races which lays the decision-theoretic groundwork for the subsequent analysis. It conceives technological racing as embedded in a stochastic process akin to optimal statistical decision problems in sequential analysis. The firm as a decision-maker decides about her strategic positioning in R&D allocations as the racing stages evolve under uncertainty toward a final stochastic destination. Along the path the firm tries to speed-optimize in reaching a winning (leadership) position under time and cost constraints

Chapter 3 follows the next step in a statistical profiling of technological evolution and innovation as it relates to competitive racing and rivalry among leading firms. Among the performance criteria to be assessed are frequency of frontier-pushing, technological domination period, innovations vs imitations in the race, innovation frequency when behind or ahead, nature of jumps, leapfrogging or frontier-sticking, inter-jump times and jump sizes, race closeness measures, and inter-frontier distance.

In Chapter 4 we develop and analyse a general model of an R&D race that takes into account the effects of firms' past R&D efforts, that is their knowledge stocks, and that timing and duration of innovation depend on them. Common behavioural models, as exposed in Chapter 3, suggest that the participants in an R&D race change their efforts as they jockey for positions against their rivals. It also implies that the currently lagging firm works harder than the leading firm in an attempt to catch up. Those models do not capture historical R&D performance as the race only relates to current actions. In contrast, we adopt the view that at each point in time, the more knowledge a firm has, a culture of knowledge, the higher her chances of making the discovery at this point. Accompanying investment in R&D, firms accumulate knowledge (learning), and some of the knowledge they have previously accumulated depreciates over time (forgetting). Net learning shapes firms' payoffs and strategies. In such a context we will find that a firm is able to reduce her R&D investment as her knowledge base increases and that the follower works harder than the leader. This would result in another pattern of competitive interaction that is more akin to a catch-up race than to increasing dominance in a frontier race.

Chapter 5 develops a framework to analyse how strategic choices are made in a sequential context when R&D competition occurs between firms, and the leadership–time tradeoffs have to be resolved in multiple stages. At issue is the way in which resources are used at each stage, i.e. are aggressive problems undertaken and solved slowly, or are quick solutions adopted in an effort to get the product to market faster? We focus on the translation between *ex-ante* asymmetries between firms in industrial settings and *ex-post* asymmetries in the equilibrium outcomes. Another focus is on understanding the implications of the tradeoff between the level of technology and product leadership and time spent on each stage in a multi-stage process.

In Chapter 6 we extend intense technological rivalry between firms to network markets in which there is uncertain technological development in product/process technologies.

Firms 'price' compete in those markets to gain market share before any of them succeeds in getting an innovation to move ahead of its rival(s). If the firms are in a technology race and the probability of innovation success is small, then a firm with a bigger network advantage is likely to attract more customers in the absence of innovation. If, however, any of those firms expect innovation with a high probability and none of them have a big (network) advantage, then they keep on sharing the market until the innovation occurs. Such a model could be extended to situations where firms are not restricted to having at most one innovation and the outcomes would result in a clustering of innovations and of different size, that is, if firms go for a major risky innovation or a series of smaller and less risky ones.

In Chapter 7 we look at the relationship between innovation and quality, investments and its strategic consequence on competition and collaboration. At the core of the explanation is the model of a sequential differential game characterized by a leader and follower interacting during a time duration in which they initiate innovation or imitation efforts and launch new products. The open-loop Stackelberg equilibrium results in explaining properties of investment trajectories, innovation and quality dynamics in different symmetric and asymmetric firm settings.

Chapter 8 focuses on a contestable market with network externalities with an incumbent and an entrant. The incumbent, unlike the entrant, already has an installed base of consumers. We look at decision situations of firms regarding how proprietary they want to make their technology, either through patent protection or through development in open-source systems. We explicitly model the direct and indirect effects of network externalities. For example, as a direct effect, consumers may prefer to use a popular word processor because they know the format of their work can be easily transported to other users' computers. An indirect effect could result if a bigger network translates into better quality. For example, more software companies are willing to produce programs for an operating system if it has a larger consumer base. This increased competition could lead to an improvement of the quality of the operating system (OS). The model predicts that using open-source technologies is likely to enhance the rate of R&D, and consequently the quality of the product. An incumbent that would choose this strategy is likely to deter entrance of a newcomer because it can play out its advantage of a larger network.

Chapter 9 deals with a specific form of asymmetries between firms embedded in market competition as, for example, in the case that one firm owns a more advanced technology. The analysis uncovers two different forces on determining the incentives to innovate. The competitive incentive is defined as the difference between a firm's profits if it innovates and the profits it would make if its rival innovated instead. The profit incentive is determined by calculating the increase in a firm's profits if it alone were investing in R&D. Given those forces, we explore how the incentives to innovate depend on the degree of product differentiation or niche building in product markets. If the products are less substitutable, the profit incentive dominates each firm's willingness to invest in R&D and it leads to the result that the currently less advanced firm engages in more R&D. If the products are very substitutable, then the competitive incentive determines the outcome of technological competition so that the currently more advanced firm increases its superiority in technology. Hence, the rate of product differentiation is shown to be an important determinant predicting the outcome of technological competition. The existence of a second period of competition largely reinforces the outcome from a single period race. An alternative outcome could be expected with technological competition when firms can share both the cost and the outcome of innovative activity. The majority of models consider the case that there is only one winner in the technological competition. However, alternative models develop technological competition in which each firm engaging in R&D can obtain a patent on a cost-reducing technology. Since innovative investments are often very expensive and knowledge has a tendency to leak to others, firms may have an incentive to conduct their innovative activities cooperatively.

In Chapter 10 we develop a general model as a rather accurate framework to describe industrial competition in an *increasing returns economy*. We refocus on the interaction between innovation, entrepreneurship, firm size and market structure in a Schumpeterian world. The factors that structurally determine an increasing returns economy, that is the resource-based loop, the Schumpeterian loop, scale economies, the different categories of learning and demand-side increasing returns (reputation) constitutes a qualitatively driven useful framework capable to explain in an endogeneous, dynamic way the number and growth of firms in a given industrial setting.

Chapter 11 opens a new dimension of competitive positioning through network competition on the basis of alliance formation (strategic alliances, joint ventures). From a strategic perspective technological competition will be refined and expanded into new markets, or new markets will be created through alliance formation. Furthermore, it could speed up competitive positioning and technological leadership in strategically important although geographically diverse markets.

Finally, Chapter 12 reviews several approaches to achieving catch-up processes taking place between leaders and followers within a group of industrialized countries (or even emerging economies) in pursuit of higher levels of productivity that could be traced to industrial racing behaviour, conduct and patterns in previous chapters.

Guide to readership

Since the reader may have diverse interests and tastes in grasping the contents of this book, it may be useful for her/him to follow a guided roadmap, as seen by the author, on how to go about skipping technical details and yet still access the main messages of the text. For first reading and learning we advise following the paths through chapter nodes indicated without an asterisk; for a deeper understanding of some technically critical chapters, the alternative path should be followed, but as an ancient Roman proverb predicts, 'if you don't know where you are going any path will lead you there'.

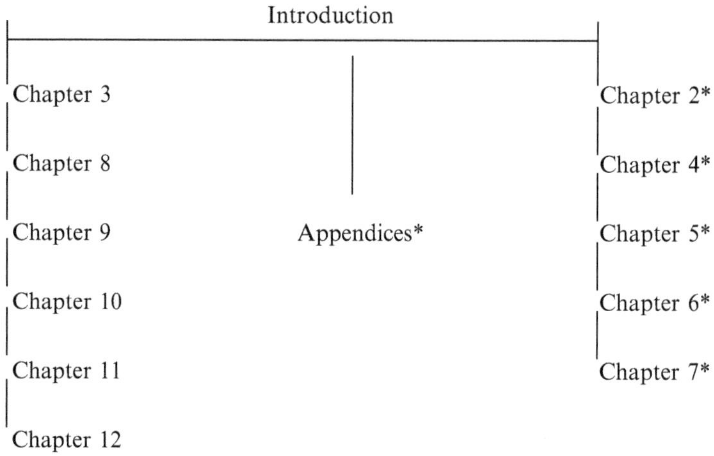

To pass through the technical (*) chapters everyone with some basic knowledge of high-school mathematics would have no difficulties. Besides, for interest in a given subject area, the chapters* are self-contained as to the focused subject area, and therefore other chapters* may be skipped without significant loss of systemic understanding.

This book proposes to be an advanced-level text for graduate students in economics, management and technology with interest in industrial economics, strategy and competition policies with special regard to high-technology industries. It strives to present an integrative approach to complex managerial and industrial leadership issues. It also attempts to back up the theoretical insights with very practical cases.

We sincerely hope that the reader will enjoy following the red line through a maze of highly interesting industrial environments and conducts.

Acknowledgments

This book emerged from lectures notes during my tenure in the School of Policy Studies at Kwansei Gakuin University, Japan which provided a unique place to devote a significant chunk of my time for writing the book. I appreciate the support and the provision of resources by the School which enabled me to complete the text without major hitches. My thanks go in particular to the former Dean of the School, Professor Amano, who in his friendly way showed much interest in the progress of the work. Several lectures based on the material of this book have been given at Kansai University, Osaka, Hitotsubashi University, Tokyo, Hong Kong University of Science and Technology, IIASA, the University of Naples, and at the World Association for Sustainable Development, Dubai.

I had interesting conversations and discussions with many participants, friends and colleagues, among them who also read and commented on parts of the text are Professors H. Eto, S. Nagaoka, Y. Ukai, M. Takashima, C. Deissenberg, A. DelMonte, D. Glynn, A. Zellner and W.G. Moss. I am grateful to all of them but, of course, the usual disclaimer applies. Some of the materials used in Chapters 6 and 12 has been previously published in the *International Journal of Technology Policy and Management* and the *World Review of Sustainable Management*, and permission has been granted by the publisher Inderscience.

Hans-Werner Gottinger

1 Taxonomy of technological racing

> Technical leadership produces high-margin products, wins competitive battles and creates new markets.
>
> *General Electric Annual Report 2003*

1.1 Introduction

This chapter tracks the evolution of a cross-section of technologies in a large span of high-technology industries embracing computers, communication, pharmaceuticals and biotechnology. It outlines that strategic interactions between the firms play a substantial role in determining firm-level and industry-level technological evolution. We can identify several races supported by selected cases across industries, each of which is the result of a subset of firms jockeying for a position either as a race leader or for a position not too far behind the leader.

The identification and interpretation of the races relies on the fact that different firms take very different technological paths to reach a common performance level. As such, the races cannot be interpreted as a free-riding situation where one firm expands resources in advancing the state of technology and the others follow closely behind. Such spillover interpretations are suspect when products are in the region of high complexity, of high risk in succeeding, and different firms typically adopt different procedural and architectural approaches.

In this chapter we present a descriptive analysis of the evolution of technology with selected case studies to recount a story of technology evolution in a given industry. The logic underlying this evolution holds in any industry in which two broad sets of conditions are satisfied.

First, it pays for a firm to have a technological lead over its rival; it also boosts its market image and enhances its reputational capital.

Second, there are various levels of technological complexity among the products introduced by various firms. Technological complexity can be represented by a multi-criteria performance measure, that is, by a vector-valued distance measure. The collection of performance indicators and parameters, being connected with each other for individual firms, form an envelope that shapes a 'technological frontier'. The technological frontier is in fact a reasonable indicator of the evolving state of knowledge (technical expertise) in the industry. At any point in time the industry technology frontier (ITF) indicates the degree of technical sophistication of the most advanced products carried by firms in that industry in view of comparable performance standards. Firm-level technology frontiers (FTF) are constructed analogously and indicate, at any point in time, the extent of technical sophistication achieved by the firm until that point in time. The evolution of firm- and industry-level frontiers is highly interactive. Groups of firm frontiers are seen to co-evolve in a manner that suggests that the respective firms are racing to catch up with, and get ahead of each other (Gottinger, 2006).

For illustrative purposes we emphasize three cases of knowledge-based industries for which we proceed to construct statistical indicators reflecting racing patterns in those industries: semiconductors/computers and telecommunications equipment (ICT), biotechnology, pharmaceutical and medical device industries. These industries are the major components of a knowledge-based network economy. Statistical indicators reflecting technology racing in those industries provide intrinsic information on knowledge leadership positions, competitive advantage and the level of welfare and wealth creation in the economies involved.

A data set could focus on a given set of products (systems) by major European, American and Asian enterprises in those industries for a sufficiently representative period of market evolution. In principle, we can identify at least two races in progress in the industries throughout a given period of duration. One comprises the world frontier race in each of those industries, the other, for example, the European frontier race which would technically constitute a subfrontier to the worldwide race. The aggregate technology frontier of the firms in a particular race (that is, ITF) is constructed in a manner similar to the individual FTFs.

Essentially, the maximal envelope of the FTFs in a particular race constitute the ITF for that race. The ITF indicates, as a function of calendar time, the best achievable performance by any firm in the race at a given date.

A statistical profiling of technological evolution and innovation is achieved for several major knowledge-based industries as they relate to competitive racing and rivalry among leading firms. Among the performance criteria to be assessed are (1) frequency of frontier pushing, (2) technological domination period, (3) innovations vs imitations in the race, (4) innovation frequency when behind or ahead, (5) nature of jumps, leapfrogging or frontier-sticking, (6) inter-jump times and jump sizes, (7) race closeness measures, (8) inter-frontier distance, (9) market leading through 'market-making' innovations and (10) leadership in 'innovation markets'.

A race may or may not have different firms in the leadership position at different times. It may be a tighter race at some times than at others, and in general may exhibit a variety of forms of interesting behaviour. While interpretation and analysis of racing behaviour is left to subsequent chapters, it is appropriate to ask at this juncture why the firms should be racing at all. As access to superior technology expands the scope of opportunities available to the firms the technology can be applied in a range of markets. However, leading-edge technology is acquired at a cost. It seems unlikely that all the firms would find it profitable to compete to be at the leading edge all the time. Also, not every firm has access to equal capabilities in leveraging a given level of technological resources. Firms may, for example, be expected to differ in their access to complementary assets that allows them to appropriately reap the benefits from their innovation. It is reasonable to assume that whatever the level of competence of a firm in exploiting its resources it will be better off the more advanced the technology.

Based on this procedure an analysis will show how dynamic competition evolved in the past.

Chapter 3 entails a novel and unique statistical profiling of industry racing behaviour for selected high-technology industry cases.

The results yield valuable, policy-relevant information on the level of technological frontiers among local, regional and national enterprises, in leading-edge, high-growth, structurally dynamic and increasing returns industries in view of major competitors on the world frontier.

Unlike other (statistical) indicators (such as patent statistics) referring to the degree of competitiveness among industries, regions and countries concerned, the proposed measures cover behavioural, dynamic movements in respective industries, and are therefore able to lend intrinsic predictive value to crucial economic variables relating to economic growth and wealth creation.

The results are likely to provide strategic support for industrial and technology policy in a regional or national context and enable policy-makers to identify strengths and weaknesses of relevant players and their environments in those markets.

The statistical indicators derived can be adapted and extended to other high-growth and fast-developing industries.

After presenting evidence of racing behaviour for some particular cases in Section 1.2, we generalize as to why multiple races might occur within an industry, and examine how lessons from these analyses influence and shape frameworks of technological evolution. Further on, in Section 1.3, we discuss frontiers and clarify why they are useful indicators of the evolving state of firm and industry knowledge. The basic pattern that emerges from the firm-level technology frontiers is indicative of racing behaviour. Section 1.4 presents a series of cases that highlight the manner in which strategic interactions between firms influence technological evolution. Section 1.5 shows the levels of conditions or constraints under which technological racing could evolve and is most likely to limit itself in the future.

1.2 Examples of technological races: old and new

In studying the evolution of high-technology industries, say over the last fifty years, one is amazed by observations on the intensity and universality of rivalry among competitors across a broad selection of industries. In many cases, developments of such industries were initiated and fostered by the interactive pattern of a continuous contest among market participants to get ahead of their rivals or not be left too far behind. We see these patterns emerging at various stages of market evolution and, at first sight, seemingly unrelated to market structures.

We identify those interactive patterns as technological races. First look at the minicomputer market in the early 1970s. At this time the Digital Equipment Corporation (DEC) was fighting its principal rival Data General (DG) which had been instantly successful because of its initial machine, the NOVA, and grew much faster than DEC did earlier in the market (Kidder, 1981). DEC's reaction was the PDP 11, a carefully orchestrated response to the challenge of DG. From DG's point of view, the most important 32-bit machine was DEC's VAX 11/780. Tracy Kidder's account portrays Tom West as one of DG's most talented engineers. He writes (Kidder, 1981, p. 29): 'It has been painful for West and for a number of engineers working with him … to watch DEC's VAX go to market, to hear it described as "breakthrough", and not have a brand-new machine of their own to show off'.

These reactions to rival product introductions form the basis for strategic interactions so crucial in determining the firm- and industry-level technology frontiers.

In the mainframe computer era, the Control Data Corporation (CDC) clearly regarded IBM as the enemy in the early stages of its history. CDC often had machines that were technologically superior to those that IBM was offering. For example, the high-end models of the CDC 6000 series, particularly the CDC 3600 and the CDC 6600, were technologically superior to even the highest end of the System 360 series that IBM introduced. Tom Watson, IBM's chairman, was concerned by CDC's activities, which became apparent in the following extract from an internal memo, dated August 1963, that emerged during an antitrust suit that the government filed against IBM (Lundstrom, 1988). 'Last week CDC had a press conference during which they officially announced their 6600 system. I understand that in the laboratory developing this system there are only 34 people, including the janitor … Contrasting this modest effort with our vast development activities, I fail to understand why we have lost our industry leadership by letting someone else offer the world's most powerful computer'. Watson wanted to have a new machine, and refused to be second best. IBM decided that a machine two-and-a-half times more powerful than CDC's machine would be an appropriate target to aim for.

Watson attributed the success of IBM to IBM's attitude to 'running scared' of the competition. We have other manifestations of intense rivalry and 'neck-and-neck' competition across industries, be it in advanced microprocessors between Intel and American Micro Devices (AMD) (see Markoff and Lohr, 2003), or specialty pharmaceuticals for medical care between Merck, Glaxo and Pfizer, and between biotech companies Amgen and Biogen. In the expanding market of software-related web services it is Microsoft against IBM, Sun Microsystems and Oracle (Lohr, 2003a,b), in consumer electronics and design it is Sony against Matsushita, Samsung or Sanyo (Belsen, 2003).

Leading in the technological race is often helped by network strategies based on network economies (Gottinger, 2003).

Flexibility and network strategy: the case of Motorola and Nokia, a case in standard-based competition

Let us take the case of Motorola and its Finnish competitor in the mobile communications markets, Nokia. During the 1980s and early 1990s, Motorola built dominance around the analogue AMPS mobile standard predominant in the United States. Motorola failed to recognize the emergence of digital standards, kept costs too high largely as a result of its dominance, and eventually lost its market position to Nokia, at the outset an obscure competitor. While there is no certainty in historical 'what ifs', had Motorola paid closer attention to developments worldwide in the mobile industry, not just the fragmented US markets, *they* might have headed the creation of substantial coalitions around emerging digital standards. In particular, they would have noticed the activities of Nokia and Ericsson, which took lead roles in attempting to drive change through coalitions, driven by a clear and consistently focused adaptable strategy (Roberts, 2004, p. 29). For a number of years, Motorola paid insufficient attention to the activities of these firms, as well as to the activities of foreign governments intent on helping to drive standards advantageous to their local firms.

Given that the mobile telecommunications industry is one of the most heavily characterized by network economics (e.g. supply and demand side economies of scale, standards, 'winner takes most'), coalitions forming around competing standards should always recommend vigilant action by competitors. Had Motorola recognized the importance of these developments much earlier, it could have responded more effectively. It is important to recognize that this framework does not suggest that network-based strategies could only be answered by network-based strategies. Radical innovations could offset and leapfrog dominating network strength in network markets (Chapter 6). Motorola in the early 1990s certainly had the resources and market power to develop its own digital offerings in-house to answer Nokia's threat. The important point of characterizing network strategies is to identify when a firm requires a competitive response, whether internally or externally focused, to other firms' network strategies, as well as how best to approach building and executing a response. Had Motorola entered the digital arena much earlier, its most effective strategy would likely have been a network-based strategy. No major mobile communications standards have prevailed in any major world markets without a coalition of varied interests. Even in cases where government regulators mandated a standard, this occurred as a result of the actions of multiple interests. The Japanese and Korean governments' mandate of specific national standards for second-generation wireless (2G wireless) occurred as a result of the interests of national firms, but in each case, multiple firms advocated for the standard. Certainly, dominant firms such as NTT in Japan exerted preponderant influence. As the wireless industry has grown globally, the creation of third-generation wireless standards (3G) has compelled all primary interest blocks – Japan, United States and Europe – to reach a global compromise.

Conversely, Nokia's approach to the global marketplace focused the firm outside its boundaries. While Nokia maintained internal control over most product development efforts, it spent considerable resources discovering customer demands through extensive direct contact with consumers. The firm also supported innovation related to its product lines through numerous coalitions and alliances with smaller firms. This collaborative approach to 'upstream' innovation included such activities as standards development and innovation on technology inputs to the company's products (e.g. improved semiconductors). Even internally, Nokia organized itself as a network of many small research teams located worldwide. Part of the motivation for this arrangement was to encourage true globalization of the Finnish firm. Finland's isolation restricted its contact with foreign markets, and therefore foreign consumers, research organizations and firms. Management made a conscious decision to globally disperse its research teams, as well as to break them into small, flexible groups.

Nokia's network strategy has provided it with a portfolio of strategic options in the midst of the brutal technology marketplace of 2000–2001, which has spread its risk while the next generation of wireless standards and behaviours evolved. Although driving marketplace standards through coalitions often fails, companies in standards-based technology industries, such as Nokia, have no choice but to either 'lead, follow, or get out of the way', to paraphrase Lee Iaccoca, the former chief of Chrysler. Nokia's broad participation in multiple standards alliances has positioned it to benefit from many different scenarios as the wireless marketplace evolves.

There are many factors that contributed to Motorola's loss of dominance in global wireless markets that do not directly pertain to our present discussion. This discussion does not propose that Motorola's lack of a network strategy led directly to its downfall. For instance, Motorola's products were much more expensive to produce, even after many impressive digital products. By 2000, Motorola's consumer wireless product line included over 5000 different components; by contrast, Nokia's wireless product line, based on modularity, required about 300, surely a better and more cost-effective way to reduce complexity (Simon, 1962). None the less, even this cost condition relates to Motorola's excessively inward-looking focus and culture. Monitoring the activities of potential competitors, as well as more effectively integrating the efforts of the company's internal R&D organization with that of its external suppliers, could have mitigated this problem. Nokia's integration of R&D efforts internally and with external partners helped to produce a modular, comparatively streamlined mix of components for their wireless products with important implications for cost and product development cycle time.

A flexible network strategy can increase the likelihood that a firm will be prepared for alternative evolutionary paths as events unfold. Strategic 'inflection points' (Grove, 1996) often occur when the future fails to cooperate with expectations and a firm lacks appropriate, efficient alternatives, as was Motorola's case. Despite setbacks during the 2000–2001 technology marketplace slowdown, if 3G proliferates, Nokia will hold a central place in a lucrative, high-growth market. Nevertheless, a superb network strategy does not guarantee marketplace success. Nokia and its partners have found it quite difficult to encourage broad adoption for 3G wireless products and services. Nokia, Ericsson and Motorola's championing of the Wireless Application Protocol (WAP) and Bluetooth has not succeeded in developing marketplace dominance for either standard. In fact, a competing standard to Bluetooth has taken an early lead. Most ominously, the diffusion of 3G has occurred more slowly than had been anticipated by most major players. Despite Nokia's masterful network strategy, based on numerous coalitions, partnerships and contact with their customers (additional nodes in the company's network!), the firm's CEO, Jorma Ollila, 'bet the company on the wireless internet' enabled by 3G. A network strategy by no means eliminates the risk of strategic commitments to a future that does not evolve as anticipated. During the first half of 2001, Nokia's market value dropped by nearly 50% (significant, even given the overall meltdown in technology equities), largely in response to a worldwide slowdown in demand for wireless products and services. Nokia will survive and eventually thrive despite a slowdown in wireless spending; however, if 3G fails to evolve as a significant global standard, Nokia's 3G-focused strategy could throw the firm into crisis.

If international coalitions increasingly prove unable to compel competing firms and consumers to adopt a particular standard, as in the 3G debate, it is unreasonable for a single firm to expect to be able to dominate a standards-driven marketplace without an effective network strategy. Motorola failed in this recognition. Because of an inward focus and the tyranny of success, Motorola failed to respond to the emergence of digital wireless until it was too late to maintain dominance. The firm continues to suffer as at this stage of observation. Nokia's broader network strategy provided the firm with a more flexible set of options; none the less, its bet-the-company approach to 3G wireless threatens the firm's long-term survival if 3G fails to materialize.

1.3 Strategic interactions and multi-stage racing

Indeed, we can observe vigorous technological races on several dimensions involving multi-product, multi-national companies, as between Nokia and Samsung on CDMA mobile handsets, as between Microsoft Windows-based systems for corporate servers and the likes of Sun, Hewlett-Packard (H-P) and IBM Linux-based software, or for the advanced consumer appliance market between Microsoft Windows and Linux-based gadgets of Sony, Matsushita, Hitachi, Toshiba and others. Further, for Unix-based blade servers there is strong rivalry between H-P, IBM, Sun and Dell Computer fighting for growing market share; for computer games it is between Sony, Microsoft and Nintendo; for web services it is between Microsoft, IBM, Bea Systems, Sun and Oracle; for microprocessors fierce battles arise between Intel and AMD; for online search services it is between Google, Microsoft, Yahoo and AOL Time Warner. When companies compete head-to-head on several technologies we say they are involved in 'multiple races'.

Overall, the recent history of high-technology industries demonstrates that dynamic competition takes place among firms in innovation-driven (Schumpeterian) industries. This has been particularly evident in the software industry. In some instances, firms race to create an entirely new product category. For example, VisiCalc defined the category of spreadsheet software and was the early market leader. But it was eventually replaced, first by Lotus 1-2-3 and subsequently, by Microsoft Excel. In other instances, dynamic competition takes the form of innovation to displace a category leader. For example, Micropro's Wordstar was the early leader in word processing software for PCs, which significantly displaced dedicated word processing systems such as those offered by Wang. But Wordstar was eventually displaced by WordPerfect. WordPerfect retained category leadership for approximately six years before being displaced by Microsoft Word, which was helped in part by the transition to graphical user interfaces, and, in particular, Windows. This pattern is not unique to computer software. It can also be observed in other industries such as (research-based) pharmaceuticals and hand-held devices (such as PDAs). For example, in 1977, SmithKline Beecham offered the first H_2-antagonist anti-ulcer drug, called Tagamet. When GlaxoWellcome entered the market in 1983 with Zantac, it quickly took market share from Tagamet. Merck (Pepcid) and Eli Lilly (Axid) also entered the market eventually. By 1988, Zantac surpassed the market share of the first mover, Tagamet. By 1993,

Zantac had 55% of the market, Tagamet had 21%, Pepcid had 15% and Axid had 9% (McKelvey, 1996).

The race to develop operating systems for personal digital assistants (PDAs) is another example of dynamic competition. Apple introduced the first hand-held PDA, called the Newton, in 1993, but that product was not a success with consumers. Following the failure of Newton, a number of firms began developing operating systems for these hand-held devices either available to consumers or in development (IDC, 2000). By 1998, the Palm OS was the clear leader in the PDA segment with a 73% share. Palm remains the category leader today, but its leadership is challenged from Microsoft's Windows CE operating system and Symbian's operating system, among others.

As we can deduce from the latter example, an essential feature of industrial racing is the prevalence of actual and potential innovative threats to leading firms, coming from inside or outside the industry, that is broadly related to the notion of 'innovation markets'. Those innovations result in competitive threats based on technologies and design approaches that differ radically from those used by the incumbent.

The examples suggest that firms adjust their R&D efforts when rivals make progress or fall behind, and, in particular, the laggard in the race gives further way to the leader. Our models in Chapter 4 assume that the past level of R&D matters on the positioning of the R&D race for the firm and that under most circumstances the follower works harder and is more likely to catch up with the leader than the leader is further advancing against the follower.

The scale of the literature can be grouped into models of symmetric R&D races and multi-stage races. In a symmetric R&D race, identical firms compete for a particular innovation by investing in R&D. A firm can increase the probability that it makes the innovation by some point in time by devoting more resources to the R&D process (Loury, 1979; Lee and Wilde, 1980; Reinganum, 1982). These models assume that the time of a successful innovation is exponentially distributed, and with it comes a so-called 'memoryless property' (MLP). That is, the knowledge bases that firms have acquired as a result of their past R&D efforts are irrelevant to firms' current R&D efforts and to the outcome of the race.

Multi-stage races can be viewed as an attempt to circumvent the MLP. In multi-stage models, firms are required to complete a number of stages or experience levels that lead up to the discovery of the innovation. A firm is ahead of a rival if it has a smaller number of stages left to complete and the competitors are head-to-head if they have the same number of stages left. Deterministic multi-stage models assume that firms transit to the next stage in a deterministic fashion, as in Fudenberg *et al.* (1983), Harris and Vickers (1985) and Lippman and McCardle (1988). The outcome of the race in these models suggests that even a small advantage by one firm causes the other to drop out of the race immediately. This strong result would be somewhat weakened when the stage-to-stage transitions are probabilistic. In the stochastic multi-stage models of Grossman and Shapiro (1987), Harris and Vickers (1987), and Lippman and McCardle (1987), the time to completion of each stage is assumed to be exponentially distributed. Consequently, while a firm's equilibrium R&D effort depends on the number of stages it and its rival have left to complete, within each stage the MLP renders firms' current R&D efforts independent of their past R&D efforts. In these models the leader devotes more resources to R&D than the follower. Thus the follower tends to fall further behind as the race progresses, whereas the leader tends to build up its advantage and the emerging pattern of strategic interactions is more like increasing dominance than a head-to-head racing-type strategic interaction.

In multi-stage races a firm has to complete a number of stages in advance of her rivals in order to win the race. A firm observes the number of stages her rival has left to complete, and adjusts her R&D efforts accordingly. A firm is ahead of a rival if she has a smaller number of stages left to complete, while she and her rival are head-to-head if they have the same number of stages left to complete, and the firm is free to alter her R&D efforts in response.

Grossman and Shapiro (1987) allow for the possibility that one firm may be ahead of the other by introducing a single intermediate step in the research programme. Thus, to win the race, a firm must complete two phases of R&D. This shows that the leader always devotes more resources to R&D than the follower. In particular, the lagging firm drops out of the race when the value of the patent is large. If the follower happens to catch up, both firms intensify their efforts. Moreover, the intensity of competition is greatest when the race is tied after the intermediate stage. Harris and Vickers (1987) extend this model to an arbitrary number of stages. They confirm that the leader always devotes more resources to R&D than the follower. Thus the follower tends to fall further behind whereas the leader tends to build up her advantage. Moreover, the follower slows down as he falls further behind whereas the leader may or may not speed up as she gets further ahead.

The model proposed in Chapter 4 hinges on observations that it is the firm's accumulated knowledge base to determine its competitive standing (and aggressiveness) *vis-à-vis* its aspiring rivals. In realistic R&D races, we are likely to meet the following 'environmental' conditions.

First, a firm is usually able to reduce its R&D spending as its knowledge base increases.

Second, as the race evolves, the follower generally works harder than the leader to keep her option for a catch-up and eventually (re)gaining leadership. The real source of this effect is that a firm's past R&D efforts have contributed to her chances in winning the R&D race.

Third, a firm can react either aggressively or submissively to an increase in her rival's knowledge base. A firm acts aggressively if it has a sufficiently large knowledge base and/or the value of the patent is large enough. While empirical explorations of the R&D race would indicate that these strategic considerations are dominated by the accumulated knowledge stock, the emerging pattern of strategic interaction is more like action–reaction than increasing dominance.

1.4 Specific examples

(a) Aircraft jet engines

In the civil aircraft jet market it has been General Electric (GE) vs Rolls Royce (RR) and United Technologies Pratt&Whitney (PW) rivalling in the supply of engines for Boeing. With Boeing's proposed new commercial aircraft, the 7E7 dreamliner, on the drawing board, the race is on among those rivals to supply the engine for the forthcoming Boeing plane. While GE had been dominating the field, and RR has been emerging in a strong second position over the past decade, PW has slipped over the past 10 years to 16% market share in the commercial engine market, a significant downturn from the 1970s when its market share was close to 90%. Thus, PW is now in a fierce catch-up race that determines its future state as a viable competitor. According to the *Wall Street Journal* (Lunsford, 2003), PW apparently missed a chance in the mid-1980s to hold onto its lead when it passed on designing a replacement engine for the Boeing 737 while instead designing an engine for the new 757 which never lived up to Boeing's expectations. In the 1990s RR grabbed market share from both GE and PW by aggressively pricing engines for both Boeing 777s and Airbus A330s. Now with the new 7E7 on the drawing board, this is a critical time for PW to make good for lost ground and move ahead.

(b) Large commercial aircraft

If one had asked the question which American company would be dominant in its global industry, say, 15 years ago, it would not have been Microsoft, Dell, IBM or even General Motors or General Electric, it most likely would have been Boeing.

Since 1958, when it first introduced the first US commercial jetliner (the 707), and later in the sixties, the 747, it had gone from success to success, grabbing global market share which topped over 60% in the commercial aircraft market in 1990, leaving McDonnell Douglas with 23% and Airbus Industries with 15%. The reversal is one of the most puzzling and stunning technological races in recent business history. While some of the factors cannot be contributed to technological racing, such as governments' policies through industrial policy and subsidization as well as business cycle swings, the dominant factors can still be explained in the framework of strategic

R&D policies. Today, in the market of large commercial aircrafts, we essentially have a duopoly of two rivals, Europe's Airbus Industries and Boeing. Just in 2003 and through 2004 Airbus has been winning more orders than Boeing, and has developed a new intercontinental wide body passenger plane (the A380) that surpasses the Boeing 747 on major payload performance criteria. It appears that the reversal in the race came about because Boeing as the leader became overconfident and assured of victory (rather than 'paranoid' as A. Grove had described it for Intel). It let its lead slip in the (manufacturing) process as well as in the product technology area. For example, on the process level, Airbus used CAD/CAM (computer-aided design/computer-aided manufacturing) productivity tools earlier and more systematically than Boeing, was more customer-value focused by having standardized common cockpits in different versions of their planes which allowed airlines to cut back in training costs for their pilots. Overall, Boeing relaxed in R&D efforts over several years, taking a product development holiday for the last eight years' (Samuelson, 2003). Heavy EU subsidization, in addition, allowed Airbus to aggressively discount the price for its new A380, allowing it to slash 30–40% off its list prices (Lunsford *et al.*, 2004). Boeing has now responded with a new type of plane, the 7E7, in different versions, that brings significant savings in operating costs and preempting environmental regulations. It remains to be seen whether this would be enough for Boeing to regain technological leadership.

(c) Internet search engines

The market for Internet search engines is growing fast. In this fast-pace development, over the past 10 years, Google has become the leader, and now occupies over 32% of all world-wide searches. Rivals are on a relentless attack which has accelerated recently. Among these are Yahoo (25%), AOL Time Warner (19%) and as a late entrant, with deep pockets, Microsoft (15%), through its MSN division. Although Yahoo and AOL Time Warner in the past have relied heavily on Google's engine (Mangalindan *et al.*, 2003) Internet search engines have become the primary starting point for Internet navigation and electronic commerce including advertising. MSN has made major investments in search engine technology over the past few years, ranking fourth in 2003, is now bound to strongly move ahead and past its rivals (Guth and Delaney, 2004) – with an enhanced content coverage over Google. Options to get ahead would also involve taking out the present leader through a takeover. Other cooperative ventures will likely slow down the race but make it much more like a marathon than a sprint. Besides, the next generation Internet searches, for example, through activating 'intelligent agents', as potent mathematical algorithms, might require a new platform that relies on new technologies which might not be obtained through acquisitions, and need proprietary development as a new platform technology. For Microsoft, Internet searches, although not foreseeable as part of its core business, might well constitute an important complementary market which aids and extends its core activities. Internet searches could be virtual showrooms for software sales, downloading and e-commerce (Clark and Richmond, 2003). To its rivals, Google is gaining a strategic position that could give it much influence on the shape of Internet commerce. Its dominance of online searches means it can reach web users from the moment they start browsing, and steer them to sellers, auctions and advertisers. Searching is the key to e-commerce. With this strategic insight in mind, MSN developed its own search engine, with the goal of using search to increase MSN subscribers and advertisers. That could be a big threat to Google, given Microsoft's history of patiently eyeing complementary markets, and then using its software clout to dominate them. Some industry experts believe that despite its prevailing position Google's technological lead is precarious, vulnerable to moves into its turf by the likes of Microsoft and Yahoo. Dominant market positions in technology tend to endure only if customers face a high cost of switching products, as with Microsoft's operating system and Intel's microprocessor chips. But with web users such considerations are of lesser concern as they can switch as easily changing channels on TV. Furthermore, there could be more potential entrants, coming from the aisle of e-commerce such as Amazon and Ebay, who have been relying previously on Google's search engines but may now opt to no longer rely on those services because of strategic control in their business.

(d) Computer servers

In the computer server market Sun Microsystems carved out a strong and for years dominant position by running against standard chip designs and software in favour of their own custom designs. This made their machines more powerful but also much more expensive. However, technology racing has now caught up with Sun when standard chips made by Intel have emerged and surpassed specialized models in performance. Also dedicated software from Microsoft or free Linux software led to a new generation of server machines rivalling those made by Sun. This allows new entrants, such as PC makers Dell, to squeeze Sun in its previously dominant market. The result is that Sun's market share is dropping fast. A natural response for Sun would be to leapfrog the competition by a significantly better price-to-performance ratio. This appears to be extremely difficult since technological advance on its own (Sun's) standard will most likely come at a significantly higher price that might not improve the price-to-performance ratio. Another strategic move could be to relinquish its own proprietary design for the mainstream standard which would mean to forego all technological advantages they previously had and which led them to success. It now appears that Sun decided to pursue the latter alternative but under difficult circumstances to re-engineer the company. Compromising on a new R&D strategy would make it harder for Sun to generate enough cash through sales of cheaper machines to plow more funds in the next round of the technology race. And this may be difficult to achieve since other major rivals, such as Hewlett-Packard and IBM, are well positioned technologically as well as endowed with the necessary cash to keep the pace. In the end, Sun will end up in a different sort of race to the one it started with, and it is less than clear that it could repeat its success. If price differentials between rivals of less-pricey machines of IBM and H-P of 30–50% exist then switching costs by present customers of Sun could be overcome. Technological races that change qualitatively because of technological standardization could eventually jeopardize product and technological leadership.

This points to the gap emerging between 'network effects' brought about by standardization against radical or 'breakthrough' innovation (Chapter 6).

In the latest tally of the world's server market (May, 2003), H-P took the first place with 28% revenue share, closely followed by IBM, Sun, Dell and at some distance Fujitsu and Siemens. However, the ranking appears different depending on types of servers: low-cost, Unix-based or advanced blade servers where IBM, Sun are leading and while Dell has the strongest momentum for low-cost servers (Clark, 2003b).

(e) Computer chips

Over the years computer experts, economists and the media have been openly wondering whether Moore's Law that the number of transistors on a chip doubles every year or two (i.e. the exponential growth of computational power in computer chips) would hit a physical or engineering limit, but so far it has held up for an astonishing quarter of a century. Only recently has the Intel Corporation announced that it has solved a problem in new chip design that replaces materials to prevent 'electrical leakage' inside chips, a growing problem as more circuits get packed onto semiconductors. If true, this event would further lift Intel in the engineering race against rivals Texas Instruments, IBM and Motorola (Clark, 2003b).

Since the development of microprocessors over three decades Intel held a commanding lead in design, architecture and market success, as clearly described by its former CEO as an obsession to forge ahead: 'Only the Paranoid Survive' (Grove, 1996). Over the past years the core of technological rivalry concerns Intel and its archrival Advanced Micro Devices (AMD). Recently, AMD unveiled its Opteron 64-bit microprocessor which squarely challenges Intel's Itanium 64-bit chip. Industry observers see the Opteron processor and its desktop version, the Athlon 64, as a 'make-or-break' test for AMD, as some sort of 'bet-your-company decision'.

Intel has developed its own 64-bit Itanium which is now in its second generation but has been slow to catch on, and now Opteron 64 offers the same performance at a lower price. Besides it promises to run existing (application) software seamlessly which improves chances for AMD to break into the lucrative market of corporate users. How fast Opteron will facilitate transition from a 32-bit to a 64-bit processor in the corporate marketplace will also depend on the upstream adoption by computer makers such as IBM and Sun. Sun and Fujitsu which make computers based on a Sun chip design called Sparc want to head off Intel's Itanium 2 in high-end server systems. This chip exploits two microprocessor brains, dubbed 'cores', which for Sun is manufactured by Texas Instruments while Fujitsu manufactures its own Sparc chip (Clark, 2003a). The struggle appears to be quite brutal, and AMD's CEO claims that 'some of our customers have woken up to find "horses heads in their beds"' (Markoff and Lohr, 2003).

Intel, the undisputed leader so far, seems to express a relaxed attitude. An Intel spokesman puts it in these terms: 'There is always competition in semiconductors, although the names of the companies can change. We continue to invest aggressively in the technology and manufacturing that help us stay ahead of the competition' (Clark, 2003a).

(f) Web services

With the advent of the Internet we see the proliferation of network services, web services, for e-commerce transactions. Web services standards seek to enable machine-to-machine communication, allowing corporate and personal databases to seamlessly transport information. The race to deliver web services will shape competition in the network service business over the next several years. There have been broadly two different paths: Microsoft on one side and a number of major software competitors on the other side, including IBM, BEA Systems, Sun and Oracle. While the Microsoft approach links web services closely to its operating system, the other camp is putting the technology in a layer of software separate from the operating system, called middleware. That lets it run on a variety of operating systems. Software competition in web services is fierce but this type of rivalry by nature is different to the one outlined in previous examples of technology companies competing with each other. While common technological competition is aiming at a 'lock-in' for the company's products, in web services a lock-in-strategy is much harder to pursue because the standards allow data to be shared easily rather than locked inside proprietary software. This is made possible by the industry creation last year of Web Services Interoperability Organization initiated and carried by the leading competitors in the field: Microsoft and IBM. As an example where despite fierce competition multiple sharing and cooperative structures emerge, one may only look at a web services project of a large US insurance company Allstate. Its data centre consists of IBM mainframes, Sun machines, other PC servers running Windows and other systems. Its databases include IBM's DB2, Oracle and Microsoft's SQL. For network software, Allstate is using Microsoft.net software tools and its Windows server software. But by sharing data with its mainframe systems it is using WebSphere, IBM's web services middleware software, which allows Microsoft and IBM systems talk to each other. The software companies competing against Microsoft are offering middleware software, based on web services standards like XML and Java, a programming language created by Sun. Will this move industry competition away from the operating system and especially from Windows? In the non-Microsoft camp, IBM, BEA, Sun and others are all competing against each other as well, and in this situation it is too early to tell who will be the winners and losers in the longer run (Lohr, 2003).

(g) Operating computer systems

Over the past decade, year after year, Microsoft successfully pushed forward its Windows operating system for server computers, gained market share and became a powerful software giant dominating the industry. In this relentless drive, Microsoft was perceived to end up as an ugly monopolist crushing its rivals along the way. However, for markets that appear typical for 'dynamic competition' or Schumpeterian rivalry, those markets could suddenly come under threat and their leaders vulnerable (Schmalensee and Evans, 2001).

For example, out of almost nowhere comes a challenge by a surprising rival: Linux, the free operating system that just a few years ago Microsoft did not even recognize as a competitor. As Linux grows, it not only stands to win lucrative parts of the server market before Microsoft, but it also threatens to lessen the value of the very software that has given Microsoft its dominant position: Windows. Linux, which got its start as a piece of free software on the Internet, has quickly moved into the mainstream of corporate computing. It is cutting into the large portion of the server market held by Unix, backed by Sun, H-P and IBM among others, which presents an increasing challenge to Microsoft. According to the International Data Corporation (IDC), the market shares in this market for 2003 for rival systems were as follows: 15.9% for Linux, 13.9% for Unix against 60.4% of Windows. The threat could enter a new dimension as businesses become increasingly comfortable using so-called 'open source' software such as Linux, and supporting software, from databases to software development tools (Chapter 8). Microsoft makes direct competitors to all those products. Linux got a significant boost in dissemination and diffusion by major players such as H-P, Dell and IBM offering server computers running on Linux, by financial service providers like Merrill Lynch using Linux-based servers, and not least by the foremost search engine Google operating their website on more than 10,000 Linux-based servers. IBM, for example, has successfully installed Linux in its newest version of its mainframe z990, as a state of the art machine for 'mission critical' application for government, corporate and scientific use (Bulkeley, 2003).

Microsoft responded. In view of public sentiments toward open-source software, it initiated a program to open some of the underlying Windows code to governments and important Microsoft customers. The move was designed to shore up costumer confidence in the reliability and security of the Windows computer code while Microsoft concentrated on strengthening its new operating system, Windows Server 2003, adding new features, such as better protection from security breaches like viruses. Another threat for Microsoft is looming from the consumer electronics market. The world's leading consumer-electronics companies agreed to promote the development of software to run appliances based on Linux. Among those are the major Japanese companies with a global reach such as Matsushita, Sony, Hitachi, NEC, Sharp, Sanyo and Toshiba. Such a move could further erode Microsoft's position in a major software application market.

(h) Autoimmune diseases

In drug development the treatment of autoimmune diseases (e.g. rheumatoid arthritis or multiple sclerosis) has been at the centre of the drug industry's fast-expanding global market. In the treatment for autoimmune disease the body's defences target its own tissues. The market is supposed to grow to $14 billion by 2010 in the US alone. About a dozen companies mostly from the biotechnology industry have launched bioengineered drugs targeting autoimmune diseases. The major rivals are mostly American companies such as Abbott Laboratories, Amgen, Biogen, Genentech and Johnson&Johnson. For years doctors treated autoimmune disorders with drugs such as cortisone that suppress the entire immune system. Often, the result was severe side effects, including infections, osteoporosis or damage to internal organs. New bioengineered drugs rely on scientific work dating back from the late 1980s when some of the biochemical 'messages' were identified which the immune system uses to coordinate its attacks on infectious disease. Success in this biotech race is anything but assured. For example, Amgen came out with Enbrel against rheumatoid arthritis in 1998, its development began in the early 1990s by Immunex, later acquired by Amgen. Judging after market introduction it had a significant impact on the disease. Drugs designed to fight autoimmune diseases are hard to make and some fail to work in human tests. Amgen itself had limited success with its own earlier arthritis drug. In 2003, anticipated biotech drugs from Biogen for Crohn's disease, an inflammatory bowel condition, and Regeneron Pharmaceuticals, for rheumatoid arthritis both failed in clinical trials. Although Enbrel was an instant success, Immunex was unable to meet the demand because of manufacturing bottlenecks starting in 1999. Manufacturing delays hindered serving would-be-patients on a waiting list, who partly defected after Johnson&Johnson introduced a rival product, Remicade, which was originally aimed at Crohn's disease but was approved for rheumatoid arthritis in 1999. This manufacturing delay cost Amgen dearly, in additional investments and lost sales, and Enbrel with a 41% market share still lags Remicade with 51% against Abbott's Humira with 8% which arrived late in the race. As it sometimes turns out for 'blockbuster' therapeutics they may also be effective for related (auto-immune) diseases. So it was for Enbrel since it proved effective against psoriasis, a dermatological disease that afflicts millions of people. Here the rivals are Biogen's Amevive, and Genentech's Raptiva. However, it appears that Amgen's Enbrel may be at least as effective, which in turn induced Johnson&Johnson and Abbott to seek similar FDA permission for Remicade and Humira, respectively, on psoriasis treatment. In this situation we may now have a multi-stage and multi-product race with uncertain outcomes.

(i) Leadership in drug R&D

As a shining example of the research-driven pharmaceutical industry, Merck was the leader of a pipeline of cutting-edge therapeutics for chronic diseases (in particular heart diseases). Unlike many of its rivals it did not build its position through mergers or mega-mergers. Instead it chose to use its research culture to advance its own product line through internally financed R&D. It was the founder's, George Merck's, dictum that 'medicine is for the people, not for the profits .... The better we have remembered this, the larger they have been'. However, in 2004 it appears that among its drug pipeline of 11 medicines, 4 of those have failed in human tests, including ambitious treatments for depression and diabetes which were cancelled. Even more so, recently, their profitable painkiller Vioxx initially had to be pulled from the market because of serious side effects.

Over many years, given Merck's obsession of profiting from scientific discovery and continuous research-driven innovation in medicines against chronic diseases, Merck was the industry model for the US and the global pharmaceutical industry. However, while Merck kept a steady course by growing from within, their rivals were busy in engaging in mega and smaller mergers to acquire the needed technologies and product lines. A real threat to the industry came when drug companies were faced with generating 'successors' to previous 'blockbuster' drugs after they faced generic competition when their patents ran out. A case in point is Schering-Plough, linked in a strategic alliance with Merck, which recently cancelled the human trial of a promising cancer drug, making a 'dry pipeline even drier' (Landers and Lubin, 2003). In the race to generate new revenues from cutting-edge medicines one promising path was pursued in mega mergers. Most of the world's top pharmaceutical companies today, Pfizer, GlaxoSmithKline (GSK) and Novartis are merger products, as are many smaller ones. Merck was rather unique in carrying out its mission of finding innovative medicines, and expanding its profits through new products rather than merger-driven cost cuts. In contrast, other pharmaceutical companies have been aggressive in pursuing smaller companies with good drugs in development, such as US biotechnology firms or specialized European or Japanese companies, either in acquiring them or forming strong alliances (Gottinger, 2004). The risk of coming up short in coming forward with innovative medicines can be seen in a particular pursuit for making a breakthrough in the race for establishing a lead in the vast antidepressant market. Merck's abandoned depression drug, known generically as Aprepitant, epitomizes the way the company set itself apart from rivals. Aprepitant was designed to block the flow of a complex brain chemical called substance P that Merck science was convinced induced depression. If it had been effective it might have offered hope to millions who were not helped by Prozac or other related drugs or suffered from side effects. Merck is now in a tight spot, but it still has the potential to come ahead in leapfrogging.

(j) Medical devices and instruments

In the multibillion dollar high-technology medical devices market catheterized stents inserted into clogged arteries form a major portion of research-intensive technology racing. Stents are metal spring-like devices used to prop open arteries. While until a few years ago, only bare metal stents were developed and surgically implanted, a change came in 2004 when Johnson&Johnson first introduced a drug-coated Cypher stent in the US, while Boston Scientific's highly promising Taxus stent already reaped sales overseas, and is scheduled to be sold in the US market after regulatory (FDA) approval. Coated stents with anti-clogging chemical entities have been proven to obviate or reduce the plaque-building process that goes with the progression of arteriosclerotic disease. The world-wide market for stents alone is estimated to be around $5 billion in 2005. In fact, the race to gain a dominating market share, between Johnson&Johnson (J&J) and Boston Scientific has become so intense that J&J teamed up but failed to acquire (with) a rival, the Guidant Corporation, to provide its stent with a superior cathetherized delivery system, supplied by Guidant, in an alliance to defeat Boston Scientific (Johannes, 2004).

(k) Patent races

Patent racing is a special segment of overall technological races that deserves special attention.

In the standard dynamic models of a patent race such as Dasgupta and Stiglitz (1980) and Reinganum (1981), the race ends with a victory after a single success. Further, Fudenberg *et al.* (1983) show that a multi-stage R&D process allows leapfrogging. Reinganum (1985) studies a market with a sequence of innovations in which a leader enjoys temporary monopoly power by holding effective patents. In her paper, firms' R&D efforts influence the time of discovery. The literature on patent policy design often considers the cumulative nature of innovation. Green and Scotchmer (1995) studied the roles of patent length and breadth when innovation occurs in two stages. Scotchmer and Green (1990) and O'Donaghue (1998) investigated the patentability (or novelty requirement). According to their models, a strong patentability requirement can stimulate R&D spending.

The model presented is closely related to that in Harris and Vickers (1987). In their model, each player is striving to reach a given finishing line before his rival. Every state is two-dimensional, one dimension for each player's current distance from the finishing line. They focus on how the efforts of competitors in a race vary with the intensity of rivalry. In contrast to their work, the finishing line here is not fixed since the prior art can be changed by pre-announcements. A main result in their paper as well as in Grossman and Shapiro (1987) is that the leader invests more than the follower, even when the cost of the effort is quadratic.

(1) Payoffs and states

Two identical firms, a and b, are engaged in a patent race. The race is a multi-stage race, an innovation is composed of discrete steps. Time is continuous and the horizon is infinite. Innovation occurs according to a Poisson process, the date at which a new step will be discovered and depends on both firms' instantaneous probabilities of success (the hazard rates). The hazard rate λ^i for each firm $i=a,b$ increases with its research effort. Firm *i*'s probability of discovering the next new step is $\lambda^i/(\lambda^i+\lambda^j)$, where $i \neq j = a, b$. Research has a flow cost $c(\lambda^i)$.

An innovation is evaluated against the prior state of the art which serves as a proxy of the current knowledge in the market regarding the innovation. The prior art is normalized to zero at the beginning of the race. To obtain a patent, a firm needs to be the first to accomplish a fixed number of innovation steps, *n*, above the state of the prior art. Firms know how many steps have been achieved by each firm and the state of the prior art. However, they cannot observe the content of their rival's research, unless the firm chooses to disclose that information. For example, a firm may choose to publish research results. Publication changes the state of the prior art. For simplicity, we assume (unless we note otherwise) that publication is free. Each successful step that was published increases the prior state of the art by one step. A simultaneous publication of one step by both firms has the same effect as a publication by one of the firms, since it is assumed to add the same information to the state of the prior art.

The states of the world are described by a pair of state variables that measure the number of (unpublished) innovation steps of each firm above the prior state of the art.

The state $s = (s^a, s^b)$ denotes the situation in which firm a had s^a innovation steps above the state of the prior art and firm b had s^b innovation steps above the prior art. Let the state space be

$$S = \{(s^a, s^b): s^i = 0, 1, 2, \ldots, n\}.$$

The number of successful innovation steps for each s^i increases by one whenever firm i experiences a success. If firm i indicates the results of k innovation steps, the state of the prior art increases by k steps. Consequently, the number of steps above the prior art for each firm decreases by k successes (or is equal to zero if a firm had less than k successes).

The value of a patent is v for the winning firm and zero for the loser. The reward at the terminal states is

$$r^i(s^a, s^b) = \begin{cases} v \text{ if } s^j < s^i = n \\ v/2 \text{ if } s^j = s^i = n. \\ 0 \text{ if } s^i < s^j = n \end{cases}$$

The tie-breaking rule can be interpreted as each firm obtaining the patent with probability ½.

The flow cost of research is $-c(\lambda)\mathrm{d}t$. The payoff of the game at state s is simply the discounted expected value of rewards for firm i.

1.5 Some caveats and constraints

The cases evolving from a cross-section of industries provide more insights about the kind, scale, dynamics and intensity of industrial racing. In a nutshell, most races appear to be multi-dimensional, multi-stage, time-critical, asymmetric, oligopolistic, diversified and unsteady with some hyperactive periods changing into more resilient modes – depending on the course of business cycles (Thurm, 2003), and overall with highly uncertain outcomes. They resemble more 'marathons' than 'sprints' in analogy to sports contests. Above all, some of the races have industry-specific characteristics. One prototype is the head-to-head rivalry in product performance and quality which looks like a 'tit-for-tat' interaction of the frontrunners (frontier race) with a relentless ('paranoid') drive to become and remain the sole and unique industry leader. (This is the Intel story as in Grove, 1996.) A timely example of a multi-dimensional race involves Sony which fends off competitors across the technology spectrum, that is, in flat-panel TV, in digital music players, in digital cameras and in game consoles (Playstation), but above all is engaged in a standard battle involving a new DVD standard (Blu-ray), the outcome of which may impact its success potential in other product categories (Economist, 2005).

Of particular interest is when knowledge accumulation of the leading company, the 'pure knowledge effect', gives it a competitive edge in R&D over the follower, thus leading to a higher probability of success in the continuing pursuit of the R&D race (Chapter 4).

At the other end of the spectrum is the mergers and acquisitions (M&A) game. Companies strive to get ahead by gobbling up other companies (with a promising research pipeline and their specific product portfolio) through which they could gain a competitive edge over their rivals or through better strategic positioning in conducting more critical R&D for larger markets or portfolio R&D. (This case relates more to pharmaceuticals, medical devices in acquiring strategic assets in biotechnology as new platform technologies.) In between, there are many mixed or overlapping forms, including the one where the race does not take place in product markets *per se*, but in 'innovation markets', that is, confined to R&D such as patent racing to build up 'patent fences' or 'cluster patents', i.e. with 'intellectual property rights' at stake but no manufacturing. Here incentives result from possible licensing fees and standard setting. The ultimate goal in innovation markets is Schumpeterian, to create new markets and for the innovator to obtain temporary monopoly positions.

While in some industries product characteristics are easily measurable in terms of welfare changes, in others they are more elusive to measurement (such as over-the-counter (OTC) drugs). Then the race might move upstream toward marketing wars or branding which is a much slower building-up effort. (Products can be leapfrogged but brands cannot!) Another dimension of racing concerns the cognitive framework or behavioural limits under which racing could emerge. Tirole (1988) established in the context of market entry games the four major firm 'incumbent' types:

- The Top Dog: big or strong to look tough or aggressive.
- The Fat Cat: big or strong to look soft or inoffensive.
- The Puppy Dog: small or weak to look soft or inoffensive.
- The Lean and Hungry Look: small or weak to look tough or aggressive.

In each of those categories asymmetries between incumbents and entrants lead to different racing intensity, duration and market outcomes, which may be even more difficult to predict as in dynamic, repeated and stochastic games the cognitive types will switch in the course. Other primarily exogeneous factors could restrict the vitality and dynamics of the race. One is given by anti-trust considerations where the likelihood of a racing dominance (through an 'essential facility' or 'killer patent'; Evans, 2004) could be considered a threat to other market participants because of the potential of monopoly creation. This is particularly true if the view of conventional static antitrust law and economics is taken (Schmalensee and Evans, 2001) in which the innovation aspect of Schumpeterian or competition dynamics of high-technology industries is not or is not sufficiently appreciated. Also, public policy events that cast doubt on the franchise or property value of successfully operating firms in dynamic industries would have severe impacts on the racing dynamics. For example, when the antitrust division of the US Justice Department initially announced its Microsoft suit and suggested remedies that many felt threatened its business model. Or if politicians declare that patent protection of research-intensive therapeutics should not apply to developing countries, or that human genome research should be made freely accessible to all mankind no matter what cost it took to complete the research. What would this mean to a value-generating race in the biotech industry? Public policy should recognize that competitive threats to the high-tech industry are numerous and by themselves sharply limit the scope for anti-competitive conduct. This is particularly true in a highly globalized market. A famous historical example from the late 1960s and early 1970s was the antitrust position that General Motors could not exceed a 60% market share of US auto production. This position was quickly made ridiculous by the invasion of Japanese and European cars. Similarly, the once-dominant position of IBM and Xerox quickly eroded. The reality is that the extreme amount of competition in the global market is the best check to anti-competitive business practices.

Another constraint is given through standard segmentation in product markets. For example, in the market for mobile cellphones there have been two major standards rivalling each other: one is the GSM standard that dominates Europe and part of the Middle East, and the CDMA standard that prevails in East Asia and the US. While Nokia of Finland clearly has a commanding lead on their products matching the GSM standard, they are also well positioned on the CDMA standard although much more threatened by the likes of Motorola, Samsung of Korea, NTT DoCoMo of Japan and emerging Chinese manufacturers such as Huawei (Bryan-Low, 2005). To the extent that the network of CDMA grows much faster than GSM, and possibly becoming dominant in the next product generation, the second race will have a significant greater impact on leadership positions, and a dominant position in the first one is no comfortable cushion in determining future outcomes.

Industrial racing in high-technology industries could result in 'spinoff' races in newly created or complementary markets where a firm avoids a race in the established market to establish a new market in which it positions itself as a 'temporary' monopolist. With increasing segmentation in technological product markets, races in crowded product markets may tend to diversify into new markets.

Government regulation has a major impact on conduct, scale and scope of technological racing. Two prominent examples include guidance through industrial policy and antitrust policies to equalize the level playing field.

The first case involves a Japanese case of technological racing in the computer industry under 'public guidance', that is the environment within which firms competed was controlled to a substantial degree by the Ministry of International Trade and Industry (MITI), now METI. There were several ways in which MITI controlled the competitive environment. First, only a handful of companies were allowed to actively participate in the computer industry. Japan's modern computer industry originated with MITI's orchestration of a 1961 agreement whereby IBM provided limited access to several technologies to 15 select Japanese firms in exchange for a limited ability to manufacture computers in Japan. Of these, only seven firms started out active in the industry: Fujitsu, Hitachi, Nippon Electric Company (NEC), Toshiba, Mitsubishi Electric, Oki Electric and Matsushita. The rationale for restricting the number of participating firms was that economies of scale dictated that a firm needed approximately 15% (domestic) market share in order to produce efficiently. MITI went further by grouping these participating firms into three separate categories. Thus, Fujitsu and Hitachi were to build the high-end mainframes, NEC and Toshiba the lower-end mainframes, while Mitsubishi and Oki were relegated to peripheral manufacture (with Matsushita dropping out of the industry). Later on, government intervention in R&D projects favoured Fujitsu, Hitachi and NEC's participation in the high-end mainframe business, with the other three manufacturing smaller, specialized machines and peripherals.

In addition, MITI severely restricted the participation of foreign firms in the Japanese market. The Foreign Capital Law required that any foreign firm participating in the Japanese market form a joint venture with a Japanese company. IBM-Japan, by virtue of already being operational before the law went into effect, was the only exception to this rule. Despite the signing of the 1961 agreement which allowed IBM-Japan to begin production in Japan, it was prevented from doing so for two years. Similarly, in 1964, it was refused permission to manufacture its computers in Japan on the grounds that Japanese companies would not be able to compete with these advanced machines. Such interference was legion. There were restrictions on how much technology could be transferred to IBM-Japan from its parent corporation and what fraction of earnings could be repatriated to the parent corporation.

Despite the technological laggardness of Japanese computers at that time, buyers were exhorted to 'Buy Japanese'. MITI controlled the ability to authorize a company to import foreign computers. Further, as late as 1982, 91% of all computers used by the government were domestic. Since government users were the ones who tended to use the large System 360 machines in which the foreign firms were especially competitive, this favouring of domestic manufacturers is all the more dramatic. Within this controlled environment, MITI used two primary techniques to foster competition among the favoured firms. First, participation in the projects funded by MITI was not guaranteed and was at least partially contingent on successful commercialization of the previous project's results. Thus, with the Japanese market insulated from foreign competition, a small number of firms raced for market share. Second, MITI used the threat of an unrestricted flood of IBM and Univac computers to make domestic firms aware of the technological level they had to achieve to compete internationally. This points out a feature that distinguishes the Japanese race from the US frontier race. Not only were the Japanese firms racing against each other, but they were all collectively striving to catch up with IBM and the other frontier firms, the industry leaders. Examples of this phenomenon are provided by the MITI-funded Super High-Performance Computer Project that tried to get the Japanese companies to catch up with the technology in IBM's 360 series of computers and by the New Series Project that organized a response to IBM's series of computers. A similar attempt was made but failed, with MITI's Fifth Generation Computer Project in the 1980s.

A severe constraint on moving ahead in technological racing is imposed by antitrust policies. Although antitrust policies aim at 'leveling the playing field' in competitive conduct, it might not be conducive to competitive rivalry for Schumpeterian industries engaging in dynamic competition. In dynamic competition products and services may not compete on a 'stand-alone basis' but rather in a connected, networked manner that will result in tying or bundling of products. Further, fierce competition in dynamic industries results from an asymmetric distribution of power (dominance) among firms instigating a strong innovation drive to move ahead rather than from an artificially symmetric power distribution imposed by equalizing antitrust policies.

A third class of constraints is facilitated through standardization. One recalls the battle of Betamax and VHS in the consumer electronics market. Rival standards would limit technological racing as long as no clear winner appears in sight.

Unlike the days of the stand-alone VCR, the next format of DVD players would be a networked product likely to affect the sales of many complementary products (McBride and Dvorak, 2004). If different industrial camps and governments cannot agree on a common standard, technological racing may not result in a significantly better product since the market may be locked into an inferior standard. Another consideration by each group is to have their standard established for which they contribute the most inputs, not easily imitated by rival groups. In that sense they would most likely gain a competitive edge.

Given those various aspects, dimensions, behavioural and time constraints, the proper methodological framework to handle racing in complex environments is through a stochastic decision and game-theoretic analysis.

References

Belsen, K. (2003) 'Sony again turns to design to lift electronics', *New York Times*, 2 February.

Bryan-Low, C. (2005) 'France's Alcatel feels pressure from the east', *The Wall Street Journal*, 10 February.

Bulkeley, W. M. (2003) 'IBM's mainframes represent bright spot in gloomy tech', *Wall Street Journal*, 13 May.

Clark, D. (2003a) 'Intel rivals will present new chips', *Wall Street Journal*, 13 October.

Clark, D. (2003b) 'Intel to report a breakthrough in chip design', *Wall Street Journal*, 5 November.

Clark, D. and Richmond, R. (2003) 'Microsoft targets Internet search, taking on Google', *Wall Street Journal*, 3 November.

Dasgupta, P. and Stiglizt, J. E. (1980) 'Uncertainty, industrial structure and the speed of R&D', *Bell Journal of Economics*, **11**, 1–28.

Evans, D. (2004) 'What's yours is mine', *Wall Street Journal*, 2 February.

The Economist (2005) 'Behind the Smiles at Sony, Special Report Sony', 12 March, 69–71.

Fudenberg, D., Gilbert, R., Stiglitz, J. and Tirole, J. (1983) 'Preemption, leapfrogging and competition in patent races', *European Economic Review*, **22**, 3–31.

Gottinger, H. W. (2003) *Economies of Network Industries*, London: Routledge.

Gottinger, H. W. (2004) 'Biotech and Pharma industries: application of network economies', *Journal of Policy Studies*, **17**, 27–34.

Gottinger, H. W. (2006) 'Strategic alliances in the global Biotech-Pharma industries', Mimeo, International Institute for Technology Management and Economics, Germany.

Green, J. R. and Scotchmer, S. (1995) 'On the division of profit in sequential innovation', *Rand Journal of Economics*, **26**, 20–33.

Grossman, G. and Shapiro, C. (1987) 'Dynamic R&D competition', *Economic Journal*, **97**, 372–387.

Grove, A. (1996) *Only the Paranoid Survive*, New York: Doubleday.

Guth, R. A. and Delaney, K. J. (2004) 'Microsoft, late to search party, seeks to capture Google's turf', *Wall Street Journal*, 11 November.

Harris, C. and Vickers, J. (1985) 'Perfect equilibrium in a model of a race', *Review of Economic Studies*, **52**, 193–209.

Harris, C. and Vickers, J. (1987) 'Racing with uncertainty', *Review of Economic Studies*, **54**, 1–21.

IDC (2000) International Data Corporation, PDA Support, New York.

Ishibashi, K. (2003) 'Japan scores big in electronics', *Wall Street Journal*, 11 December.

Johannes, L. (2004) 'Boston Scientific cites stent gains over rival J&J', *Wall Street Journal*, 3 February.

Kidder, T. (1981) *The Soul of a New Machine*, New York: Avon Books.

Landers, P. and Lublin, J. S. (2003) 'Merck's big bet on research by its scientists comes up short', *Wall Street Journal*, 28 November.

Lee, T. and Wilde, L. (1980) 'Market structure and innovation: a reformulation', *Quarterly Journal of Economics*, **94**, 429–438.

Lippman, S. and McCardle, K. (1987) 'Dropout behavior in R&D races with learning', *The Rand Journal of Economics*, **18**, 287–298.

Lippman, S. and McCardle, K. (1988) 'Preemption in R&D races', *European Economic Review*, **32**, 1661–1669.

Loury, G. (1979) 'Market structure and innovation', *Quarterly Journal of Economics*, **93**, 395–410.

Lohr, St. (2003a) 'Competitors shape strategy to gain edge in web services', *The New York Times*, 3 February.

Lohr, St. (2003b) 'In software, still testy after all these years', *The New York Times*, 9 December.

Lundstrom, D. (1988) *A Few Good Men from Univac*, Cambridge, MA: MIT Press.

Lunsford, J. L. (2003) 'For Pratt, a cutting-edge plane offers a shot at a big comeback', *Wall Street Journal*, 12 December.

Lunsford, J. L. (2004) 'New friction puts Airbus, Boeing on course for fresh trade battle', *Wall Street Journal*, 1 June.

Mangalindan (2003) 'Rising clout of Google prompts rush by internal rivals to adapt', *Wall Street Journal*, 16 July.

Markoff, J. and Lohr, St. (2003) 'Advanced micro issues challenge to rivals', *The New York Times*, 19 April.

McBride, S. and Dvorak, P. (2004) 'Toshiba says four studios back its HD-DVD format', *Wall Street Journal*, 24 November.

McKelvey, M. (1996) *Evolutionary Innovations: The Business of Biotech*, New York: Oxford University Press.

O'Donaghue, T. (1998) 'Patent protection when innovation is cumulative'. Mimeo, Department of Economics, University of California, Berkely.

Reinganum, J. (1982) 'A dynamic game of R&D; patent protection and competitive behavior', *Econometrica*, **50**, 671–688.

Reinganum, J. F. (1981) 'Dynamic games of innovation', *Journal of Economic Theory*, **25**, 21–41.

Reinganum, J. F. (1985) 'Innovation and industry evolution', *Quarterly journal of Economics,* **100**, 81–99.

Roberts, J. (2004) *The Modern Firm*, Oxford: Oxford University Press.

Samuelson, R. J. (2003) 'How Airbus eclipsed Boeing', *Wall Street Journal*, 26 June.

Schmalensee, R. and Evans, D. S. (2001) 'Some economic aspects of antitrust analysis in dynamically competitive industries', National Bureau of Economic Research, Cambridge, MA, WP 8268 (http://www.nber.org/papers/w8268).

Scotchmer, S. and Green, J. R. (1990) 'Novelty and disclosure in patent law', *Rand Journal of Economics*, **21**, 131–146.

Simon, H. (1962) *The Sciences of the Artificial: The Architecture of Complexity*, Cambridge, MA: MIT Press.

Tirole, J. (1988) *The Theory of Industrial Organization*, Cambridge, MA: MIT Press.

Thurm, S. (2003) 'After the boom, Cisco is learning to go slow', *Wall Street Journal*, 7 May.

2 Modelling stochastic innovation races

The process of innovative competition is like a race in which it is necessary to run and run well to endure and to prosper, but the race is more like a marathon than a sprint.

Fisher *et al.*, *Folded, Spindled and Mutilated: Economic Analysis and* US vs IBM (1983)

2.1 Introduction

Firms' strategic decisions may sometimes change from stage to stage. At any given stage the next stage may be known to be tentative in nature even when the firm reveals its strategy in private or in public. We may say then that the firm is actually headed towards a stochastic destination. To simplify the analysis, we first follow a decision-theoretic (and later a game-theoretic) line of reasoning, in the sense that we look at the individual firm's options against any or all of their rivals. It relates to the exploration of Kamien and Schwartz (1982), where they showed that the intensity of rivalry between market participants leads to an increased speed of R&D which is the main characteristic of a frontier race. To motivate the problem we first give a characterization of racing behaviour. Since the firm faces extensive uncertainty in its strategic positioning, we assume the decision period is stochastic. However, to formulate the problem we let the decision period be deterministic (in finite time) but the destination be stochastic. At a further step we also assume the decision period to be stochastic. If the decision period is stochastic, the problem of identifying the optimal trajectory in terms of strategic positioning, i.e. in the plane-based innovation race, is a dynamic optimization problem which can be solved by dynamic programming, or other calculus of variations or numerical methods. Along this line we deal with special issues of racing behaviour among firms. By complementation in Section 2.7 we draw the dynamic decision problem of innovation racing in suitable game-theoretic terms.

In Section 2.2 we identify a formal architecture of innovation races. Section 2.3 displays from a decision-theoretic view a special class of stochastic racing for a given fixed decision period, which is extended to the case when the decision period is also stochastic in Section 2.4. Section 2.5 suggests that through strategic interactions two major forms of races emanate, which are further extended as multi-stage races in Section 2.7. In Section 2.8 we draw the simple structure of innovation games that serve as paradigms for more specialized classes of games in later chapters. Conclusions follow in Section 2.9.

2.2 Characteristics of innovation races

The concept of a race is intrinsic to sports events, crossing the finishing line first is everything to a racer, and the rewards may be immense by reaching for the gold. In general, if such a race evolves, the race looks like a collection of separately running sequential machines (finite automata) each acting under resource and time constraints, until a winning outcome clearly emerges. A winner may establish himself at the first trials or runs, leaving very little chance for those left behind to catch up. The situation of competitive rivalry among firms or businesses in high-technology industries may resemble more complex paradigms of a race that appear more difficult to describe than a sports event. First of all, the finishing line may not be sharply defined. It could be a greater market share than any of the rivals attain, it may be a higher profitability given the share, or a higher growth potential. In terms of process, it could be even a slow race at the beginning which might accelerate to whatever the finishing line constitutes. It may be a race that is open to new entrants along the way, in a dormant, low innovation-driven industry that brings changes to this industry. It may allow 'disruptive' moves among rivals, unheard of in conventional races, such as 'leapfrogging', 'taking a breath and recharge' or redefining a new race through mergers and acquisitions, in the course of the given one. Races may be endogeneous, induced by changes in innovation patterns, market structures and productivity cycles. All these issues of complexity may justify to set up a racing model that captures many of the essential features. This would be a model of a stochastic race which is proposed in the sequel. Let us describe the characteristics of such a race on achieving technological and market supremacy (Gottinger, 1996). A finishing line would be ahead of a present technological frontier which would be the common ground for starting the race.

Let TF(C) be each racing company's technological knowledge frontier while TF(I) would be the respective industry's frontier represented by the most advanced company as a benchmark. All firms engage in pushing their own frontier forward which determines the extent to which movements in the individual TF(C) of the racing firms translate into movements of the TF(I). While a variety of situations may emerge, the extremal cases involve: either one firm may push the frontier at all times, with the others following closely behind, or all firms share more or less equally in the task of advancing the TF(I). The first situation corresponds to the existence of a unique technological leader for a particular race, and a number of quick followers. The other situation corresponds to the existence of multiple technological leaders. In some industries, firms share the task for pushing the frontier forward more equally than in other industries. This is usually the case the more highly paced and dynamic is the race in an industry. In any race of this sort 'closeness' is an important but relative attribute. Some races, such as frontier races, are all close by construction; however, some might be closer than others. As a closeness measure of the race at any particular time one could define $c(t) = \sum_0^N [\mathrm{TF}(\mathrm{C}_i) - \mathrm{TF}(\mathrm{I})]^2 / N(t)$ where $N(t)$ is the number of active firms in that industry at time t. The measure thus constructed has a lowest value of 0, which corresponds to a 'perfectly close' race. Higher values of the measure correspond to races that are less close. Unlike other characteristics, such as the domination period length during a race, innovation when ahead vs when behind, leapfrogging vs frontier sticking, which describe the behaviour of a particular feature of the race and of a particular firm in relation to the frontier, the closeness measure is more of an aggregate statistic of how close the various racing parties are at a point in time. The closeness measure is simply an indication of the distance to approach a benchmark, and it does not say anything about the evolution of the technological frontier. To see this, note that if none of the frontiers were evolving, the closeness measure would be 0, as it would if all the frontiers were advancing in perfect lock-step with one another.

On the basis of previous theoretical works on these issues, e.g. Fudenberg *et al.* (1983), Harris and Vickers (1985, 1987), Reinganum (1989), etc., there have been attempts to categorize similarities and differences among various races due to a range of behaviour rules in races although very little empirical work has been done to substantiate these claims (Lerner, 1997). The very robust feature that appears to be common to all races is that there is a pronounced tendency for a firm to innovate more when it falls behind in the race. In less dynamic industries the race seems most prone to catchup behaviour rather than frontier-pushing behaviour. Further, even the catch-up behaviour evidenced by firms in this race is less aggressive in that it seldom tries to leapfrog the frontier. Rather, the firms tend to exhibit less frontier-sticking behaviour than the firms in high-technology, fast-paced industries. Overall, these facts seem to suggest that the incremental returns to a firm that occupies the race leader position seem to be lower in the first than in the second category. The first category also tends to be occupied by firms with the most unequal frontier-pushing activity.

By taking the (large mainframe) computer industry as an example, an interesting point to note is that each frontier advance is embodied in a machine that is frequently the product of years of planning. More often, the performance target of a computer mainframe is set when the project begins, although there is some uncertainty in the time taken to achieve this target performance. This, in conjunction with the racing behaviour, implies that firms must constantly anticipate their rivals' actions when deciding on their technology strategy (Fisher *et al.*, 1983). This is marvellously described in the novel by Tracy Kidder (1981). Empirical studies confirm that such anticipation does, in fact, crucially impact the targeting decision. This validates the emphasis of strategic interaction placed by the 'racing behaviour' perspective.

2.3 Stochastic race in a deterministic decision period

The problem: on a Euclidean plane let N be a set of n points (x_i,y_i); $i=1,\ldots n$; let n probabilities p_i; $i=1,\ldots,\ n$ be given such that $\sum p_i=1$. We use the Euclidean distance on a plane because innovation characteristics are at least two-dimensional, that is, it would apply to so-called system products that consist of at least two components. The probabilities will most likely be subjective probabilities determined by the individual firm's chances to position itself, endogeneously determined by its distance to the finishing line or its proximity to the next rival in the race. They may be formed by considering the firm's own position in the race as well as depending on the stochasticity of the rivals' efforts. As a first approximation we may let the firm's R&D investment x_i, in relation to the total investment of its rivals $\sum x_j$, determine the probability $p_i=x_i/\Sigma x_j$. Let a starting point, point $(x_0,\ y_0)$ or (point 0) also be given; let $f(S)$: $S\geq 0$ be a function such that:

$$f(0)=0, \tag{2.1}$$

$$f(S)>0;\ \forall\ S>0, \tag{2.2}$$

$$f(S+\varepsilon)\geq f(S);\ \forall\ S,\varepsilon>0, \tag{2.3}$$

and such that except for $S=0$, $f(S)$ is (not necessarily strictly) convex and represents the cost of racing at speed S; let $F>0$ be given (the fixed time value); and finally let $T>0$ be given (the decision period). It is required to minimize the following function by choosing $t\equiv(x_t,y_t)$ and S (i.e. choose a point t, to be at T time units from now, and a speed S with which to proceed afterwards, so that the expected total cost to cross the 'finishing line' will be minimized:

$$Z(t,S)=\mathrm{FT}+f(d(0,t)/T)d(0,t)+(f(S)+F/S)\sum_{i=1}^{n}p_i d(t,i) \tag{2.4}$$

where $d(i,j)$ is the Euclidean distance between points i and j. We denote the optimal S by S^*, and similarly we have t^* and $Z^*=Z(t^*,S^*)$. Note that FT is a constant, so we can actually neglect it; the second term is the cost of getting to t during T time units, i.e. at a speed of $d(0,t)/T$. Now, clearly the problem of finding S^* can be solved separately, and indeed we start by solving it.

The speed race problem

If we look at the list of stipulations for $f(S)$, equation (2.1) just means that we can stop and wait at zero marginal cost (which we first keep as a strict assumption to justify the flexibility of the race). Equation (2.2) indicates some moving forward or standing still, and (2.3) assumes some minimal speed, since if f is not monotone for $S > 0$, then it has a global minimum for that region at some S, say S_{min}, where the function assumes the value $f_{min} \leq f(S);\ \forall\ S > 0$. Now suppose we wish to move at a speed of λS_{min}; $\lambda \varepsilon$ (0,1], during T time units, thus covering a distance of λTS_{min}; then who is to prevent us from waiting $(1 - \lambda)T$ time units, and then going at S_{min} during the remaining λT time units, at a variable cost of f_{min} per distance unit? As for the convexity requirement, which we actually need from S_{min} and up only, this is not a restriction at all! Not only do all the firms we mentioned behave this way in practice generally, but even if they did not, we could use the convex support function of f as our 'real' f, by a policy, similar to that discussed above, of moving part-time at a low speed and part-time at a higher one at a cost which is a linear convex combination of the respective f's. Figure 2.1 'sums' our treatment of an ill-behaved function (in dotted lines where almost never used). We will also assume that f is continuously

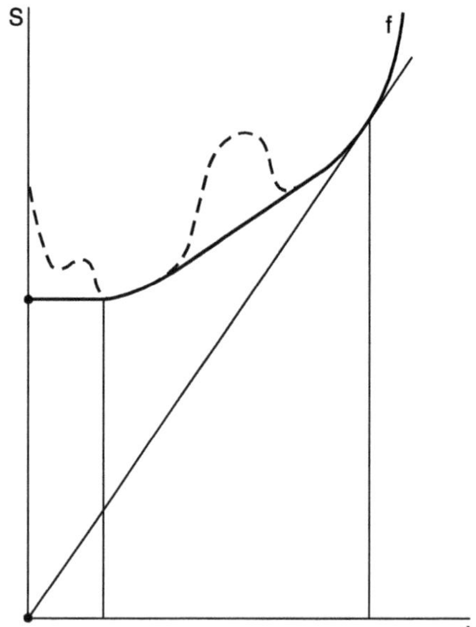

Figure 2.1 A well-behaved (bold) and an ill-behaved (dotted) speed function.

differentiable – see Zang (1981) as to why this would not be restrictive in practice.

Lemma 2.1: *Let $\bar{f}$ (S); S > 0 be any positive cost function associated with moving at speed* S *continuously and let (2.1) hold, then by allowing mixed speed strategies, we can obtain a function f* (S)*; S > 0 such that f is positive, monotone non-decreasing and convex, and reflects the real variable costs.*

Now, since each time unit cost is F, and we can go S distance units during it, each distance unit's 'fair share' is F/S. To this add $f(S)$, to obtain the cost of a distance unit at a speed of S when the firm knows where it is going, and requires their fixed costs to be covered. (On the other hand, not knowing what it wants to do means that the firm has to lose the F money, or part of it.) Denote the total cost as above by $TC(S)$, or

$$TC(S) = f(S) + F/S. \tag{2.5}$$

But, F/S is strictly convex in S, and $f(S)$ is convex too, so $TC(S)$ is strictly convex. Further, $\lim_{\varepsilon \to 0^+} TC(\varepsilon) = \infty$, so $TC(S)$ has a unique minimum, S^*. Since we practically assume differentiability, then

$$S^* = \arg\{f'(S) = F/S^2\} \tag{2.6}$$

and we can obtain it numerically. (S^* is also depicted in Figure 2.1, where a ray from the origin supports f.)

Choosing t *optimally*

Our problem is to find the point t, or the 'decision point', where we elect to be at the end of the decision period. Then, we will know with certainty what we have to do, so we will proceed at S^* to whichever point i chosen, at a cost of $TC(S^*)d(t,i)$. Denoting $TC(S^*) = TC^*$, we may rewrite (2.4) as follows:

$$Z(t) = \mathrm{FT} + f(d(0,t)/T)d(0,t) + TC^* \sum_{i=1}^{n} p_i d(t,i). \tag{2.7}$$

Theorem 2.1: *Z(t) is strictly convex in* t.

Proof. Clearly FT is a constant so it is convex. Let $h(w) = f(w/T)w$, hence our second term, $f(d(0,t)/T)d(0,t)$ is $h(d(0,t))$. By differentiation we can show that $h(w)$ is strictly convex, monotone increasing and non-negative. $d(0,t)$ is convex (being a norm), and it follows that $h(d(0,t))$ is strictly convex as well (see Theorem 5.1 in Rockafellar (1970), for instance). As for the third term, it is clearly convex (since $\{p_i\}_{i=1,\ldots,n}$ are non-negative probabilities), and our result follows for the sum.

It follows that a unique minimum Z^* exists for Z, within the convex hull of the $n+1$ points $0,1,\ldots,n$. In order to find this minimum we look for a point such that the gradient ∇Z is zero. We now examine the two components of ∇Z, by x_t and y_t.

$$\frac{\delta Z}{\delta x_t} = \frac{x_t - x_0}{d(0,t)}(f(S) + Sf'(S)) + TC^* \sum_{i=1}^{n} p_i \frac{x_t - x_i}{d(t,i)}, \tag{2.8}$$

$$\frac{\delta Z}{\delta y_t} = \frac{y_t - y_0}{d(0,t)}(f(S) + Sf'(S)) + TC^* \sum_{i=1}^{n} p_i \frac{y_t - y_i}{d(t,i)} \tag{2.9}$$

where

$$S = d(0,t)/T. \tag{2.10}$$

The 'length' of the gradient L_t is

$$L_t = [(\delta Z/dx_t)^2 + (\delta Z/\delta y_t)^2]^{1/2}. \tag{2.11}$$

We can gain some more insight into the problem if we consider two limiting cases: (i) $T \to \infty$; and (ii) $T \to 0$.

(i) The T $\to \infty$ *Case*

Here we assume $S_{\min} > 0$ (and not 0^+). Under this assumption, at a cost of $f_{\min}$ per distance unit, the firm can arrive anywhere during the decision time. Hence we have $p_0 = f_{\min}/TC^*$, and our problem is solved. As usual, denote the solution point by t^*, and clearly for T large enough we are not going to move during the whole decision period, but rather only during T^* time units of it, where

$$T^* = d(0,t^*)/S_{\min}. \tag{2.12}$$

Hence the same solution is obtained for any $T > T^*$.

It may be advisable to try solving under the assumption that $T > T^*$, and then check the assumption. This way, even if $f'(s)$ jumps at $S_{\min}$, we will not have any problems with it. If $T > T^*$, we are through, and in any case we know that $S > S_{\min}$.

(ii) The T $\to 0^+$ *Case*

Recall (2.5), with S^* as per (2.6) we have $TC^* = f(S^*) + F/S^*$, and from (2.6) we easily obtain

$$F/S^* = S^* f'(S^*). \tag{2.13}$$

Now substitute (2.12) to TC^*, and we have

$$TC^* = f(S^*) + S^*f'(S^*). \tag{2.14}$$

Let $W(S)$, as defined below

$$W(S) \equiv f(S) + Sf'(S), \tag{2.15}$$

be the relative weight of the starting point 0 in (2.8) and (2.9). We observe that $W(S)$ is a monotone increasing function (since f, f' and $f'' > 0$), and that $W(S^*) = TC^*$. But at t^* (2.8) and (2.9) are zero, hence

$$\frac{x_0 - x_t^*}{d(0, t^*)} W(S) = TC^* \sum_{i=1}^{n} p_i \frac{x_t^* - x_i}{d(t^*, i)} \tag{2.16}$$

$$\frac{y_0 - y_t^*}{d(0, t^*)} W(S) = TC^* \sum_{i=1}^{n} p_i \frac{y_t^* - y_i}{d(t^*, i)}. \tag{2.17}$$

Squaring (2.16) and (2.17), adding them and taking the square root again, we obtain

$$W(S) = TC^* \left[\left(\sum_{i=1}^{n} p_i \frac{x_t^* - x_i}{d(t^*, i)} \right)^2 + \left(\sum_{i=1}^{n} p_i \frac{y_t^* - y_i}{d(t^*, i)} \right) \right]^{1/2}. \tag{2.18}$$

Clearly $W(S) \leq TC^*$ (the magnitude of a vector sum is less than the sum of the magnitudes), with equality only in the special case where all the points, including the starting point, are colinear, and both 0 and t are to the same side of all the rest (in which case the firm can behave as if it knows where it is going, since it has to reach the first point at least, and it knows the decision will be made by the time it gets there). But $W(S)$ is monotone, hence if $W(S) < TC^*$ then $S < S^*$, and

$$d(0, t^*) \leq TS^*. \tag{2.19}$$

Following (2.18) we define $G(t)$ for any $t \varepsilon \{E^2 - N\}$ (i.e. any point on the plane and 0, but not $i \varepsilon N$)

$$G(t) = TC^* \left[\left(\sum_{i=1}^{n} p_i \frac{x_t^* - x_i}{d(t^*, i)} \right)^2 + \left(\sum_{i=1}^{n} p_i \frac{y_t^* - y_i}{d(t^*, i)} \right) \right]^{1/2}. \tag{2.20}$$

For $t = t^*$, and S chosen optimally (2.18) plus (2.20) yield

$$G(t^*) = W(S). \tag{2.21}$$

Now (for the first time) we use the data $T \to 0$, and by (2.18) we have

$$\lim_{T \to 0^+} d(0, t^*) = 0. \tag{2.22}$$

That is, we only have to determine in which direction and at what speed to proceed, but we will not get very far. The direction we choose is that of $-\nabla Z(t^*)$, as we always have to; but now we can take 0 instead of t^*, using equation (2.22) so we do not have to search for this value. As for the speed, we choose S^∇ (the 'gradient' speed), such that

$$S^\nabla \triangleq \arg\{W(S) = G(0)\}, \tag{2.23}$$

since by (2.21) this is the value for $t^* = 0$.

Since the speed is one of the parameters we are interested in, we present a theorem which will also hold for the stochastic decision period case.

Theorem 2.2: *The gradient speed* S^∇ *as defined at any point, is an upper bound for the optimal speed at that point, and* S^* *is an upper bound for* S^∇.

Proof. By Theorem 2.1, $Z(t)$ is strictly convex, hence along the segment $\overline{0, t^*}$ it is also strictly convex, and since $Z(t^*) \leq Z(t)$; $\forall t$, it is monotone decreasing along the segment. Let $g(t)$ be the absolute value of the directed derivative along $\overline{0, t^*}$. Clearly $g(t)$ is monotone decreasing for λ when $t = \lambda_0 + (1 - \lambda)t^*$ (i.e. $x_t = \lambda x_o + (1 - \lambda)x_t^*$, $y_t = \lambda y_o + (1 - \lambda)y_t^*$, o being an index and not a number here). For $\lambda = 0$ the slope $g(0)$ is bounded from above by $G(0)$, since $G(0)$ reflects the steepest descent (in the direction of $-\nabla$Z). At t^*, the direction of $\overline{0, t^*}$ is the steepest descent direction itself, by (2.18). Summing these assertions we have

$$G(0) \geq g(0) > g(t^*) = G(t); 0 \neq t^*. \tag{2.24}$$

It follows that the gradient speed S^∇ is an upper bound on the speed for any movement from 0, and we can designate any point as 0. That is,

$$S \leq S^\nabla \leq S^*. \tag{2.25}$$

So S^∇, which is rather easy to compute, is an upper bound on our speed anywhere, and it would be easy to extend the proof to the stochastic decision period case, using the basic attributes of the expectation.

The stopping line and the waiting region

For $T \geq T^*$, we obtain $S = S_{\min}$, and by (2.15) it follows that $W(S) = f_{\min}$. Using (2.21) we have

$$G(t^*) = f_{\min}. \tag{2.26}$$

Now, starting at different points, but such that $G(0) > f_{\min}$ and $T > T^*$ as defined for them we should stop at different decision points, respectively (unless we start from colinear points, on $\overline{0, t^*}$), each satisfying (2.25). Actually there is a locus of points satisfying (2.26), which we call D as follows

$$D = \{t \,\varepsilon\, E^2 / G(t) = f_{\min}\}. \tag{2.27}$$

We call D the stopping line (although it may happen to be a point). Now denote the area within D, inclusive, as C, or

$$C = \{t \,\varepsilon\, E^2 / G(t) \leq f_{\min}\}. \tag{2.28}$$

C is also called the waiting area, since being there during the decision period would imply waiting. Clearly $C \subseteq D$, with $C = D$ for the special case where one of the points $N \cup 0$ is the only solution for a large T. In case $C \neq D$, however, we have a non-empty set E as follows

$$E = C - D \,(\text{or } C/D). \tag{2.29}$$

Specifically, there is a point in C, and in E if $E \neq \emptyset$, for which $G = 0$ (if $C = D$, we have to define G as 0 there, since it includes 0/0 terms); we denote this point by $t_{\min}$, i.e.

$$G(t_{\min}) = 0. \tag{2.30}$$

Clearly, in order to identify $t_{\min}$, we do not need any information about the starting point or any of the costs we carry, but just the information on N and $\{p_i\}$.

We are now ready to discuss the stochastic decision period case. In that connection, note that some of our results so far, such as Theorem 2.1, the stopping line, etc., are not dependent upon T, hence they can serve us for the stochastic decision period case as well.

2.4 The stochastic decision period case

Our problem is exactly as before, except that T is a random variable (RV) now. Conceivably the p_i values could be influenced by information such as 'the decision has not yet been made', but we do not consider this case in detail (i.e. we assume statistical independence between T and the choice). Our RV may be discrete (contact with management is at predetermined times), continuous or mixed. We discuss the discrete case in detail, and show how to accommodate the continuous case by the discrete one. We assume that the distribution of T is given. (Bayesians will find no fault with that assumption, hopefully. Others will have to take it at face value.) Let

$$P(T = h_j) = q_j; j \varepsilon J = \{1,2,\ldots.\}, \tag{2.31}$$

where $q_j \geq 0$; $\forall j$ and they sum to one, of course; J may be finite or not, the index o is maintained for the start as before, and we may assume $h_o = q_o = 0$ for it. Our problem is to find the best set of decision points t_j (or t_j^* when optimality is assumed), such that as long as the decision is not yet made by $T = h_j$ we proceed from t_j^* to t_{j+1}^* (starting at $t_o = t_o^*$). Let v_j be the conditional probability of decision at h_j, given it has not been made yet, i.e.

$$v_j = P(T - h_j/T > h_{j-1}) = q_j / \left(1 - \sum\nolimits_{k=1}^{j-1} q_k\right); \tag{2.32}$$

and, following (2.7) we define $Z(t_{j-1},t_j)$:

$$\begin{aligned} Z(t_{j-1},t_j) = {} & F(h_j - h_{j-1}) + f(d(t_{j-1},t_j)/(h_j - h_{j-1}))d(t_{j-1},t_j) \\ & + v_j TC^* \sum_{i=1}^{n} p_i d(t_j,i) + (1 - v_j) Z_{j+1}(t_j,t_{j+1}^*). \end{aligned} \tag{2.33}$$

This formulation lends itself to dynamic programming very naturally, and assuming optimality we define $Z_j^*(t_{j-1})$:

$$Z_j^*(t_{j-1}) = {}^{\min}_{tj}\{Z_j(t_{j-1},t_j)\} = Z_j(t_{j-1},t_j^*). \tag{2.34}$$

Before proceeding further with the general solution, two limiting cases will help us to confine our search to a manageable area. These are analogues of cases we discussed above, and here is the payoff for the effort there.

The $P(T < T^*) \rightarrow 0$ Case: this is the analogue of the $T \geq T^*$ case, so we proceed at $S_{\min}$ to the correct spot along the stopping line. We refer to the solution as the 'slow' trajectory.

The $P(T < \varepsilon) \rightarrow 1$; $\forall\ \varepsilon > 0$ Case: this case to which we refer as the 'gradient' case, which is analogous to the $T \rightarrow 0^+$ case, since it stipulates that with probability approaching 1 this is indeed anticipated. Therefore we move at a speed of S^Δ in the $-\nabla Z(t)$ direction. Now, under the stipulation, the probability that we will go far before the decision is negligible, but this does not deter us from defining the steepest descent, or (minus) gradient trajectory all the way until the stopping line. The 'gradient' speed we use is a function of t, which may be obtained by

$$S^{\nabla}(t) = \arg\{f(S^{\nabla}(t)) + S^{\nabla}(t)f'(S^{\nabla}(t)) - G(t) = 0\}, \tag{2.35}$$

which is a direct extension of (2.23).

Since by Theorem 2.2 (which extends almost directly to the stochastic decision period case) $S^{\nabla}(t)$ is an upper bound on S, we also refer to this as the fast trajectory. It is interesting (although intuitively clear) to note that S^{∇} is decreasing along the fast trajectory.

Lemma 2.2: *When moving along the gradient trajectory, which we denote by* X(t), *in the* $-\nabla Z(\mathrm{t})$ *direction,* $\mathrm{S}^{\nabla}(\mathrm{t})$ *is monotone non-increasing.*

Proof. Let $z(t)$ be the expected distance to the final destination from t given a decision (recall, a decision is due immediately), i.e.

$$z(X(t)) = TC^* \sum_{i=1}^{n} p_i d(t,i). \tag{2.36}$$

Then $G(t)$ as per (2.20), is z's directional derivative along $X(t)$, i.e.

$$G(t) = |z'(X(t))|. \tag{2.37}$$

We want to show that $G(t)$ is monotone non-increasing (which will imply our lemma by (2.35) and the monotonicity of $W(S)$). We know that $G(t)$ decreases from $G(0)$ to $f_{\min}$ without changing signs along $X(t)$, so it will suffice to show that $z''(X(t)) \geq 0$. But $X(t)$ is a trajectory in the steepest descent direction, hence if we differentiate it by t, twice, we get

$$\dot{X}(t) = -\nabla_z(X(t)), \tag{2.38}$$

$$\ddot{X}(t) = -\nabla^2 z(X(t))\,\dot{X}(t) = \nabla^2 z(X(t))\nabla_z(X(t)). \tag{2.39}$$

We continue and differentiate $z(X(t))$, twice again, to obtain

$$z'(X(t)) = \nabla z(X(t))^T\,\dot{X}(t), \tag{2.40}$$

$$z''(X(t)) = \dot{X}(t)^T\nabla^2 z(X(t))\,\dot{X}(t) + \nabla z(X(t))^T\,\ddot{X}(t). \tag{2.41}$$

Finally, by substituting (2.39) and (2.41) we have

$$z''(X(t)) = \left(\dot{X}(t) + \nabla z(X(t))\right)^T \nabla^2 z(X(t))\left(\dot{X}(t) + \nabla z(X(t))\right), \tag{2.42}$$

which is a bilinear form ($Y^T\nabla^2 zY$). Now, z is clearly a convex function, hence $\nabla^2 z$ is positive semidefinite (at least), and our result follows.

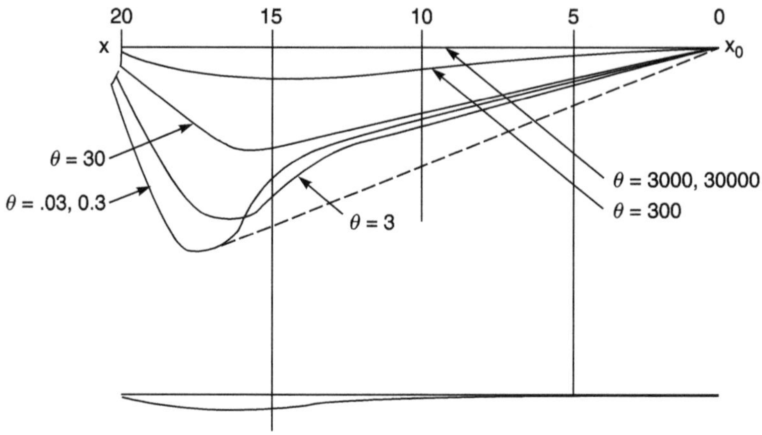

Figure 2.2 Exponentially distributed decision period, for expectations and for randomly chosen points.

Figure 2.2 depicts the results of a program run for an exponentially distributed decision period, for seven expectations θ, and for seven randomly chosen points of randomly chosen weights (probabilities). For large θs, the trajectories were virtually the same as the slow trajectory. For small θs, a similar behaviour was observed relative to the fast trajectory. Interestingly, although all the trajectories, were within the convex hull of the area between these two trajectories, one of them actually intersected the fast trajectory. The speed race obeyed Theorem 2.2. In general, one might say that the faster the decision is due, the faster we should move, and the longer our total trajectory may be – since we do not expect to stick to it for a long time; the slower the decision is due, the more we tend to go slowly and along a 'mildly curving' trajectory (if not exactly straight). In any case, as far as the physical location of the trajectory is concerned, we should not find ourselves out of the area defined by the convex hull of the extreme case trajectories, –*E* (which we should never enter even if it is in that convex hull). If we do, we can always find a better trajectory for immediate or delayed advantage, within this area. We denote this result in Theorem 2.3.

Theorem 2.3: *One should never leave the convex hull of the slow and fast trajectories*, –E.

Proof. By negation, as described above.

Incidentally, the results in Figure 2.2 even show that the convex hull of a 'medium' speed trajectory and the slow one contains all the 'slower' speed trajectories. However, this may not be extendable to more general distributions, where the relative 'speeds' may not be uniquely implied.

Note, however, that if we do not use an exaggerated vertical scale, all trajectories seem rather straight! Practically, there is no doubt that location-wise we should pick some straight trajectory, even the slow one, and just optimize the speed race. This would yield most of the potential gain, with the additional benefit of a less complex race problem.

If we return now to equations (2.33) or (2.34), we can see that the stopping line and the search area are what we need practically to obtain a working dynamic programming model. We may start by assuming the slow trajectory, which would make our decision variable univariate, and we can fold back satisfactorily by assuming that the, say, *k*th step will bring us to the stopping line. It may happen that we overshoot the starting point, but it should not be difficult to adjust. However, we do not suggest using this method here, since it does not seem to justify the programming effort, and we can simply use a multivariable library search method instead. A problem might be if local minima exist besides the global minimum; the next theorem removes this obstacle.

Theorem 2.4: *The problem of locating the decision points* t_j^* *so as to minimize* $(Z_1^* t_0)$ *is convex.*

Proof. By iterative application of Theorem 2.1.

2.5 Strategic interactions in an innovation race

In tracking the evolution of part of the Japanese telecommunications industry over several years, it shows that strategic interactions between the firms play a substantial role in determining the firm level and the industry level of technological evolution. In particular, we identify several 'races', each of which is the result of a subset of firms jockeying for a position either as a race leader or for a position not too far behind the race leader (see Chapter 3). The identification and interpretation of the races relies on the fact that different firms take very different technological paths to reach a common 'cycle time' level. In view of the previous description of a stochastic race it is pertinent to distinguish between two kinds of racing behaviour. A lagging firm might simply try to close the gap between itself and the technological leader at any point in time ('frontier-sticking' behaviour or catch-up race), or it might try to actually usurp the position of the leader by 'leapfrogging' it (frontier race). When there are disproportionately large payoffs to being in the technical lead (relative to the payoffs that a firm can realize if it is simply close enough to the technical frontier), then one would expect that leapfrogging behaviour would occur more frequently than frontier-sticking behaviour. All attempts to leapfrog the current technological leader might not be successful since many lagging firms might be attempting to leapfrog the leader simultaneously, and the leader might be trying to get further ahead simultaneously. In this regard, one should both report the attempted leapfroggings and the realized leapfroggings. Thus, we may distinguish between two-layer races – in the first, leapfroggings may occur stochastically; in the second, followers or imitators may just try to catch up with the frontier, by frontier-sticking behaviour.

2.6 Multi-stage races

All the existing work on multi-stage races is in the patent race framework. Harris and Vickers (1987) show that the leader invests more than the follower in a multi-stage patent race scenario. Their result generalizes a similar result due to Grossman and Shapiro (1987) for two-stage games. In contrast, instead of analysing aggregate resource allocation, we discuss how given resources are allocated. In line with the stochastic race model suggested in Sections 2.3 and 2.4, there could be an important strategic advantage of being aggressive or fast in each stage of a multi-stage race. Our focus will be on characterizing the differences in the expected payoff functions of firms as they get ahead of their rivals (or fall behind) and closer to the finishing line. Our explanation here will be intuitive, and analytical treatment is given in the appendix to this Chapter. We can speak of the monopoly (or duopoly) term becoming more important in a payoff expression as the ratio of its coefficient to that of the duopoly (or monopoly) term rises. It can be established that the monopoly term in the expected payoff expression of the leading firm in a two-firm multi-stage race becomes progressively more important as it gets further ahead of its rival, provided the lead meets a minimum threshold. The threshold lead is smaller the closer the lead firm is to the finishing line. Conversely, the duopoly term in the expected payoff expression of the lagging firm becomes more important as it falls further behind, subject to the same threshold lead considerations as the leading firm. We assume the leading firm's payoff, when it has finished all stages and is reaping monopoly profits, as a function of the lead it has over its rival. Let the lead firm receive a payoff when it has finished all n stages, and the rival is in the first stage. Then the coefficient on the monopoly term rises faster than that on the duopoly term as the lead increases. Then, using this property of the coefficients, we consider the leading firm's payoff as a function of its lead, when it is in the last stage of the n-stage race. Once again, we show that as long as the lead exceeds a threshold lead (which may be 0), the coefficient on the monopoly term rises faster than that on the duopoly term as the lead increases. A method of recursion can be applied, where the relationship of the coefficients when the lead firm is at stage s of the race is used to derive similar relationships when the lead firm is at stage s–1. The procedure is similar for the lagging firm. In this case we can show that the duopoly coefficients rise faster than the monopoly ones as the lead increases, subject to the threshold lead considerations. The same is shown to be true recursively when the lead position is at position n–1, n–2, etc. This characterization highlights two forces that influence a firm's choices in various stages: proximity to the finishing line and distance between the firms. The probability of reaping monopoly profits is higher the farther ahead a firm is of its rival, and even more so the closer the firm is to the finishing line. If the lead firm is far from the finishing line, even a sizeable lead may not translate into the dominance of the monopoly profit term, since there is plenty of time for the lead situation to be reversed, and failure to finish first remains a probable outcome. In contrast, the probability that the lagging firm will get to be a monopolist becomes smaller as it falls behind the lead firm. This raises the following question: what kinds of actions cause a firm to get ahead? Intuitively, one would expect that a firm that is ahead of its rival at any time t, in the sense of having completed more stages by time t, is likely to have chosen the faster, less-aggressive strategy more often. The monopoly term is increasingly important to a firm that falls behind. Further simple calculations suggest that the firm that is ahead is likely to have made less-aggressive choices than the firm that is behind in the race.

A further interesting question is whether a lead results in a greater likelihood of increasing lead and then in an increased chance of leapfrogging (as in a frontier race) or in more catch-up behaviour (as in a catch-up). The existing literature (Grossman and Shapiro, 1987; Harris and Vickers, 1987) has suggested that a firm that surges ahead of its rival increases its investment in R&D and speeds up while a lagging firm reduces its investment in R&D and slows down. Consequently, these papers suggest that the lead continues to increase. However, when duopoly competition and dichotomy of the race (in frontier-pushing and frontier-sticking behaviour) are accounted for, the acceleration of a leading firm occurs only under special circumstances. In high-tech industries, such as computers and telecommunications, it could be expected that monopoly profits do not change substantially with increased aggressiveness, but duopoly profits do change substantially with increased aggressiveness. Then a firm getting far enough ahead such that the monopoly term dominates its payoff expression will always choose the fast strategy, while a firm that gets far enough behind will always choose the slow and aggressive approach. Then the lead is likely to continue to increase. If, on the other hand, both monopoly and duopoly profits increase substantially with increased aggressiveness, then even large leads can vanish with significant probabilities.

2.7 Stochastic innovation games

Some generalization of the stochastic innovation race is given through 'simple timing games' (Fudenberg and Tirole, 1991). In that type of game a player has the choice either to 'stop' in the innovation race or to move ahead. For example, in two-player timing games, one can restrict the attention to sub-game perfect equilibria. In terms of definition, any dynamic game that has a sub-game perfect equilibrium has a Nash Equlibrium, in any of its sub-games.

That is, in the context of simple timing games, if one player has stopped, the remaining player faces a maximization problem to be solved. The player's payoff functions can be easily cast as functions of time. If only one player stops at some stage, then he is the 'leader', and receives some payoff, and his opponent receives a follower payoff. An example where 'stopping times' in a timing game plays a central role is the 'war of attrition'. In a simple version, two opponents are fighting for the price over time with given fighting costs per period. If one stops fighting in a given period t, his opponent wins the prize without incurring a fighting cost in that period. If both stop simultaneously then neither wins the prize.

More generally, two immediate conclusions can be reached from this type of game. One is that each player i prefers his opponent stopping first at any time t to any outcome where player i stops first at some later t. And since fighting forever is costly, each player would rather quit immediately than fight forever. Two non-stationary variants emerging from these properties have emerged in the literature: (1) 'eventual continuation' and (2) 'eventual stopping'.

1. Eventual continuation

This example is due to Fudenberg *et al.* (1983). Two firms are engaged in a patent race, and 'stopping' means abandoning the race. The expected productivity of research is initially low; that is, if both firms conduct R&D until one makes a discovery then both firms have a negative expected value. However, if the productivity of R&D increases over time so that there are threshold levels at which both firms are active, then it will be a dominant strategy for any firm not to stop (R&D). Suppose the patent has the value v, and the cost of obtaining it at every stage is c. If firm i has started to conduct R&D until date t, it pays $c_i dt$ and makes a discovery with probability $p_i dt$ between t and $t+dt$. The net flow profit as expected is $[p_i(t)v-c_i]\,dt$. Thus, in this game, we have the leader's payoff

$$L_i(t) = \int_0^t [p_i(\tau)v - c_i]\exp\left(-\int_0^\tau [p_1(s) + p_2(s)]ds\right)\exp(-r\tau)d\tau$$

where r is the rate of interest. The probability that no one has discovered at date τ conditional on both players having stayed in the race is $\exp\left(-\int_o^\tau [p_1(S)+p_2(S)]\,ds\right)$. We assume that an R&D monopoly is viable

$$0 < \int_o^t [p_i(\tau)v - c_i]\,dt.\exp\left(-\int_o^\tau [p_1(s) + p_2(s)]\,ds\right)\exp(-r\tau)\,d\tau = F(0)$$

and that a duopoly is not viable at date 0. Because $p_i(.)$ is increasing, if a monopoly is viable at date 0, then it is viable from any date $t > 0$ on. Therefore, it is optimal for each player to stay in until discovery once his opponent has quit. The follower's payoff is thus

$$F_i(t) = \int_o^t [p_i(\tau)v - c_i].\exp\left(-\int_o^\tau [p_1(s) + p_2(s)]\,ds\right)\exp(-r\tau)\,d\tau$$
$$+ \int_o^\tau [p_i(\tau)v - c_i].\exp\left(-\left(\int_o^\tau \left[p_i(s)\,ds + -\int_o^t [p(s)\,ds\right]\right)\right)\exp(-r\tau)\,d\tau.$$

The leader's and the follower's payoff paths in a continuous-time innovation race are pursued in Chapter 7.

2. Eventual stopping

In this case, two firms wage duopoly competition in a market. If one quits, the other becomes a monopoly. Demand is declining, and until a threshold time, firm *i* stops making a profit as a duopolist, and after that time, firm *i* is no longer viable as a monopolist. Examples where eventual stopping fits are in declining industries, where exit from the industry is an 'all-or-nothing' choice. In such a case a big firm will become unprofitable faster than a smaller one, and so the big firm will be forced to exit first. Foreseeing this eventual exit, the small(er) firm will stay in, and backward induction implies that the big firm exits one the market shrinks enough that staying in earns negative profit flows.

In studying multi-stage racing we will focus in Chapter 6 on 'Markov' or 'state space' strategies in which the past influences current play only through its effect on a state variable that summarizes the direct effect of the past on the current environment. Since the state captures the influence of past play on the strategies and payoff functions for each sub-game, if a player's opponents use Markov strategies, that player has a best response that is Markov as well. Markov-type games relate to a class of differential games, more closely treated in Chapter 7, which are (continuous-time analogues of) stochastic games in which the state evolves according to a (deterministic) differential equation.

2.8 Conclusions

Stochastic models of racing embrace several features that resemble moving objects towards a stochastic final destination. Together with game-theoretic treatments of racing behaviour we look into racing patterns of individual firms in view of their strategic responses to their racing environment. Among those features we identified the speed race problem, the selection of an optimal decision point t^*, to optimize a gradient trajectory (of technological evolution) and to determine the 'stopping line and the waiting region'. Such a model would be compatible with observations on innovation races in high-technology industries, in particular, with race-type behaviours such as leapfrogging and catch-up, striking a balance between moving ahead and waiting. The model can be improved by incorporating constraints into it. For example, constraints on an innovation path could be given by road blocks such as a bankruptcy constraint or an R&D uncertain payoff constraint. Some of these constraints may be conceptually easy to introduce, others may be tougher, such as an investment constraint if the total innovation effort en route to t^* plus the worst case would violate it. In such a case one may want to weight the distant finishing line unproportionately.

It is interesting to speculate on the implications of the way the firms in major high-tech markets, such as telecommunications, split clearly into the two technology races, with one set of firms clearly lagging the other technologically. The trajectories of technological evolution certainly seem to suggest that firms from one frontier cannot simply jump to another trajectory. Witness, in this regard, the gradual process necessary for the firm in the catchup race to approach those in the frontier race. There appears to be a frontier 'lock-in' in that once a firm is part of a race, the group of rivals within that same race are the those whose actions influence the firm's strategy the most. Advancing technological capability is a cumulative process. The ability to advance to a given level of technical capability appears to be a function of existing technical capability. Given this path dependence, the question remains: why do some firms apparently choose a path of technological evolution that is less rapid than others? Two sets of possible explanations could be derived from our case analysis, which need not be mutually exclusive. The first explanation lingers primarily on the expensive nature of R&D in industries like telecommunications and computers which rely on novel discovery for their advancement. Firms choosing the catch-up race will gain access to a particular technical level later than those choosing the frontier, but will do so at a lower cost.

References

Fisher, F., McGowan, J. and Greenwood, J. (1983) *Folded, Spindled, and Mutilated: Economic Analysis and* US vs IBM, Cambridge, MA: MIT Press.

Fudenberg, D., Gilbert, R., Stiglitz, J. and Tirole, J. (1983) 'Preemption, leapfrogging and competition in patent races', *European Economic Review* **22**, 3–30.

Fudenberg, D. and Tirole, J. (1991) *Game Theory*, Cambridge, MA: MIT Press.

Kamien, M.J. and Schwartz, N. (1982) *Market Structure and Innovation*, Cambridge: Cambridge University Press.

Gottinger, H.W. (1996) 'Technological races', *Tonan Ajia Kenkyn Nenpo, Annual Review of Economics, Japan*, **38**, 1–9.

Grossman, G. and Shapiro, C. (1987) 'Dynamic R&D competition', *Economic Journal*, **97**, 372–387.

Harris, C. and Vickers, J. (1985), 'Perfect equilibrium in a model of a race', *Review of Economic Studies*, **52**, 193–209.

Harris, C. and Vickers, J. (1987), 'Racing with uncertainty', *Review of Economic Studies*, **54**, 305–321.

Kidder, T. (1981), *The Soul of a New Machine*, Boston: Little, Brown and Company.

Lerner, J. (1997), 'An empirical exploration of a technology race', *The Rand Journal of Economics*, **28**(2), 228–247.

Reinganum, J. (1989), 'The timing of innovation', in *The Handbook of Industrial Organization*, Volume 1, R. Schmalensee and R. Willig (eds.), Amsterdam: North Holland, Chapter 14.

Rockafellar, R.T. (1970), *Convex Analysis*, Princeton, NJ: Princeton University Press.

Zang, I. (1981), 'Discontinuous optimization by smoothing', *Mathematics of Operations Research*, **6**, 140–152.

Appendix

On the expected gain by the model

Our model is based on the fact that the decision period is going to be completely wasted unless we utilize it. This places an obvious upper bound on our expected gain, V, namely

$$V \leq FT. \tag{A1}$$

In the stochastic case, similarly

$$V \leq FE(T). \tag{A2}$$

Clearly, the only way we can approach this upper bound is if $G(t)$ approaches TC^* along the trajectory, throughout the decision period; for instance, if the points are close to each other and far from the start. In this case we behave as if the destination is known. However, in both the deterministic and the stochastic decision period case, if T is large, we cannot do anything at least part of the time. It is obvious that in the deterministic case the gain cannot exceed FT^*, which makes us rewrite $G(t)$, and similarly (A2):

$$V \leq F_{\min}\{T, T^*\} \tag{A3}$$

$$V \leq F_{\min}\{E(T), T^*\}. \tag{A4}$$

But, suppose now that $T \geq T^*$, can we really expect to gain even FT^*? The answer of course is no. In this case $G(t)$ is rather low, at least towards the stopping line where it reaches $f_{\min}$. We may compute V for this case by the following formula

$$V = TC^* \left(\sum_{i=1}^{n} p(d(0,i) - d(t^*, i)) \right) - f_{\min} d(0, t^*), \tag{A5}$$

where the gross gain is the improvement in the expected 'future' total costs to reach the final destination, but we have to subtract the 'present' variable costs, in this case $f_{\min}\, d\,(0,t^*)$. By substituting $f(d(0,t^*)/T)d(0,t^*)$ for these costs, we obtain the expected gain in the deterministic case:

$$V = TC^* \left(\sum_{i=1}^{n} p(d(0,i)) - d(t_i^*\, i)) \right) - f(d(0,t^*)/T)d(0,t^*). \tag{A6}$$

In the stochastic case, similarly, we have the following result:

$$V = FE(T) + TC^* \left(\sum_{i=0}^{n} p(d(0,i)) \right)^N - z_1^*(0). \tag{A7}$$

A similar result can be obtained at any stage, given that we reached it without decision, but we omit it. Note, however, that this expected gain is monotone non-increasing. For instance, once we reach the stopping line it drops to zero, since there is nothing useful we can do any more. Note, that if we start anywhere in C, all the formulas above, including the bounds (A3) and (A4), yield $V = 0$ (e.g. $T^* = 0$ in this case).

3 Statistical indicators of technological hypercompetition

The real competitive problem is laggards versus challengers, incumbents versus innovators, the inertial and imitative versus the imaginative.

Hamel and Prahalad, *Competing for the Future* (1994)

3.1 Introduction

Racing behaviour is a dynamic story of how technology unfolds in an industry. In contrast to any existing way of looking at the evolution of technology, racing behaviour recognizes the fundamental importance of strategic interactions between competing firms, and therefore of dynamic competition. Thus firms take their rivals' actions into account when formulating their own decisions. The importance of this characterization is at least twofold. At one level, racing behaviour has implications for understanding technology strategy at the level of the individual firm and for understanding the impact of policies that aim to spur technological innovation. At another level, racing behaviour embodies both traditions that previous writings have attempted to synthesize: the 'demand–pull' side emphasized by economic theorists and the 'technology–push' side emphasized by the autonomous technical evolution school. It remains an open problem how technological races can be induced endogenously, e.g. by changes in economic variables (such as costs, prices and profitability).

Our stochastic model of a race in Chapter 2 embraces several features that resemble moving objects towards a stochastic final destination. By exploring 'hypercompetitive' patterns of racing behaviour, in respective industries, we look into racing patterns of individual firms in view of their strategic responses to their racing environment. Among those features we identify is the *speed race problem*, the selection of an *optimal decision point*, to optimize a gradient trajectory (of technological evolution) and to determine the *stopping line and the waiting region*. The corresponding empirical framework lies in the prescription and description of industrial racing patterns which have to be viewed as identifying objectives or benchmarks for performance evaluation of firms, industries, and expanded to regions and national economies (Chapter 12).

(a) A key objective is to explore and explain which type of 'racing behaviour' is prevalent in global high-technology industries, as exemplified by information technology (computer and telecommunications) industries. The pattern evolving from such racing behaviour would be benchmarked against the frontier racing type of the global technological leaders.
(b) Another objective is to draw policy inferences on market structure, entrepreneurship, innovation activity, industrial policy and regulatory framework in promoting and hindering industry frontier races in a global industrial context.
(c) Given the statistical profile of technological evolution and innovation for respective global industries, how does it relate to competitive racing and rivalry among the leading firms? Among the performance criteria to be assessed are frequency of frontier pushing, technological domination period, innovations vs imitations in the race, innovation frequency when behind or ahead, nature of jumps, leapfrogging or frontier-sticking, inter-jump times and jump sizes and race closeness measures.
(d) An empirical profiling and proliferation of racing in these global industries can be explored, comprising data sets identifying 'relationship between technological positions (ranks) of firms in successive years' (10–25-year period).
(e) Do observed racing patterns in respective industries contribute to equilibrium and stable outcomes in the world economy? To what extent are cooperative ventures (global governance) between states justified to intervene? In particular, we investigate the claim, as put forward by the Group of Lisbon (1995) that as a likely future scenario triadization will remain 'the prevailing form of economic globalization', in view of observations that increased intensity in racing patterns within key industries could lead to instability and welfare losses in triadization.
(f) More specifically, in an era of ongoing deregulation, privatization, liberalization and lifting of trade barriers, we explore whether industry racing patterns are sufficiently controlled by open world-wide markets or whether complementary international agreements (regulations, controls) are needed to eliminate or mitigate negative externalities (without compromising the positive externalities that come with industry racing).

The effects of racing patterns on the industrial organization of particular industries are assessed: how does the behaviour of leading firms influence the subcontracting relation between the purchasing firms and their subcontractors, that is, which racing pattern induces a strengthening of their vertical links and what behaviour of the parent firm in technological racing encourages their subcontractors to be technologically (and thus managerially) independent?

This chapter proceeds as follows. We first review the state of research in the industrial economics of technological racing. Then the model framework identifies the essential elements under which racing patterns will occur. This will be complemented by empirical considerations of the diversity and complexity of racing patterns, and the measurement problems that result.

In Section 3.2 we provide a focused review of up-to-date economic research into the evolution of industry and competitive performance of firms within the industry. In Section 3.3 we present a selection of statistical indicators reflecting and measuring patterns of racing behaviour. Section 3.4 presents a case of the Japanese telecommunications industry to complement and support the discussion of measurement in the previous section. Section 3.5 provides a summary of results and a discussion in the context of the catch-up literature. Conclusions follow in Section 3.6.

3.2 Catch-up or leapfrogging

It was Schumpeter (1942) who observed that it is the expectation of supernormal profits from a temporary monopoly position following an innovation that is the chief driver of R&D investment. Along this line, the simplest technology race model would be as follows. A number of firms invest in R&D. Their investment results in an innovation with the time spent in R&D subject to some uncertainty (Gottinger, 1989). However, a greater investment reduces the expected time to completion of R&D. The model investigates how many firms will choose to enter such a contest, and how much they will invest.

Despite some extensive theoretical examination of technological races there have been very few empirical studies on the subject (Lerner, 1997), and virtually none in the context of major global industries, and on a comparative basis. This will be one major focus in this chapter.

Technological frontiers at the firm and industry race levels offer a powerful tool through which to view evolving technologies within an industry. By providing a benchmarking roadmap that shows where an individual firm is relative to the other firms in the industry, they highlight the importance of strategic interactions in the firm's technology decisions.

Does lagging behind one's closest technological rivals cause a firm to increase its innovative effort? The term 'race' suggests that no single firm would want to fall too far behind, and that every firm would like to get ahead. If a firm tries to innovate more when it is behind than when it is ahead, 'catch-up' behaviour will be the dominant effect. Once a firm gets ahead of its rivals noticeably, its rivals will step up their efforts to catch up. The leading firm will slow down its innovative efforts until its rivals have drawn uncomfortably close or have surpassed it. This process repeats itself every time a firm gets far enough ahead of its rivals. An alternative behaviour pattern would correspond to a firm increasing its innovative effort if it gets far enough ahead, thus making catch-up by the lagging firms increasingly difficult. For any of these forms there appears to be a clear link to market and industry structure, as termed 'intensity of rivalry' by Kamien and Schwarz (1982). We investigate two different kinds of races: one that is a frontier race among leaders and 'would-be' leaders, and another that is a catch-up race among laggards and imitators.

Another aspect of innovation speed has recently been addressed by Kessler and Bierly (2002). As a general rule they found that speed to racing ahead may be less significant the more 'radical' (drastic) the innovation appears to be and the more likely it leads to a dominant design.

These two forms have been applied empirically to the development of the Japanese computer industry (Gottinger, 1998), that is, a frontier race model regarding the struggle for technological leadership in the global industry between IBM and 'Japan Inc.' guided by MITI, and a catch-up race model relating to competition among the leading Japanese mainframe manufacturers as laggards.

Furthermore, it is interesting to distinguish between two kinds of catch-up behaviour. A lagging firm might simply try to close the gap between itself and the technological leader at any point in time ('frontier-sticking' behaviour), or it might try to actually usurp the position of the leader by 'leapfrogging' it. When there are disproportionately large payoffs to being in the technical lead (relative to the payoffs that a firm can realize if it is simply close enough to the technical frontier), then one would expect that leapfrogging would occur more frequently than frontier-sticking (Owen and Ulph, 1994). Alternatively, racing toward the frontier creates the 'reputation' of being an innovation leader facilitating to maintain and increase market share in the future (Albach, 1997). All attempts to leapfrog the current technological leader might not be successful since many lagging firms might be attempting to leapfrog the leader simultaneously and the leader might be trying to get further ahead simultaneously.

Correspondingly, one should distinguish between attempted leapfroggings and realized leapfroggings. The leapfrogging phenomenon (although dependent on industry structure) appears as the predominant behaviour pattern in the US and Japan frontier races (Brezis *et al.*, 1991). Albach (1994) cites studies for Germany that show otherwise.

Leapfrogging behaviour influenced by the expected size of payoffs as suggested by Owen and Ulph (1994) might be revised in compliance with the characteristics of industrial structure of the local (regional) markets, the amount of R&D efforts for leapfrogging and the extent of globalization of the industry. Even in the case where the payoffs of being in the technological lead are expected to be disproportionately large, the lagging firms might be satisfied to remain close enough to the leader so as to gain or maintain a share in the local market. This could occur when the amount of

R&D efforts (expenditures) required for leapfrogging would be too large for a lagging firm to be viable in the industry and when the local market has not been open enough for global competition: the local market might be protected for the lagging local firms under the auspices of measures of regulation by the government (e.g. government purchasing, controls on foreign capital) and the conditions preferable for these firms (e.g. language, marketing practices). When the industrial structure is composed of multi-product firms, as, for example, in the Japanese computer industry, sub-frontier firms may derive spillover benefits in developing new products in other technologically related fields (e.g. communications equipment, consumer electronic products). These firms may prefer an R&D strategy just to keep up with the technological frontier level (catch-up) through realizing a greater profit stream over a whole range of products.

What are the implications of the way the firms split cleanly into the two technology races, with one set of firms clearly lagging the other technologically? The trajectories of technological evolution certainly seem to suggest that firms from one frontier cannot simply jump to another trajectory. Witness, in this regard, the gradual process necessary for the firms in the Japanese frontier to catch up with the global frontier firms. There appears to be a frontier 'lock-in', in that once a firm is part of a race, the group of rivals within that same race are those whose actions influence the firm's strategy the most. Advancing technological capability is a cumulative process. The ability to advance to a given level of technical capability appears to be a function of existing technical capability. Given this 'path dependence', the question remains: why do some firms apparently choose a path of technological evolution that is less rapid than others? We propose two sets of possible explanations, which need not be mutually exclusive. The first explanation hinges primarily on the expensive nature of R&D in industries such as the computer industry, which rely on novel scientific discovery for their advancement. Firms choosing the sub-frontier will gain access to a particular technical level later than those choosing the frontier, but will do so at a lower cost. Expending fewer resources on R&D ensures a slower rate of technical evolution. The second explanation relates mainly to technological spillovers. Following the success of the frontier firms in achieving a certain performance level, this fact becomes known to the sub-frontier firms. In fact, leading-edge research in the computer industry is usually reported in scientific journals and is widely disseminated throughout the industry. The hypothesis is that partial spillover of knowledge occurs to the sub-frontier firms, whose task is then simplified to some extent. Notice that the sub-frontier firms still need to race to be technological leaders, as evidenced by the analysis above. This implies that the spillovers are nowhere near perfect. Firm-specific learning is still the norm. However, it is possible that knowing something about which research avenues have proved successful (for the frontier firms) could greatly ease the task for the firms that follow and try to match the technical level of the frontier firm.

3.3 Statistical measurements of industrial racing patterns

We can establish statistically descriptive measures of racing behaviour that support the richness of the model outcomes as prescribed by the optimal process of the model outlined in Chapter 2.

The point of departure for a statistical analysis of industrial racing patterns is that the technological frontier is in fact a reasonable indicator of the evolving state of the knowledge (technical expertise) in the industry. At any point in time the industry frontier (ITF) indicates the degree of technical sophistication of the most advanced product in the industry, in a sense described below. Firm-level technology frontiers (FTF) are constructed analogously and indicate, at any point in time, the extent of the technical sophistication achieved by any firm until that point in time.

In this context we define 'race' as a continual contest for technological superiority among some subset of firms within a well-defined industry (classification). Under this conceptualization a race is characterized by a number of firms whose FTFs remain 'close' together over a period (T) of, say, 10 to 25 years. The distinctive element is that firms engaging in a race have FTFs substantially closer together than the FTFs of any firms not in the race. A statistical analysis should reflect that a race, as defined, may or may not have different firms in the leadership position at different times. It may be a tighter race at some times than at others, and, in general, may exhibit a variety of forms of industrial behaviour.

We look for clusters of firms whose FTFs remain close enough throughout the 25-year period (formal measures of closeness are defined and measured). We identify at least two races in progress in the industries with up to 25 years' duration. One comprises the world frontier race in each of those industries, the other a sub-frontier race (say, North America, Europe, East Asia) which would technically constitute a sub-frontier to the world, also allowing for the sub-frontier to be the frontier. Since the data set by no means exhausts the firms in the industry, it is certainly easier to accept that these are the significant technological races in progress. The technology frontier of the firms in a particular race (that is, the ITF) is constructed in a manner similar to the individual FTFs. Essentially, the maximal envelope of the FTFs in a particular race constitute the ITF for that race. So the ITF indicates, as a function of calendar time, the best achievable performance by any firm in the race.

Characterization of statistical indicators of industrial racing

The empirical explorations examine the features of the innovative process that are common to all the races, and those that distinguish between them. This will help us to understand some of the similarities and differences between different technology strategies that the races appear to represent. A frontier is 'pushed' forward when the performance level of the technology (for the firm in case of FTF and for the racing group of firms in the case of ITF) is being enhanced. For example, to what extent are different firms in each race responsible for pushing the frontier forward (i.e. to what extent are movements in the individual FTFs of the racing firms translated into movements of the ITF)?

While a variety of situations are possible, the extremes are the following: (a) one firm may push the frontier at all times, with the others following closely behind, and (b) all firms share more or less equally in the task of advancing the ITF. Extreme situation (a) corresponds to the existence of a unique technological leader for a particular race, and a number of quick followers. Situation (b), on the other hand, corresponds to the existence of multiple technological leaders.

We first demonstrate those cases by using the data of major players in the global telecommunications equipment industry (TIS, 2000), and then we look at a special situation that describes the racing pattern of the Japanese telecommunications (services) industry (InfoCom Research, 2000).

a. Assessment of frontier pushing

The relevant statistics for the three races are given in Table 3.1 in illustrative terms.

b. Domination period statistics

Accepting the view that a firm has greater potential to earn rents from its technological position if it is ahead of its race suggests that it would be interesting to examine the duration of time for which a firm can expect to remain ahead once it finds itself pushing its ITF. We define statistically the 'domination period' to be the duration of time for which a firm leads its particular race. It is interesting to note that the mean domination period is virtually indistinguishable for the three races, and lies between three and four years. A difference of means test cannot reject the hypothesis that the means are identical. So firms in each of the races can expect to remain ahead for approximately the same amount of time after they have propelled themselves to the front of their respective races. However, the domination period tends to be a more uncertain quantity in the world frontier race

Table 3.1 Pushing the frontier

	Number of times each firms pushes the frontier						
Firm #	*1*	*2*	*3*	*4*	*5*	*6*	*Total*
World frontier	6	3	2	–	–	–	11
EU frontier	4	1	0	2	1	2	10
Japan frontier	3	3	5	–	–	–	11

and in the EU frontier race than in the Japan frontier race (as evidenced by the lower domination period standard deviation in Table 3.2).

c. Catch-up statistics

If a firm tries to innovate more when it is behind then when it is ahead, then catch-up behaviour will be the dominant effect. (Evidence that catch-up behaviour is the norm is also provided by data from the US and Japanese computer industry.) Extending this evidence to illustrate our innovation race statistics, we make up Table 3.3A.

Table 3.2 Domination period statistics

Frontier	*Mean (years)*	*S.D. (years)*	*n*
World frontier	3.44	5.19	9
EU frontier	3.88	3.14	8
Japan frontier	3.86	2.20	8

One-tailed difference of means t tests: World frontier and EU frontier: $t = 0.2$, d.f. $= 15$; World frontier and Japan frontier: $t = 0.19$, d.f. $= 15$.

Table 3.3A More innovations when behind or ahead

World frontier *Firms #*	*1*	*2*	*3*				*Total*
Total innovations	7	8	3				18
Number when ahead	3	2	0				5
% when ahead*	43	25	0				28
% of time ahead**	81	10					20
EU frontier *Firms #*	*1*	*2*	*3*	*4*	*5*	*6*	*Total*
Total innovations	9	3	6	5	2	2	27
Number when ahead	2	0	1	1	0	1	5
% when ahead*	22	0	17	20	0	50	19
% of time ahead**	29	47	3	36	18	75	
Japan frontier *Firms #*	*1*	*2*	*3*				*Total*
Total innovations	8	6	10				24
Number when ahead	2	2	3				7
% when ahead*	25	33	30				29
% of time ahead**	16	45	52				

One-tailed difference of means t test: percent when ahead vs percent of time ahead: $t = 1.62$, d.f. $= 22$, $p < 0.06$.
*Percentage of innovations occurring when firm leads its race.
**Percentage of time that a firm leads its race.

Table 3.3B Firm jump sizes larger behind or ahead?

	When ahead		*When behind*	
	Mean number	*Number of jumps*	*Mean number*	*Number of jumps*
World frontier	2.37	5	2.92	12
EU frontier	2.84	4	2.82	22
Japan frontier	6.41	7	8.41	17

For each firm, this table compares the fraction of the total innovations carried out by the firms (i.e. the fraction of the total number of times that the FTF advanced) when the firm in question was leading its race with the fraction of time that the firm actually led its race. In the absence of catch-up behaviour, or behaviour leading to a firm increasingly dominating its rivals, we would expect to see no difference in these fractions. Then the fraction of time that a firm is ahead of its race could be an unbiased estimator of the fraction of innovations that it engages in when it is ahead.

The data, however, suggest that this is not the case. Difference of means tests indicate that the fraction of time that a firm leads its race is larger than the fraction of innovations that occur when the firm is ahead, i.e. more innovations occur when the firm is lagging than would be expected in the absence of catch-up or increasing dominance behaviour. Catch-up behaviour is supported by additional observations, as in Table 3.3B, that the firms make larger jumps (i.e. the FTF advances more) when they are behind than when they are leading the race.

d. Leapfrogging statistics

From this, the distinction emerges between two kinds of catch-up. A lagging firm might simply try to close the gap between itself and the technological leader at any point in time (frontier-sticking behaviour), on it might try to actually usurp the position of the leader by 'leapfrogging' it when there are disproportionally larger payoffs in being in the technical lead (relative to the payoffs that a firm can realize if it is simply close enough to the technical frontier), then one would expect that leapfrogging behaviour would occur more frequently than frontier-sticking behaviour.

Tables 3.4 and 3.5 describe the results of some analyses of this leapfrogging/frontier-sticking phenomenon. All attempts to leapfrog the

Table 3.4 Nature of jumps: leapfrogging or frontier-sticking

	Total jumps	*Attempted leapfrogs*		*Realized leapfrogs*	
World frontier	18	15	83%	11	61%
EU frontier	27	13	48%	10	37%
Japan frontier	24	15	63%	11	46%

Table 3.5 Inter-jump times and jump sizes

	Inter-jump times		*Jump sizes*	
	Mean	*S.D.*	*Mean*	*S.D.*
Time and jump statistics summarizing all FTFs				
World frontier	3.87	3.42	2.87	2.52
EU frontier	3.59	2.76	2.95	1.94
Japan frontier	3.81	1.50	7.82	14.14
Time and jump statistics summarizing ITFs				
World frontier	2.90	2.99	2.02	1.02
EU frontier	3.11	2.02	2.14	1.36
Japan frontier	2.90	1.70	6.04	2.15

current technological leader might not be successful since many lagging firms might be attempting to leapfrog the leader simultaneously. Correspondingly, we report both the attempted leapfroggings and the realized leapfroggings. It appears likely that the leapfrogging phenomenon would be more predominant in world frontier than in the EU frontier races.

e. Inter-frontier distance

How long does 'knowledge' take to spillover from frontier firms to sub-frontier firms? This requires investigating 'interfrontier distance'. One measure of how much sub-frontier firms' technology lags the frontier firms' technology could be graphed as 'sub-frontier lag' in terms of calendar time. At each point in time, this is simply the absolute difference in the sub-frontier performance time and the frontier performance time. The graph would clearly indicate that this measure has been declining or increasing more or less monotonically over the past 25 years to the extent that the sub-frontier firms have been able/unable to catch up with the frontier firms. A complementary measure would be to assess the difficulty of bridging the lag. That is, how much longer does it take the sub-frontier to reach a certain level of technical achievement after the frontier has reached that level. Thus it might very well turn out that the inter-frontier distance may be decreasing although the difficulty in bridging the gap is increasing.

f. Race closeness measure (RCM)

None of the previous analyses tell us how close any of the overall races are over a period of time. The races are all close by construction, however, some might be closer than others. We define 'a measure of closeness' of a race (RCM) at a particular time as follows: $RCM(t)=\sum_0^N [f_i(t)-F(t)]^2/N(t)$ where $f_i(t)$ is the firm's FTF at time t, $F(t)$ is the ITF at time $t=\max \text{FTF}(t)$ and $N(t)$ is the number of active firms at time t. The measure thus constructed has a lowest value of 0, which corresponds to a 'perfectly close' race. Higher values of the measure correspond to races that are less close. Unlike the earlier characteristics (domination period length, innovation when ahead versus when behind, leapfrogging versus frontier-sticking) which investigate the behaviour of a particular feature of the race and of a particular firm in relation to the race frontier, the RCM is more of an aggregate statistic of how close the various racing parties are at a point in time. The closeness measure is simply an indication of parity, and not one that says anything *per se* about the evolution of the technological frontier. To see this, note that if none of the frontiers were evolving, the closeness measure would be 0, as it would if all the frontiers were advancing in perfect lock-step with one another.

3.4 The Japanese race

We present here a demonstration case of a race in the Japanese mobile communications service sector which is a further (although more limited) application of industrial racing patterns.

In Japan, cellular telephone service used to be provided by 9 regional NTT DoCoMo companies and 15 other so-called Type I telecommunications carriers, as by the end of 1995 (MPT, 2000). From 1990 to 1996 there had been a dramatic increase in subscriptions for cellular telephone services, on average over 100% annual increase, far exceeding the number of standard telephone subscriptions (InfoCom Research, 2000). This process has even accelerated by the introduction of personal handy-phone systems (PHS), a low-cost mobile communication system for limited urban use. This remarkable growth was driven by intense competition causing a rapid lowering and diversification of service rates, the introduction of a user-ownership system, resulting in lower prices for cellular phones, and significant technological advances – resulting in the availability of smaller, lighter sets, digital services and longer-life batteries. All mobile operations experienced a dramatic increase in contracts during this period, most notably NTT DoCoMo, the DDI Cellular Group, the Digital Phone Group, the Tuka and the Astel Group.

From a market domination standpoint, the sample includes all the major firms, as Table 3.6 shows. Given Table 3.6, Tables 3.7–3.13 show that diversity and intensity of the Japanese innovation race for that particular industry. In the context of this case we identify an innovation as an improved product performance at the same price. It is useful to attempt a summary of the similarities and points of difference among races on the basis of the above case.

The very robust feature that appears to be common to all races is that there is a pronounced tendency for a firm to innovate (imitate) more when it falls behind in the race. Firms in the catch-up race seem to be most prone to imitative behaviour. Then frontier-pushing behaviour evidenced by firms in this race was less aggressive in that it seldom tried to leapfrog the frontier. Rather, the firms tend to exhibit more incremental/frontier-sticking behaviour than the firms in the frontier race. Overall, these facts seem to suggest that the incremental returns to a firm that occupies the race leadership position seem lower in the catch-up race than in the frontier race.

The catch-up race also differed from the other in that it was the one with the most unequal frontier-pushing behaviour. The catch-up race does appear to generate behaviour that is somewhat steadier than that exhibited

Table 3.6 Market shares of major Japanese mobile communications carriers (1996)

Firm	*Market share (%)*
NTT DoCoMo	32
DDI Cellular Group	12
IDO	8
Digital Phone Group	7
Tuka	10
Astel	12
Rest (nine firms)	19

Table 3.7 Innovation frontier race and catch-up

	Number of times each firm attempts to push the frontier or attempts a catch-up						
Firm #	*1*	*2*	*3*	*4*	*5*	*6*	*Total*
Frontier	15	8	5	–	–	–	28
Catch-up	25	20	15	10	8	5	83

Table 3.8 Domination period statistics

	Mean (years)	*Standard deviation (years)*	*n*
Frontier	1.2	1.6	24
Catch-up	2.2	0.8	83

Table 3.9A Frontier: innovations in the race

Firm #	*1*	*2*	*3*	*Total*
Total innovations	8	6	3	17
Innovations when ahead	3	2	0	5
Percentage when ahead*	37.5	33	0	33
Percentage of time ahead**	78	20	25	

Table 3.9B Catch-up: innovations (imitations) in the race

Firm #	*1*	*2*	*3*	*4*	*5*	*6*	*Total*
Total innovations/imitations	15	8	6	5	2	2	38
Numbers when ahead	2	0	1	1	0	1	5
Percentage when ahead*	22	0	17	20	0	50	19
Percentage of time ahead**	29	47	3	36	18	75	

*Percentage of innovations occurring when firm leads its race.
**Percentage of time that a firm leads its race.

Table 3.10 Firm jump sizes when ahead or behind in a race

	When ahead		*When behind*	
	Mean jump size	*# of jumps*	*Mean jump size*	*# of jumps*
Frontier	2.35	5	2.95	14
Catch-up	1.56	3	3.85	26

Table 3.11 Nature of jumps: leapfrogging or frontier-sticking

	Total jumps	*Attempted leapfrogging*	*Realized leapfrogging*
Frontier	22	20 91%	14 63%
Catch-up	30	12 40%	10 33%

Table 3.12 Inter-jump times and jump sizes

	Inter-jump times		*Jump sizes*	
	Mean	*S.D.*	*Mean*	*S.D.*
Time and jump statistics summarizing all firm technology frontiers				
Frontier	4.86	3.82	5.84	2.52
Catch-up	4.54	2.24	5.45	1.95

Table 3.13 Race closeness measures ($0 \leq c(t) \leq 0.5$; $c(t)=0$ 'very close', $c(t)=0.5$ 'lagging')

	Mean	*S.D.*
Frontier	0.05	–
Catch-up	0.25	0.18

by the frontier race. While the domination period distribution has a higher mean for the catch-up race than it does for the frontier race, its covariance is by far the smallest (Table 3.8). Related to this is the fact that the distributions of inter-jump times (i.e. the times between frontier advancements) have similar means for the two races, but the catch-up race exhibits the smallest variance (Table 3.12).

Turning to the race closeness measures, they indicate by the higher means that in the catch-up race jump sizes are larger than those of the frontier race because of the higher degree of 'laggardness' among the racers.

3.5 Further discussion

This chapter sets out to examine and measure racing behaviour on technological positions among firms in high-technology industries, as exemplified by the globally operating telecommunications and computer industries. In measuring the patterns of technological evolution in these industries we attempt to answer questions about whether and to what extent their racing patterns differ from those firms in respective industries that do not operate on a global scale. Among the key issues we want to address is the apparent inability of technology-oriented corporations to maintain leadership in fields that they pioneered. There is a presumption that firms fail to remain competitive because of agency problems or other suboptimal managerial behaviour within these organizations. An alternative hypothesis is that technologically trailing firms, in symmetric competitive situations, will devote greater effort to innovation, so that a failure of technological leaders to maintain their position is an appropriate response to the competitive environment. In asymmetric situations, with entrants challenging incumbents, research could demonstrate whether startup firms show a stronger endeavour to close up to or leapfrog the competitors. Such issues would highlight the dynamics of the race within the given market structure in any of the areas concerned. We observe two different kinds of market asymmetries bearing on racing behaviour: (a) risk-driven and (b) resource-based asymmetries.

When the incumbents' profits are large enough and do not vary much with the product characteristics, the entrant is likely to choose the faster, less-aggressive option in each stage as long as he has not fallen behind in the race. The incumbent's behaviour is influenced by what is known as the 'replacement effect' (Tirole, 1988). The conventional replacement effect says that, in an effort to maximize the discounted value of its existing profit stream, the incumbent (monopolist) invests less in R&D than an entrant, and thus expects to be replaced by the entrant (in the case where the innovation is drastic enough that the firm with the older technology would not find it profitable to compete with the newer technology). In one of our models, when the incumbent's flow profit is large enough, the same replacement effect causes the incumbent to be replaced only temporarily (if the innovation is drastic). Subsequently, she is likely to regain a dominant position in the market since she has a superior version of the new technology.

In view of resource-based asymmetries, we observe, as a firm's stage resource endowment increases, it could use the additional resources to either choose more aggressive targets or to attempt to finish the stage sooner, or both. This hypothesis suggests two interpretations, suitable for empirical exploration: (a) if the demand for new products displays different elasticities for different local/regional markets, then we might expect there to be only imperfect correlation between aggressiveness and resource richness when products from different markets are grouped together; (b) if, however, demand for these products is not inelastic enough, then we would expect resource-rich firms to aim for both higher speed in R&D and greater aggressiveness.

A further point of exploration is whether chance leads result in greater likelihood of increasing lead, or in more catch-up behaviour. Previous work in this regard (Grossman and Shapiro, 1987; Harris and Vickers, 1987) has suggested that a firm that surges ahead of its rival increases its investment in R&D and speeds up while a lagging firm reduces its investment in R&D and slows down. Consequently, previous work suggests that the lead continues to increase. However, based on related work for the US and Japanese telecommunications industry (Gottinger, 1998), when duopoly and monopolistic competition and product system complexity for new products are accounted for, the speeding up of a leading firm occurs only under rare circumstances. For example, a firm getting far enough ahead such that the (temporary) monopoly term dominates its payoff expression will always choose the fast strategy, while a firm that gets far enough behind will always choose the slow and aggressive approach. Then the lead is likely to continue to increase. If, on the other hand, both monopoly and duopoly profits increase substantially with increased aggressiveness then even large leads can vanish with significant probability.

Overall, this characterization highlights two forces that influence a firm's choices in the various stages: proximity to the finish line and distance between the firms. This probability of reaping monopoly profits is higher the farther ahead a firm is of its rival, and even more so the closer the firm is to the finish line. If the lead firm is far from the finish line, even a sizeable lead may not translate into the dominance of the monopoly profit term, since there is plenty of time for the lead situation to be reversed and failure to finish first remains a probable outcome. In contrast, the probability that the lagging firm will get to be a monopolist becomes smaller as it falls behind the lead firm. This raises the following question: what kind of actions cause a firm to get ahead? Intuitively, one would expect that a firm that is ahead of its rival at any time t, in the sense of having completed more stages by time t, is likely to have chosen the faster, less-aggressive (that is, more incremental) strategy more often. We will construct numerical estimates of the probability that a leading firm is more likely to have chosen a strategy less aggressively (faster) to verify this intuition.

Moving away from the firm-led race patterns revolving in a particular industry to a clustering of racing on an industry level is putting industry in different geoeconomic zones against each other and becoming dominant in strategic product/process technologies. Here racing patterns among industries in a relatively free-trade environment could lead to competitive advantages and more wealth creating and accumulating dominance in key product/process technologies in one region at the expense of others. The question is whether individual races on the firm level induce such like races on the industry level and, if so, what controlling effects may be rendered by regional or multilateral policies on regulatory, trade and investment matters.

The point is that racing behaviour in leading high-technology industries by generating frontier positions create cluster and network externalities pipelining through other sectors of the economy and creating competitive advantages elsewhere, as supported by the 'increasing returns' debate (Arthur, 1996). In this sense we can speak of positive externalities endogenizing growth of these economies and contributing to competitive advantage.

It is interesting to speculate on the implications of the way the firms in major high-technology markets, such as telecommunications, split clearly into the two major technology races, with one set of firms clearly lagging the other technologically. The trajectories of technological evolution certainly seem to suggest that firms from one frontier cannot simply jump to another trajectory. Witness, in this regard, the gradual process necessary for the firm in the catch-up race to approach those in the frontier race. There appears to be a frontier 'lock-in', in that once a firm is part of a race, the group of rivals within that same race are those whose actions influence the firm's strategy the most. Advancing technological capability is a cumulative process. The ability to advance to a given level of technical capability appears to be a function of existing technical capability. Given this path dependence, the question remains: why do some firms apparently choose a path of technological evolution that is less rapid than others. Two sets of possible explanations could be derived from our case analysis, which need not be mutually exclusive. The first explanation lingers primarily on the expensive nature of R&D in industries such as telecommunications and computers, which rely on novel discovery for their advancement. Firms choosing the catch-up race will gain access to a particular technical level later than those choosing the frontier, but will do so at a lower cost.

3.6 Summary and conclusion

We show how dynamic competition and hypercompetition evolved in the past and what competition in the future would look like, whether the rate of technological advancement in the industry has changed over the last 25 years, and whether divergence or convergence of frontier and catch-up races deserves attention. Furthermore, there are at least two interesting issues regarding the rate of technological advancement. The first relates to the efforts of the firms over time, and the second relates to the translation of these efforts into results:

1 We exhibit a statistical profiling of industry racing behaviour for selected high-technology industry cases. Such statistical profiling on an industrial economy level would aid policy-makers in improving industrial and microeconomic structural policies.
2 The results yield valuable, policy-relevant information on the level of regional technological frontiers of North American, European or Japanese corporations, in leading-edge, high-growth and structurally dynamic industries in view of major competitors on the world frontier.
3 Unlike other (statistical) indicators (such as patent statistics) referring to the degree of competitiveness among industries, regions and countries concerned, the proposed measures cover behavioural, dynamic movements in respective industries, and are therefore able to lend intrinsic predictive value to crucial economic outcomes relating to economic growth and wealth creation.
4 The results are likely to provide strategic support for industrial and technology policy in a European context and enable policy-makers to identify strengths and weaknesses of relevant players and their environments in those markets.

The statistical indicators derived can be adapted and extended to other high-growth and fast-developing industries.

References

Abramovitz, M. (1986) 'Catching up, forging ahead, and falling behind', *Journal of Economic History*, **66**, 385–406.

Albach, H. (1994) Information, Zeit und Wettbewerb (Information, Time and Competition), J. H. von Thünen-Vorlesung, in *Jahrestagung des Vereins für Sozialpolitik*, Münster 1993, Berlin: Duncker und Humblot, 113–154.

Albach, H. (1997) 'Global competition among the few', The Ehrenrooth Lectures, Swedish School of Economics and Business Administration, Res. Reports 40. Helsingfors.

Arthur, B. (1996) 'Increasing returns and the new world of business', *Harvard Business Review*, July–August, 100–109.

Brezis, E., Krugman, P. and Tsiddon, D. (1991) 'Leapfrogging: a theory of cycles in national technological leadership', National Bureau of Economic Research (NBER), Working Paper # 3886.

Gottinger, H. W. (1989) 'Stochastics of innovation processes', *Journal of Economics*, **49**, 123–138.

Gottinger, H. W. (1998) 'Technological races', *Annual Review of Economics (Japan)*, **38**, 1–9.

Grossman, G. and Shapiro, C. (1987) 'Dynamic R&D Competition', *Economic Journal*, **97**, 372–387.

The Group of Lisbon (1995) *Limits to Competition*, Cambridge, MA: MIT Press.

Harris, C. and Vickers, J. (1987) 'Racing with uncertainty', *Review of Economic Studies*, **54**, 305–321.

InfoCom Research (2000) *Information and Communications in Japan*, Tokyo.

Kessler, E. H. and Bierly, P. E. (2002) 'Is faster really better? An empirical test of the implications of innovation speed', *IEEE Transactions on Engineering Management*, **49**, 2–12.

Kamien, M. I. and Schwarz, N. L. (1982) *Market Structure and Innovation*, Cambridge: Cambridge University Press.

Lerner, J (1997) 'An empirical exploration of a technology race', *The Rand Journal of Economics*, **28**(2), 228–224.

Ministry of Post and Telecommunications, MPT (2000) *Communications in Japan*, White Paper, Tokyo.

Owen, R. and Ulph, D. (1994) 'Racing in two dimensions', *Journal of Evolutionary Economics*, **4**, 185–206.

TIS (2000) Telecommunications Industry Statistics, www.fcc.gov./bureaus/common_carrier/reports.

Tirole, J. (1988) *Theory of Industrial Organization*, Cambridge, MA: MIT Press.

4 Technological racing and competitive structures

> Microsoft has been caught off guard by several new technologies. In nearly every case, Microsoft first waffled, then readjusted and finally dragged its challenger into a battle of relentless product improvements.
>
> Robert A. Guth, *Wall Street Journal*, 19 May, 2003

4.1 Introduction

In studying the evolution of high-technology industries, say over the last fifty years, one is amazed by observations on the intensity and universality of rivalry among competitors across a broad selection of industries. In many cases the development of such industries were initiated and fostered by the interactive pattern of a continuous contest among market participants to get ahead of their rivals or not left too far behind. We see these patterns emerging at various stages of market evolution and seemingly unrelated to market structures.

The reactions to rival product introductions form the basis for strategic interactions so crucial to determining the firm- and industry-level technology frontiers.

Chapter 1 showed various manifestations of intense rivalry and 'neck-and-neck' competition across industries.

Overall, the recent history of high-technology industries demonstrates that dynamic competition takes place among firms in innovation-driven (Schumpeterian) industries. As we can deduce from those examples, an essential feature of industrial racing is the prevalence of actual and potential innovative threats to leading firms, coming from inside or outside the industry, that is broadly related to the notion of 'innovation markets'. Those innovations result in competitive threats based on technologies and design approaches that differ radically from those used by the incumbent.

Some descriptions of races suggest that firms adjust their R&D efforts when rivals make progress or fall behind, and, in particular, the laggard in the race gives further way to the leader. Our model assumes that the past level of R&D matters on the positioning of the R&D race for the firm, and that under most circumstances the follower works harder and is more likely to catch up with the leader than the leader is further advancing against the follower.

The given literature can be grouped into models of symmetric R&D races and multi-stage races. In a symmetric R&D race, identical firms compete for a particular innovation by investing in R&D. A firm can increase the probability that it makes the innovation by some point in time by devoting more resources to the R&D process (Loury, 1979; Lee and Wilde, 1980; Reinganum, 1982). These models assume that the time of a successful innovation is exponentially distributed, and with it comes a so-called 'memoryless property' (MLP). That is, the knowledge bases that firms have acquired as a result of their past R&D efforts are irrelevant to firms' current R&D efforts and to the outcome of the race.

Multi-stage races can be viewed as an attempt to circumvent the MLP. In multi-stage models firms are required to complete a number of stages or experience levels that lead up to the discovery of the innovation. A firm is ahead of a rival if it has a smaller number of stages left to complete and the competitors are head-to-head if they have the same number of stages left. Deterministic multi-stage models assume that firms transit to the next stage in a deterministic fashion, as in Fudenberg *et al.* (1983), Harris and Vickers (1985), and Lippman and McCardle (1988). The outcome of the race in these models suggests that even a small advantage by one firm causes the other to drop out of the race immediately. This strong result would be somewhat weakened when the stage-to-stage transitions are probabilistic. In the stochastic multi-stage models of Grossman and Shapiro (1987), Harris and Vickers (1987), and Lippman and McCardle (1987), the time to completion of each stage is assumed to be exponentially distributed. Consequently, while a firm's equilibrium R&D effort depends on the number of stages it and its rival have left to complete, within each stage an MLP renders firms' current R&D efforts independent of their past R&D efforts. In these models the leader devotes more resources to R&D than does the follower. Thus the follower tends to fall further behind as the race progresses whereas the leader tends to build up its advantage and the emerging pattern of strategic interactions is more like increasing dominance than a head-to-head racing-type strategic interaction.

In multi-stage races a firm has to complete a number of stages in advance of her rivals in order to win the race. A firm observes the number of stages her rival has left to complete and adjusts her R&D efforts accordingly. A firm is ahead of a rival if she has a smaller number of stages left to complete while she and her rival are head-to-head if they have the same number of stages left to complete, and the firm is free to alter her R&D efforts in response.

Grossman and Shapiro (1987) allow for the possibility that one firm may be ahead of the other by introducing a single intermediate step in the research programme. Thus, to win the race, a firm must complete two phases of R&D. This shows that the leader always devotes more resources to R&D than the follower. In particular, the lagging firm drops out of the race when the value of the patent is large. If the follower happens to catch up, both firms intensify their efforts. Moreover, the intensity of competition is greatest when the race is tied after the intermediate stage. Harris and Vickers (1987) extend this model to an arbitrary number of stages. They confirm that the leader always devotes more resources to R&D than does the follower. Thus the follower tends to fall further behind whereas the leader tends to build up her advantage. Moreover, the follower slows down as he falls further behind whereas the leader may or may not speed up as she gets further ahead.

The model proposed hereafter hinges on observations that it is the firm's accumulated knowledge base to determine its competitive standing (and aggressiveness) *vis-à-vis* its aspiring rivals. In realistic R&D races, we are likely to meet the following 'environmental' conditions.

First, a firm is usually able to reduce its R&D spending as its knowledge base increases. Second, as the race evolves, the follower generally works harder than the leader to keep her option for a catch-up and eventually (re)gaining leadership. The real source of this effect is that a firm's past R&D efforts have contributed to her chances in winning the R&D race. Third, a firm can react either aggressively or submissively to an increase in her rival's knowledge base. A firm acts aggressively if it has a sufficiently large knowledge base and/or the value of the patent is large enough. While empirical explorations of the R&D race would indicate that these strategic considerations are dominated by the accumulated knowledge stock, the emerging pattern of strategic interaction is more like action–reaction than increasing dominance.

In Section 4.2 we present a dynamic model of knowledge acquisition with the MLP whose sensitivities to computational parameters are explored in Section 4.3. For some basic scenarios equilibrium strategies and the payoffs derived are the subject of Section 4.4. This is followed by a social welfare analysis of the racing processes in Section 4.5, with conclusions drawn in Section 4.6.

In the next sections the modelling broadly follows previous work by Dorazelski (2003) and Reinganum (1989) and makes taxonomic use of accumulative learning dynamics in knowledge acquisition and utilization generating diverse computational paradigms of competitive dynamics. Section 4.2 explains the model construction being implemented into a few parameter variations in Section 4.3. Section 4.4 looks at computationally feasible equilibrium strategies and payoffs. Section 4.5 conducts some welfare analysis of the path-dependent outcomes, while Section 4.6 concludes the chapter.

4.2 The model

Consider an R&D race in which two firms are simultaneously seeking a particular innovation. Firms compete to be the first to make the discovery by investing in R&D. As a firm invests in R&D, its chances to make the discovery at this point in time increase and, in addition, the firm accumulates knowledge which might help it to make the discovery later on. Firms' knowledge stocks depreciate over time. Firms may differ in the knowledge they possess at the outset of the R&D race, but are identical in every other respect. Time is continuous and the horizon is infinite.

Let $z_i(t)$ denote a firm's accumulated knowledge and $u_i(t)$ its rate of knowledge acquisition. t denotes continuous time and a dot above a variable denotes its derivative with respect to time. For simplicity, we write z_i and u_i instead of $z_i(t)$ and $u_i(t)$, respectively. Firm 1's accumulated knowledge evolves according to

$$\dot{z}_1 = u_1 - \delta z_1,\ z_1(0) = z_1^0 \geq 0,\ \delta \geq 0.$$

If $\delta > 0$, firm 1's knowledge base depreciates over time. Firm 1's hazard rate of successful innovation is given by

$$h_1 = \lambda u_1 + \gamma z_1^{\varphi}$$

where λ measures the effectiveness of current R&D efforts in making the discovery and γ the effectiveness of past R&D efforts. The hazard rate might be thought of as the rate at which the discovery is made at a certain point in time given that it has not been made before. An exponential distribution of success times implies that the hazard rate is independent of past R&D efforts. Hence, if $\gamma = 0$, the model inherits the MLP of the exponential distribution. Indeed, the special case of $\gamma = 0$ corresponds to the models analyzed by Reinganum (1981, 1982).

On the other hand, the model does not suffer from the MLP if $\gamma > 0$. We allow for a nonlinear influence of the firm's accumulated knowledge on its hazard rate. In particular, the marginal impact of past R&D efforts is increasing or decreasing depending on whether the hazard rate is convex ($\varphi > 1$) or concave ($\varphi < 1$).

The firm which makes the innovation first is awarded a patent of positive value $\overline{P} > 0$, whereas its rival receives nothing if patent protection is assumed to be perfect. On the other hand, if patent protection is imperfect as the empirical evidence suggests (see for example Cohen *et al.* (2000)), the loser receives a positive payoff $\underline{P}$, where $\overline{P} > \underline{P} > 0$. $\overline{P}$ is understood to be the expected net present value (NPV) of all future revenues from marketing the innovation net of any costs the firm incurs in doing so. Similarly $\underline{P}$ is the expected NPV of the all future cash flows including costs of imitation.

Let $V_i(z_1, z_2)$ denote the value of the race to firm 1 when firm 1 has accumulated $z_1 \geq 0$ units of knowledge and firm 2 has accumulated $z_2 \geq 0$ units of knowledge. In what follows, we characterize the value function under the presumption that firms behave optimally (Bertsekas, 1987).

Consider a short interval of time of length dt. From t to $t + dt$, firm 1's accumulated knowledge changes from z_1 to $z_1' = z_1 + \dot{z}_1\, dt$. The Bellman equation is

$$V_i(z_1, z_2) = \max h_1 dt\overline{P} + h_2 dt\underline{P} - c(u_1)dt \\ + (1 - h_1 dt - h_2 dt)1/(1 + rdt)V_i(z_1', z_2'),$$

where we ignore terms of order. $(dt)^2$. $\overline{P}$ is the benefit to innovation, $\underline{P}$ the benefit to imitation and r the interest rate. The cost incurred to acquire knowledge at rate u_1 is $c(u_1) = (1/\eta)\, u_1^{\eta}$, $\eta > 1$, where η is the elasticity of the cost function. The RHS of the Bellman equation is composed of four terms. First, during a time interval of length dt, firm 1 wins the race with probability $h_1 dt$ and receives a prize $\overline{P}$. Second, firm 2 wins the race with probability $h_2 dt$ and firm 1 receives a prize $\underline{P}$. Third, firm 1 spends $c(u_1)dt$ on R&D. Fourth, if neither firm 1 nor firm 2 wins the race, an event that has probability $(1 - h_1 dt - h_2 dt)$, the race continues in state (z_1', z_2) at time $t+dt$. From the definition of the value function, this is worth $[1/(1 + rdt)]V_1(z_1', z_2')$ to firm i at time t.

Rearranging, adding and subtracting $V_1(z_1', z_2')$, and dividing by dt yields

$$0 = \max_{u_1 \geq 0} \frac{V_1(z_1', z_2') - V_1(z_1, z_2)}{dt} - \frac{V_1(z_1', z_2') - [1/(1 + rdt)]V_1(z_1', z_2')}{dt} \\ + h_1\overline{P} + h_2\underline{P} - c(u_1) - (h_1 + h_2)\frac{1}{1 + rdt}V_1(z_1', z_2').$$

Taking the limit as $dt \to 0$ and rearranging yields

$$rV_1(z_1, z_2) = \max_{u_1 \geq 0} h_1(\overline{P} - V_1(z_1, z_2)) + h_2(\underline{P} - V_1(z_1, z_2) - c(u_1) \\ + \frac{\partial}{\partial z_1}V_1(z_1, z_2)\dot{z}_1 + \frac{\partial}{\partial z_2}V_1(z_1,z_2)\dot{z}_2.$$

Hence, $V_1(z_1, z_2)$ can be interpreted as the asset or option value (to firm 1) of participating in the race. This option is priced by requiring that the opportunity cost of holding it, $rV_1(z_1, z_2)$, equals the current cash flow, $-c(u_1)$, plus the expected capital or loss flow. The latter is composed of three parts, namely the capital gain from the race, $\overline{P} - V_1(z_1, z_2)$, multiplied by the likelihood of doing so, h_1, the capital cost from losing the race, $\underline{P} - V_1(z_1, z_2)$, multiplied by the likelihood of doing so, h_2, and the capital gain or loss flow attributable to changes in the knowledge stocks $\frac{\partial}{(\partial z_1)}V_1(z_1, z_2)\dot{z}_1 + \frac{\partial}{(\partial z_2)}V_1(z_1, z_2)\dot{z}_2$.

Substituting for h_1, h_2, $\dot{z}_1$ and $\dot{z}_2$ and rearranging, the Bellman equation becomes

$$\begin{aligned}0 = \max_{u_1 \geq 0}(\lambda u_1 + \gamma z_1^{\varphi})\overline{P} + (\lambda u_2 + \gamma z_2^{\varphi})\underline{P} - \frac{1}{\eta}u_1^{\eta} \\ - (r + \lambda u_1 + \gamma z_1^{\varphi} + \lambda u_2 + \gamma z_2^{\varphi})V_1(z_1, z_2) \\ + \frac{\partial}{\partial z_1}V_1(z_1, z_2)(u_1 - \delta z_1) + \frac{\partial}{\partial z_2}V_1(z_1, z_2)(u_2 - \delta z_2).\end{aligned}$$

We focus on symmetric stationary Nash equilibria in feedback strategies. Hence, firm i's strategy $u_1 : [0, \infty)^2 \rightarrow [0, \infty)$ maps the accumulated knowledge of firms 1 and 2 into a rate of knowledge acquisition.

Let $u_2^*(z_1, z_2)$ denote firm 2's equilibrium strategy. Carrying out the indicated maximization yields firm 1's first-order condition (FOC) for an interior solution $u_2^*(z_1, z_2) > 0$. Since the objective function is strictly concave due to $\eta > 1$, the FOC is also sufficient for an interior solution. Hence,

$$u_1^*(z_1, z_2) = (\lambda(\overline{P} - V_1(z_1, z_2)) + \frac{\partial}{\partial z_1}V_1(z_1, z_2))^{\frac{1}{\eta - 1}}. \tag{4.1}$$

The firm's incentives to engage in R&D are governed by two considerations. First, as the firm invests an additional dollar in R&D, its chances to make the discovery at this point in time increase by λ. Since the capital gain from winning the race is $\overline{P} - V_1(z_1, z_2)$, the marginal benefit accruing to the firm therefore is $\lambda(\overline{P} - V_1(z_1, z_2))$. Second, the firm expands its own knowledge base which carries a marginal benefit of $\frac{\partial}{(\partial z_1)}V_1(z_1, z_2)$.

Define $V_1(z_1, z_2) = V(z_1, z_2)$ and $u_1^*(z_1, z_2) = u^*(z_1, z_2)$. Then, using symmetry, the value of the race to firm 2 when firm 1 has accumulated z_1 units of knowledge and firm 2 has accumulated z_2 units of knowledge is given by $V_2(z_1, z_2) = V(z_2, z_1)$ and firm 2's equilibrium strategy is $u_2(z_1, z_2) = u(z_2, z_1)$. The Bellman equation at $u_1(z_1, z_2) = u(z_1, z_2)$ and $u_2(z_1, z_2) = u(z_2, z_1)$ becomes

$$\begin{aligned}0 = \lambda u(z_1, z_2) + \gamma z_1^{\varphi})\overline{P} + (\lambda u(z_2, z_1) + \gamma z_2^{\varphi})\underline{P} - \frac{1}{\eta}u^*(z_1, z_2)^{\eta} \\ - (r + \lambda u^*(z_1, z_2) + \gamma z_1^{\varphi} + \lambda u^*(z_1, z_2) + \gamma z_2^{\varphi})V_1(z_1, z_2) \\ + \frac{\partial}{\partial z_1}V(z_1, z_2)(u^*(z_2, z_1) - \delta z_1) \\ + \frac{\partial}{\partial z_2}V(z_1, z_2)(u^*(z_2, z_1) - \delta z_2).\end{aligned} \tag{4.2}$$

Equation (4.1) can be substituted into equation (4.2) to obtain one functional equation in one unknown function.

Proposition 4.1: *Let* $\gamma = 0$. *Then the partial differential equation (PDE) in equation (4.2) admits a unique constant solution* V, $0 < \mathrm{V} < \overline{\mathrm{P}}$, characterized by

$$0 = \lambda u^*(\overline{P} + \underline{P}) - \frac{1}{\eta}(u^*) - (r + 2\lambda u^*)V \tag{4.3}$$

where

$$u^* = (\lambda(\overline{P} - V))^{\frac{1}{\eta-1}}. \tag{4.4}$$

Proof. At a constant solution V, equations (4.2) and (4.1) reduce to equations (4.3) and (4.4), respectively. Define

$$\Delta(V) = (\lambda(\overline{P} - V))^{\frac{\eta}{\eta-1}}\left(1 - \frac{1}{\eta}\right) + \lambda(\lambda\overline{P} - V)^{\frac{1}{\eta-1}}(\underline{P} - V) - rV.$$

At $V = \overline{P}$, we have

$$\Delta(\overline{P}) = -r\overline{P} < 0.$$

At $V = 0$, we have

$$\Delta(0) = (\lambda\overline{P})^{\frac{\eta}{\eta-1}}\left(1 - \frac{1}{\eta}\right) + \lambda(\lambda\overline{P})^{\frac{1}{\eta-1}}\underline{P} > 0$$

since $\eta > 1$. Since $\Delta(V)$ is continuous in V, there exists a solution to $\Delta(V) = 0$ by the intermediate value theorem.

It remains to establish uniqueness of the solution. We have

$$\Delta'(V) = -2\lambda(\lambda(\overline{P} - V))^{\frac{1}{\eta-1}} - \frac{\lambda^2}{\eta - 1}(\lambda(\overline{P} - V))^{\frac{2-\eta}{\eta-1}}(\underline{P} - V) - r,$$

$$\Delta''(V) = \frac{3\lambda^2}{\eta - 1}(\lambda(\overline{P} - V))^{\frac{2-\eta}{\eta-1}} + \frac{\lambda^3(2 - \eta)}{(\eta - 1)^2}(\lambda(\overline{P} - V))^{\frac{3-2\eta}{\eta-1}}(\underline{P} - V).$$

Rearranging yields

$$\Delta''(V) = \frac{\lambda^2}{\eta - 1}(\lambda(\overline{P} - V))^{\frac{2-\eta}{\eta-1}} + \left(3 + \frac{2 - \eta}{\eta - 1}\frac{\underline{P} - V}{\overline{P} - V}\right).$$

Note that the term in the second parenthesis governs the sign of $\Delta''(V)$. Differentiating it yields

$$-\frac{2-\eta}{\eta-1}\frac{\overline{P}-\underline{P}}{(\overline{P}-V)^2}$$

which is non-negative if $\eta \geq 2$ and non-positive if $1<\eta<2$. Consider the case of $\eta \geq 2$ first. Then the term in parentheses is non-decreasing in V and achieves its minimum of

$$3+\frac{2-\eta}{\eta-1}\frac{\underline{P}}{\overline{P}} \geq 2$$

at $V=0$, where the last inequality uses the facts that $0 \leq \underline{P}/\overline{P} \leq 1$ and $-1 \leq (2-\eta)/(\eta-1) \leq 0$ whenever $\eta \geq 2$. Hence, $\Delta''(V) \geq 0$ and the claim follows. Consider the case of $1<\eta<2$ next. Then the term in parenthesis is non-increasing in V, achieves its maximum of

$$3+\frac{2-\eta}{\eta-1}\frac{\underline{P}}{\overline{P}} \geq 3$$

at $V=0$, and approaches $-\infty$ as V approaches $\overline{P}$. By continuity it follows that $\Delta''(V) \geq 0$ around $V=0$ and $\Delta''(V) \leq 0$ around $V=\overline{P}$. Since the term in parenthesis changes sign at most once, so does $\Delta''(V)$, and the claim follows.

4.3 Parameter scenarios

The numerical techniques to solve the PDE in equation (4.2) require specific parameter values.

The model has eight parameters, the interest rate r, the rate of depreciation δ, the elasticity of the cost function η, the effectiveness of past R&D efforts γ, the curvature with respect to past R&D efforts ϕ, the value of innovation $\overline{P}$, and the value of imitation $\underline{P}$. While we are able to fix some parameters by normalizing the units of measurement, we choose the remaining ones to match empirical observations from the biotech/pharma industry (Gottinger, 2004).

By setting $c=1$ in the cost function $c(u_1)=(c/\eta)u_1^{\eta}$, $c>0$, we have already normalized monetary units. The time scale is determined once we choose the interest rate r. The following Lemma shows that we are free to impose an additional normalization by choosing the unit of measurement of knowledge.

Lemma 4.2: *Suppose that the value and policy functions* $\mathrm{V}(z_1, z_2)$ *and* $\mathrm{u}(z_1, z_2)$ *satisfy the first-order condition and Bellman equation.*

Rescale knowledge by defining $\bar{z} = cz$, $c > 0$. Define the value and policy functions on the rescaled state space as

$$\overline{V}(\bar{z}_1, \bar{z}_2) = c^{\eta} V(z_1, z_2),$$
$$\bar{u}(\bar{z}_1, \bar{z}_2) = cu^*(z_1, z_2).$$

Then $\overline{V}(\bar{z}_1, \bar{z}_2)$ and $\bar{u}^*(\bar{z}_1, \bar{z}_2)$ satisfy the first-order condition and Bellman equation defined on the rescaled state space provided that $\overline{P}$, $\underline{P}$, λ, and γ are rescaled appropriately.

Proof. Rescale the parameters by replacing $\overline{P}$ with $c^{\eta}\overline{P}$, $\underline{P}$ with $c^{\eta}\underline{P}$, λ with λ/c, and γ with γ/c^{φ}. These definitions imply that the hazard rate remains the same across parameterizations, i.e. $\bar{h}_i = h_i$. Moreover, $\dot{\bar{z}}_i = c\dot{z}_i$, $(\partial\bar{z}_1/\partial\bar{u}_1) = 1$, $(\partial\overline{V}/\partial\bar{z}_1)(\bar{z}_1, \bar{z}_2) = c^{\eta-1}(\partial V/\partial z_1)(z_1, z_2)$, and $(\partial\overline{V}/\partial\bar{z}_2)(\bar{z}_1, \bar{z}_2) = c^{\eta-1}(\partial V/\partial z_2)(z_1, z_2)$. Hence,

$$\frac{\lambda}{c}c^{\eta}\overline{P} - \overline{u^*}(\bar{z}_1, \bar{z}_2)^{\eta-1} - \frac{\lambda}{c}\overline{V}(\bar{z}_1, \bar{z}_2) + \frac{\partial\overline{V}}{\partial\bar{z}_1}(\bar{z}_1, \bar{z}_2) = 0$$
$$\Leftrightarrow \lambda c^{\eta-1}\overline{P} - c^{\eta-1}u^*(z_1, z_2)^{\eta-1} - \lambda c^{\eta-1}V(z_1, z_2) + c^{\eta-1}\frac{\partial V}{\partial z_1}(z_1, z_2) = 0$$
$$\Leftrightarrow \lambda\overline{P} - u^*(z_1, z_2)^{\eta-1} - \lambda V(z_1, z_2) + \frac{\partial V}{\partial z_1}(z_1, z_2) = 0$$

which is the same as the first-order condition defined on the original state space. Similarly,

$$0 = \bar{h}_1 c^{\eta}\overline{P} + \bar{h}_2 c^{\eta}\underline{P} - \frac{1}{\eta}\overline{u^*}(\bar{z}_1, \bar{z}_2)^n - (r + \bar{h}_1 + \bar{h}_2)\overline{V}(\bar{z}_1, \bar{z}_2)$$
$$+ \frac{\partial\overline{V}}{\partial\bar{z}_1}(\bar{z}_1, \bar{z}_2)\dot{\bar{z}}_1 + \frac{\partial\overline{V}}{\partial\bar{z}_2}(\bar{z}_1, \bar{z}_2)\dot{\bar{z}}$$
$$\Leftrightarrow 0 = h_1 c^{\eta}\overline{P} + h_2 c^{\eta}\underline{P} - c^{\eta}\frac{1}{\eta}u(z_1, z_2)^n - (r + h_1 + h_2)c^{\eta}V(z_1, z_2)$$
$$+ c^{\eta}\frac{\partial V}{\partial z_1}(z_1, z_2)\dot{z}_1 + c^{\eta}\frac{\partial V}{\partial z_2}(z_1, z_2)\dot{z}_2$$
$$\Leftrightarrow 0 = h_1\overline{P} + h_2\underline{P} - \frac{1}{\eta}u^*(z_1, z_2)^n - (r + h_1 + h_2)V(z_1, z_2)$$
$$+ \frac{\partial V}{\partial z_1}(z_1, z_2)\dot{z}_1 + \frac{\partial V}{\partial z_2}(z_1, z_2)\dot{z}_2$$

which is again the same as the Bellman equation defined on the original state space.

In light of Lemma 4.2, we set γ=1. The effectiveness of a firm's knowledge base in making the discovery is determined by both γ and φ. In particular, the marginal impact of past R&D efforts is increasing or decreasing depending on whether the hazard rate is convex ($\varphi > 1$) or concave ($\varphi < 1$). We explore the effect of a non-linear influence of the firm's accumulated knowledge on its hazard rate by contrasting a model with $\varphi = 1$ to models with $\varphi \neq 1$.

Reasonable values for the remaining parameters can be obtained using data from the pharmaceutical industry. For example, Henderson and Cockburn (1996) examine the determinants of research productivity in the pharmaceutical industry. Estimating the impact of research expenditures on the number of patent grants, they find an elasticity of current and past expenditures of 0.030 and 0.035, respectively. While there is no analogue to the number of patent grants in this model, their results suggest that current and past R&D efforts are roughly of the same importance, and accordingly we set $\lambda = \gamma$.

The rate at which knowledge depreciates is notoriously hard to estimate. Griliches (1980), for example, indicates that his estimate of R&D productivity in manufacturing industries is consistent with a wide range of depreciation rate ranging from $\delta = 0$ to $\delta = 0.3$. To construct a proxy for the stock of knowledge, Henderson and Cockburn (1994, 1996) use a depreciation rate of $\delta = 0.2$ in their study of the pharmaceutical industry, and we assume likewise.

The returns to R&D in the pharmaceutical industry have been studied thoroughly. Based on a sample of 67 New Chemical Entities (NCEs) that were introduced in the US between 1980 and 1984, Grabowski and Vernon (1994) estimate an internal rate of return of 11.1% which compares to a real cost-of-capital estimate of 10.5%. The latter estimate is based on all business activities of pharmaceutical firms and reflects the real after-tax cost of capital on equity plus debt for the pharmaceutical industry. If capital markets are perfect, the discount rate equals the cost of capital. We therefore set $\gamma = 0.105$. Finally, the average time until the discovery of a NCE is 3 years followed by another 9 years of testing prior to marketing. Since the model of an R&D race applies to the discovery phase, we implicitly assume that $\overline{P}$ and $\underline{P}$ captures all costs incurred after the first 3 years. Nevertheless, we maintain that $\underline{P} = 0.2\overline{P}$ in line with the market shares for innovative drugs since imitation tends to be less costly than innovation (see Mansfield *et al.* (1981)). The remaining parameters, $\overline{P}$ and η, could in principle be chosen such that the expected duration of the race equals 3 years and the expected internal rate of return equals 11.1%. There are two difficulties, however.

First, the expected duration of the race and the expected internal rate of return depend on the initial knowledge bases of the competing firms, z_1^0 and z_2^0. Differences in the available knowledge at the outset of the race can be interpreted as distinctive competencies of the participating firms. Henderson and Cockburn (1994, 1996) have shown that research productivity varies greatly among pharmaceutical firms and depends on their past experience with similar research programme as well as the organizational characteristics of their research divisions. The initial knowledge bases could in part be chosen by specifying the ratio of the number of races won by the 'technologically superior' firm to the number of races won by the 'technologically inferior' firm. This would, however, require at least one additional data point. Hence, I set $z_1^0 = z_2^0 = 0$ for now and later explore the effects of differences in the initial knowledge bases.

Second, as the following Lemma shows for the special case of γ=0, the expected internal rate of return may fail to be finite.

Lemma 4.3: *Let $\gamma = 0$. Then the expected internal rate of return is infinite.*

Proof. By proposition 4.1, the value function $V(z_1, z_2)$ and R&D expenditures $u^*(z_1, z_2)$ are constants. Let $\gamma = \min\{\tau_1, \tau_2\}$, where τ_1 is the random date of a successful innovation by firm *i*. The expected payoff of firm 1 given by

$$V = E\left\{e^{-r\tau}(\overline{P}1(\tau_1 = \tau) + \underline{P}1(\tau_2 = \tau)) - \int_0^{\tau} e^{-r\tau}c(u^*)dt\right\}$$

$$= \int_0^{\infty} e^{-r\tau - \int_0^t h_1(s) + h_2(s)ds} (\overline{P}h_1(t) + \underline{P}h_2(t) - c'(u^*))dt$$

$$= \frac{\lambda u^*(\overline{P} + \underline{P}) - \frac{1}{\eta}(u^*)^{\eta}}{r + 2\lambda u^*}$$

where 1(.) denotes the indicator function and $h_1(t) = \lambda u^*$ is the hazard rate of firm *i*. Proposition 1 gives $V > 0$. Since the expected internal rate of return is defined as the value of γ that makes the above expression zero, it follows that the expected internal rate of return is infinite.

It is of course possible to restore the finite expected internal rate of return by introducing a fixed set up cost into the model. However, such a fixed set up cost does not affect the behaviour of firms during the race (albeit it governs their decision of whether or not to participate in the race) and hence seems ill-suited for our purposes. Absent a constraint on the expected internal rate of return (*In*), there is a host of values of $\overline{P}$ and η such that the expected duration of the race equals 3 years. Fortunately, firms' behaviour is qualitatively similar over a wide range of such values. In particular, we have compared models with $\eta = 2$ to models with $\eta = (In3/In2)$ and $\eta = 3$, respectively. If $\eta = (In3/In2)$, a doubling of R&D efforts leads to a tripling of costs, where as if η=3, a doubling of R&D efforts increases costs eightfold. Despite this substantial variation in the cost function, the value and policy functions remain qualitatively similar. We thus set η=2 in what follows.

We explore five scenarios. The special case of γ=0 corresponds to the models analysed by Reinganum (1981a, 1982) in that the model inherits the MLP of the exponential distribution. The polar case to γ=0 is λ=0, which gives rise to a model in which a firm's current R&D efforts do not directly aid it in winning the race, but indirectly aid it by adding to the firm's knowledge base. Since $\gamma > 0$, the model does not suffer from the MLP.

Between the polar cases of $\gamma=0$ and $\lambda=0$ are models in which both current and past R&D efforts have an impact on a firm's hazard rate. The starting point is $\varphi=1$, leading to a model in which the hazard rate is linear in the firm's knowledge base. We next allow for a non-linear influence of the firm's accumulated knowledge on its hazard rate. In particular, the marginal impact of past R&D efforts as increasing or decreasing depends on whether the hazard rate is convex ($\varphi > 1$) or concave ($\varphi < 1$). We obtain a concave or a convex hazard rate by setting $\varphi = 1/2$ and $\varphi = 2$.

Table 4.1 summarizes the parameter values for five scenarios.

4.4 Equilibrium strategies and payoffs

We explore the five scenarios listed in Table 4.1. In each case, we report the value function $V(z_1, z_2)$ and R&D expenditures $c(u^*(z_1, z_2))$, an increasing and convex transformation of R&D effort $u^*(z_1, z_2)$. We consider R&D expenditures because they are measured in terms of monetary units divided by time units rather than 'knowledge units' per unit of time. In addition, we measure the intensity of competition as the sum of R&D expenditures $c(u^*(z_1, z_2)) + c(u^*(z_2, z_1))$ and we look at the sum of the value functions $V(z_1, z_2) + V(z_2, z_1)$to gauge the impact of competition on the joint payoff of the participating firms.

1. Current R&D effort alone

Proposition 4.1 guarantees that there exists a unique constant solution for the value and policy functions whenever $\gamma = 0$. The model inherits the MLP of the exponential distribution and, although firms accumulate knowledge during the course of the R&D race, the equilibrium payoffs and strategies are independent of the knowledge bases and are given by $V = 0.0815$ and $c(u^*) = 0.0139$. It follows that the sum of R&D expenditures is $2c(u^*) = 0.0227$ and the combined value of the race to firms 1 and 2 is

Table 4.1 Parameter values

	λ	γ	φ	$\overline{P}$
Polar cases:				
Current R&D effort alone	1	0	–	0.248
Past R&D effort alone	0	1	1	0.137
Intermediate cases:				
Linear hazard rate	1	1	1	0.0435
Concave hazard rate	1	1	½	0.0350
Convex hazard rate	1	1	2	0.1035

$2V = 0.1630$. Since the knowledge bases that firms have acquired as a result of their past R&D efforts are irrelevant to firms' current R&D efforts and therefore to the outcome of the race, there is no sense in which one can properly speak of one competitor being ahead of another, or the two competitors being neck-and-neck.

2. *Past R&D effort alone*

In this case the value equilibrium payoffs and strategies are no longer independent of firms' knowledge bases. Obviously learning (and forgetting) shape firms' equilibrium strategies and payoffs. First, R&D expenditures $c(u^*(z_1, z_2))$ are decreasing in z_1, i.e. a firm decreases its current R&D effort as its knowledge base becomes larger. Second, $c(u^*(z_1, z_2))$ can be either decreasing or increasing in z_2. Inspection of the partial derivatives of $c(u^*(z_1, z_2))$ reveals that $\frac{\partial}{\partial z_2} c(u(z_1, z_2)) > 0$ whenever $z_1 \geq 0.6886 + 0.4856 z_2$. Hence, a firm sometimes increases its own R&D effort as its rival accumulates knowledge. This 'aggressive' behaviour is confined to the leading firm whereas the firm with the smaller knowledge base always reacts 'submissively'.

Third, the follower devotes less resources to R&D than does the leader. The reason lies in the *pure progress effect* which, independently of strategic considerations, causes a firm to increase its R&D effort as it gets closer to the finishing line (Grossman and Shapiro, 1987). Intuitively, the gain from winning the race from the intermediate stage of a two-stage race is larger than the gain from making the intermediate discovery from the initial stage. Consequently, the R&D investment of the leading firm exceeds that of the lagging firm.

Underlying these results is what might be called the *pure knowledge effect*. The pure knowledge effect is independent of strategic considerations. The source of this effect is that the firm's past R&D efforts contribute to its chances of winning the R&D race because the firm's knowledge base enters its hazard rate. Hence, knowledge has a productive function in the model. In contrast, the stages of a multi-stage R&D race are mere labels: due to the exponential distribution of success times in each stage, the expected time until a stage is completed is *ceteris paribus* the same in all stages of a multi-stage R&D race. Due to the pure knowledge effect, a firm is able to conserve on its R&D investment as its knowledge base increases, i.e. $c(u^*(z_1, z_2))$ is decreasing in z_1. Moreover, the firm with the larger knowledge base is able to conserve more on its investment in R&D than is the firm with the smaller knowledge base. The follower thus devotes more resources to R&D than the leader, i.e. $c(u^*(z_1, z_2)) > c(u^*(z_2, z_1))$ if and only if $z_1 < z_2$. Since the leader's chances of winning the race *ceteris paribus* exceed those of the follower, the value of the race to the follower is, therefore, less than the value to the leader, i.e. $V(z_1, z_2) < V(z_2, z_1)$ if and only if $z_1 < z_2$.

Fourth, $V(z_1, z_2)$ is increasing in z_1 and decreasing in z_2, i.e. an increase in its own knowledge base benefits the firm but an increase in the rival's knowledge base hurts it. The reason is again that knowledge is productive. Indeed, as the following proposition shows, a firm benefits in the limit as the size of its own knowledge base approaches infinity since it wins the race for certain, whereas the firm loses the race for sure as the size of its rival's knowledge base approaches infinity.

Proposition 4.4

$$\lim_{z_1 \to \infty} V(z_1, z_2) = \overline{P} \textit{ and } \lim_{z_2 \to \infty} V(z_1, z_2) = \underline{P}$$

Proof. Consider an R&D race that starts from the initial knowledge bases z_1^0 and z_2^0. Let $\gamma = \min\{\tau_1,\tau_2\}$, where τ_1 is the random date of a successful innovation by firm i, and pick an arbitrary time path of firm i's current R&D effort $u_i(t)$. The expected payoff of firm 1 is given by

$$\overline{J}_1(u_1(.), u_2(.), z_1^0, z_2^0)$$

$$= E\{e^{-r\tau}(\overline{P}1(\tau_1 = \tau) + \underline{P}1(\tau_2 = \tau) - \int_0^{\tau} e^{-rt}c(u_1(t))dt\}$$

where 1(.) denotes the indicator function.

The cumulative distribution function (cdf) of τ_i is given by $F_i(t) = 1 - \exp(-\int_0^{\tau} h_i(s)ds)$ and the cdf of τ by $F(t) = 1 - \exp(-\int_0^{\tau} h_1(s) + h_2(s)ds)$, where $h_i(t)$ is the underlying hazard rate of firm i. To see how the expected payoff depends on the initial knowledge of firms 1 and 2 recall that firm 1's hazard rate of successful innovation is

$$h_1(t) = \lambda u_1(t) + \gamma z_1(t)^{\varphi}$$

and the accumulation equation implies

$$z_1(t) = z_1^0 e^{-\delta t} + \int_0^t u_1(s)e^{-\delta(t-s)}ds,$$

where we have made the dependence of the knowledge stock on time explicit. It follows that $h_i(t) \to \infty$ as $z_i^0 \to \infty$ for all t close to zero. This implies $F_i(t) \to 1(t \geq 0)$ and $F_i(t) \to 1(t \geq 0)$ as $z_i^0 \to \infty$. Hence, the limiting distributions of τ_i and τ are that of a mass point at zero. Consequently,

$$\lim_{z_1 \to \infty} \overline{J}_1(u_1(.), u_2(.), z_1^0, z_2^0) = \overline{P},$$

$$\lim_{z_2 \to \infty} \overline{J}_1(u_1(.), u_2(.), z_1^0, z_2^0) = \underline{P},$$

provided that $u_1(t)$ and $u_2(t)$ are finite.

Since the above argument holds for arbitrary time paths $u_1(.)$ and $u_2(.)$, it holds for the time paths induced by the equilibrium strategies. But $\bar{J}_1(u_1(.), u_2(.), z_1^0, z_2^0)$ evaluated at the equilibrium strategies equals $V(z_1^0, z_2^0)$which proves the claim.

Fifth, the measure of the intensity of competition, $c(u^*(z_1, z_2)) + c(u^*(z_2, z_1))$, is convex and the combined value of the race to firms 1 and 2, $V(z_1, z_2) + V((z_2, z_1)$, is concave. Hence, holding the combined amount of knowledge constant, competition is more intense among firms with unequal knowledge stocks than among firms with equal knowledge stocks. This contrasts with the literature on multi-stage races, where competition tends to be fiercest when firms are neck-and-neck. The concavity of $V(z_1, z_2) + V((z_2, z_1)$ mirrors the convexity of $c(u^*(z_1, z_2)) + c(u^*(z_2, z_1))$. As we discuss in Section 4.5, the non-cooperative game involves too much competition as compared to the planner's solution whenever the gap between the payoff to the winner $\overline{P}$ and the payoff to the loser $\underline{P}$ is large. Hence, since competition is less intensive when firms are neck-and-neck than when one firm is ahead of the other, there is less rent dissipation when firms are neck-and-neck than when one firm is ahead of the other.

3. Intermediate cases: linear hazard rate

There are important similarities with the case of past R&D effort. In particular, $c(u^*(z_1, z_2))$ is decreasing in z_1 but can either be decreasing or increasing in z_2 and, while the follower works harder than the leader, the value of the race is greater to the leader than to the follower.

However, there are also important differences. First, the difference between the follower's and the leader's current R&D effort is larger. The follower works harder than the leader if and only if $c(u^*(z_1, z_2)) > c(u^*(z_2, z_1))$ for all $z_1 < z_2$ which is equivalent to

$$-\lambda V(z_1, z_2) + \frac{\partial}{\partial z_1} V(z_1, z_2) > -\lambda V(z_2, z_1) + \frac{\partial}{\partial z_1} V(z_2, z_1)$$

for all $z_1 < z_2$. In case $\lambda = 0$, a necessary and sufficient condition is thus $\frac{\partial}{\partial z_1} V(z_1, z_2) > \frac{\partial}{\partial z_1} V(z_1, z_2)$. In case of $\lambda > 0$, $\frac{\partial}{\partial z_1} V(z_1, z_2) > \frac{\partial}{\partial z_1} V(z_1, z_2)$ and $V(z_1, z_2) < V(z_2, z_1)$ are jointly sufficient. But since the value of the race to the leader exceeds its value to the follower, the first effect is reinforced by the second.

Second, $c(u^*(z_1, z_2))$ decreases less rapidly with z_1. The reason is that a firm has an additional incentive to invest in R&D since its current R&D effort not only adds to the firm's knowledge base but also impacts its hazard rate.

Third, $c(u^*(z_1, z_2))$ changes more slowly with z_2 since the firm compensates for the additional knowledge of its rival by investing more in R&D. Nevertheless, $c(u^*(z_1, z_2))$ decreases in z_2 in particular around the origin.

Intuitively, one might feel that there is no point in trying even harder as one's rival advances faster. On the other hand, firm 1 should invest in R&D up to a point where the expected marginal benefit equals the marginal cost. Since the probability that firm 2 wins the R&D race in the next short interval of time dt is increasing in z_2, the probability that firm 1 has to sustain whatever current R&D effort it chooses beyond time $t+dt$ is decreasing in z_2. Hence, the point where the expected marginal benefit of investment in R&D equals its marginal cost is reached for an increasing level of R&D investment. Phrased differently, as firm 2 advances, firm 1 takes its chances and invests heavily but briefly in R&D. Moreover, since the expected duration of the race is decreasing in z_1, aggressive behaviour on behalf of firm 1 becomes more likely as it accumulates knowledge.

4. Concave hazard rate

The main features of the case of a linear hazard rate are preserved. One minor but noteworthy difference is that the intensity of competition is by far the greatest along the axes of the state space and declines rapidly towards the diagonal of accumulated knowledge. Holding the combined amount of knowledge constant, the difference in firms' knowledge stocks is maximal at the axes and minimal at the diagonal. This feature is due to a combination of two facts. First, due to the decreasing returns nature of its own hazard rate, a firm's incentive to accumulate knowledge deteriorates quickly. Second, along the axes $c(u^*(z_1, z_2))$ decreases slowly with z_2 since, due to the decreasing returns nature of its rival's hazard rate, the firm is easily able to compensate for the additional knowledge of its rival by investing more in R&D. The concavity of $V(z_1, z_2) + V(z_2, z_1)$ again mirrors the convexity of $c(u^*(z_1, z_2)) + c(u^*(z_2, z_1))$.

5. Convex hazard rate

Unlike the case of a concave hazard rate, this case differs greatly from the case of a linear hazard rate. First, $c(u^*(z_1, z_2))$ first increases in z_1, then decreases. While the knowledge effect eventually dominates, the increasing returns nature of the hazard rate initially gives a firm a strong incentive to invest in R&D. Again, it can be either decreasing or increasing in z_2.

Second, as a consequence, the follower may work either more or less than the leader. In particular, the follower initially devotes less resources to R&D than the leader, i.e. $c(u^*(z_1, z_2)) < c(u^*(z_2, z_1))$ if z_1 is small and $z_1 < z_2$, but eventually devotes more resources to R&D than the leader, i.e. $c(u^*(z_1, z_2)) > c(u^*(z_2, z_1))$ if z_1 is large and $z_1 < z_2$. A closer inspection shows that a sufficient condition for $c(u^*(z_1, z_2)) > c(u^*(z_2, z_1))$ is $z_1 \geq 0.6446 - z_2$ and $z_1 < z_2$. Nevertheless, the value of the race to the leader continues to exceed its value to the follower.

Third, $c(u^*(z_1, z_2)) > c(u^*(z_2, z_1))$, the measure of the intensity of competition is quasi-concave and $V(z_1, z_2) + V(z_2, z_1)$ is quasi-convex. Competition is equally intense on an ellipsis with center on the diagonal. Holding the combined amount of knowledge constant, this implies that competition is most intense when firms are neck-and-neck. The quasi-convexity of $V(z_1, z_2) + V(z_2, z_1)$ again mirrors the quasi-concavity of $c(u^*(z_1, z_2)) > c(u^*(z_2, z_1))$.

Discussion

The results appear at least partially to fly in the face of the existing literature. Contrary to the literature on symmetric R&D races, where the knowledge bases that firms have acquired as a result of their past R&D efforts are irrelevant to firms' current R&D efforts and therefore to the outcome of the race, we show that learning and forgetting shape firms' equilibrium payoffs and strategies. Opposite to the common wisdom on multi-stage races, we find that a firm is able to reduce its R&D investment as its knowledge base increases and that the follower works harder than the leader. In the case of a convex hazard rate, the increasing returns nature of the hazard rate gives rise to a counteracting force because it provides the firm with a strong incentive to invest in R&D.

Underlying the results is the *pure knowledge effect*. This effect arises because a firm's past R&D efforts contribute to its chances of winning the R&D race as the firm's knowledge base enters its hazard rate. In contrast, the expected time until a stage is completed is *ceteris paribus* the same in all stages of a multi-stage R&D race due to the exponential distribution of success times in each stage. In other words, the stages of a multi-stage R&D race are mere labels, whereas knowledge plays a productive role in the model.

A firm can either react aggressively or submissively to an increase in its rival's knowledge base. In multi-stage models, the follower slows down as he falls further behind whereas the leader may or may not speed up as he gets further ahead. In the model, aggressive or submissive behaviour is not tied to a firm's relative position. Rather, a firm reacts aggressively if it has a sufficiently large knowledge base and reacts submissively otherwise.

Lastly, we find in contrast to some literature on multi-stage races that competition is not necessarily fiercest when firms are neck-and-neck. Competition is generally most intense along the axes and least intense along the diagonal. In case of a convex hazard rate, however, competition is equally intense on an ellipsis with its centre on the diagonal.

Despite the abundance of anecdotal evidence, empirical research on R&D races is sparse. Most empirical research on R&D races has used firm-level investment data to test the reactions between the leader and the follower in an industry. In line with this model, these studies observe patterns of strategic interactions that are similar to action–reaction rather than increasing dominance. Grabowski and Baxter (1973) found that in the chemical industry firms increase R&D expenditures in response to rivals' outlays and that the two largest firms respond quickly to changes in each other's R&D policies. Exploratory research indicates that leaders and followers react positively to each other's increases in R&D expenditures, while fringe firms set R&D expenditures independently. Scherer (1992) found that firms with greater domestic sales in more concentrated U.S. markets were likely to react much more aggressively to increasing import competition than smaller firms or firms in less concentrated markets.

A major problem in empirical work is that it has to disentangle movements along firms' reaction curves for given knowledge bases from shifts in these reaction curves due to changes in the knowledge bases. More recent studies thus attempt to operationalize the notion of a knowledge base. These studies again lend more support to action–reaction than to increasing dominance. In Chapter 2 we proposed statistical measures on a firm's technological position relative to its rivals by constructing a technological frontier for the high-end computer industry. We identify national or regional clusters of firms and find that within each cluster firms race against each other for the technological leadership position. We find that firms that fall behind the technological frontier may engage in catch-up behaviour. Focusing on the outputs of the R&D process rather than on its inputs, Lerner (1997) attempts to directly measure a firm's technological position relative to its rivals. In his study of the disk drive industry, he finds that firms that trail the industry leader display a greater propensity to innovate. In the present context, however, this finding must be interpreted with some care. Since the current generation of disk drives is superseded by the next generation, it may well be the case that the rewards from winning the race are lower for the leader than for a follower because the leader replaces itself, which in turn would give the leader an incentive to decrease its current efforts, i.e. product market competition allows the replacement effect to operate.

4.5 Welfare analysis

We now study the welfare implications by comparing the outcome of the non-cooperative game to the planner's solution. The planner strives to maximize the benefits of the innovation to society by prescribing firms' investment in R&D. Let $W(z_1, z_2)$ denote the value of the race to the planner when firm 1 has accumulated $z_1 \geq 0$ units of knowledge and firm 2 has accumulated $z_2 \geq 0$ units of knowledge. Consider a short interval of time of length dt. From t to $t+dt$, firm 1's accumulated knowledge changes from z_1 to $z_1' = z_1 + \dot{z}_1 dt$. The Bellman equation is

$$W(z_1, z_2) = \max_{u_1 \geq 0,\, u_2 \geq 0} (h_1 dt + h_2 dt)Q - c(u_1)dt - c(u_2)dt + (1 - h_1 dt - h_2 dt)\frac{1}{1 + rdt} W(z_1', z_2')$$

where terms of order $(dt)^2$ are ignored. Q is the benefit to society from making the innovation. Note that in general, $Q \neq \overline{P} + \underline{P}$. But if $Q = \overline{P} + \underline{P}$. Then the planner's problem coincides with the problem of firms that are able to collude on setting their R&D efforts.

Letting $dt \to 0$ along the lines of section 4.3 yields

$$\begin{aligned} 0 = \max_{u_1 \geq 0,\, u_2 \geq 0} & (\lambda u_1 + \gamma z_1^{\phi} + \lambda u_2 + \gamma z_2^{\phi})Q - \frac{1}{\eta} u_1^{\eta} - \frac{1}{\eta} u_2^{\eta} \\ & - (r + \lambda u_1 + \gamma z_1^{\phi} + \lambda u_2 + \gamma z_2^{\phi}) W(z_1, z_2) \\ & + \frac{\partial}{\partial z_1} W(z_1, z_2)(u_1 - \delta z_1) + \frac{\partial}{\partial z_2} W(z_1, z_2)(u_2 - \delta z_2). \end{aligned}$$

Carrying out the indicated maximization yields the FOCs for an interior solution $u_1^{**}(z_1, z_2) > 0$ and $u_2^{**}(z_1, z_2) > 0$. Since the objective function is strictly concave due to $\eta > 1$, the FOCs are also sufficient for an interior solution. Hence,

$$u_1^{**}(z_1, z_2) = (\lambda(Q - W(z_1, z_2)) + \frac{\partial}{\partial z_1} W(z_1, z_2))^{\frac{1}{\eta - 1}}, \tag{4.11}$$

$$u_2^{**}(z_1, z_2) = (\lambda(Q - W(z_1, z_2)) + \frac{\partial}{\partial z_2} W(z_1, z_2))^{\frac{1}{\eta - 1}}. \tag{4.12}$$

Due to symmetry, $W(z_1, z_2) = W(z_2, z_1)$ and therefore $\frac{\partial}{\partial z_1} W(z_1, z_2) = \frac{\partial}{\partial z_2} W(z_2, z_1)$. Hence, if $u_1^{**}(z_1, z_2) = u_1^{**}(z_1, z_2)$ is the optimal policy for firm 1, then $u_2^{**}(z_1, z_2) = u_2^{**}(z_1, z_2)$ is the optimal policy for firm 2. The Bellman equation at $u_1 = u^{**}(z_1, z_2)$ and $u_2 = u^{**}(z_1, z_2)$ becomes

$$\begin{aligned} 0 = & (\lambda u^{**}(z_1, z_2) + \gamma z_1^{\phi} + \lambda u^{**}(z_2, z_1) + \gamma z_2^{\phi})Q \\ & - \frac{1}{\eta} u^{**}(z_1, z_2)^{\eta} - \frac{1}{\eta} u^{**}(z_2, z_1)^{\eta} \\ & - (r + \lambda u^{**}(z_1, z_2) + \gamma z_1^{\phi} + \gamma z_2^{\phi} + \lambda u^{**}(z_2, z_1)) W(z_1, z_2) \\ & + \frac{\partial}{\partial z_1} W(z_1, z_2)(u^{**}(z_1, z_2) - \delta z_1) \\ & + \frac{\partial}{\partial z_2} W(z_1, z_2)(u^{**}(z_2, z_1) - \delta z_2). \end{aligned} \tag{4.13}$$

Equation (4.11) can be substituted into equation (4.13) to obtain one functional equation in one unknown function.

As in the case of the non-cooperative game, we establish the existence of a unique solution for the special case of $\gamma = 0$. This special case corresponds to the models analysed by Reinganum (1981, 1982) in that learning and forgetting have no bearing on the outcome of the race.

Proposition 4.5: *Let $\gamma = 0$. Then the PDE in equation (4.13) admits a unique constant solution* W, $0 < W < Q$, characterized by

$$0 = 2\lambda u^{**}Q - \frac{2}{\eta}(u^{**})^{\eta} - (\gamma + 2\lambda u^{**})W, \tag{4.14}$$

where

$$u^{**} = (\lambda(Q - W))^{\frac{1}{\eta - 1}}. \tag{4.15}$$

Proof. At a constant solution W equations (4.13) and (4.11) reduce to equations (4.14) and (4.15). Define

$$\Delta(W) = 2(\lambda(Q - W))^{\frac{\eta}{\eta-1}}\left(1 - \frac{1}{\eta}\right) - rW.$$

At $W = Q$, we have

$$\Delta(Q) = -rQ < 0.$$

At $W = 0$, we have

$$\Delta(0) = 2(\lambda Q)^{\frac{\eta}{\eta-1}}\left(1 - \frac{1}{\eta}\right) > 0$$

since $\eta > 1$. Since $\Delta(W)$ is continuous in W, there exists a solution $\Delta(W) = 0$ by intermediate value theorem. Uniqueness of the solution follows from noting that $\Delta(W)$ is decreasing in W.

Welfare comparisons are complicated by the fact that they hinge on the social benefits Q as well as on the private benefits $\overline{P}$ and$\underline{P}$. It has long been argued that $Q \gg \overline{P} + \underline{P}$ for several reasons. First, an important part of the value of an innovation are the benefits accruing to consumers. Second, firms' R&D efforts generate spillovers that benefit firms in other industries or even the firms themselves at a later time. Both consumer surplus and spillovers are neglected in $\overline{P} + P$. Indeed, based on case studies of 17 industrial innovations, Mansfield *et al.* (1977) estimate a median social rate of return of 56% compared to a median private rate of return of 25%.

More recent evidence suggests that patent races reflect excessive patenting from a social perspective. In particular, the building of patent fences around some core innovation and the amassing of large patent portfolios are indicative of socially wasteful investment in R&D. Patent fences may not only preclude innovations that substitute for the core innovation, but also innovations that improve upon it. Similarly, the amassing of large patent portfolios may impede entry into the industry and the spur to innovative activity that usually accompanies it (Cohen *et al.*, 2000).

For these reasons, in what follows we focus on the case where social and private benefits are equal, $Q = \overline{P} + \underline{P}$. This implies that the planner's solution coincides with the collusive outcome in which firms coordinate their R&D efforts to maximize the value of making the innovation and receiving $\overline{P} + \underline{P}$.

If firms behave non-cooperatively, then there is in general a misallocation of resources. In particular, since the planner maximizes the sum of the individual payoffs and is free to replicate the non-cooperative outcome, the value of the race to the planner must be at least as big as the combined value of the race to firms 1 and 2 whenever $Q = \overline{P} + \underline{P}$, i.e. $V(z_1, z_2) + V(z_2, z_1) \leq W(z_1, z_2)$.

On the one side we can depict the difference between the value of the race to the social planner and the combined value of the race to firms 1 and 2,

$$\Delta^{W,\Sigma V}(z_1, z_2) = W(z_1, z_2) - (V(z_1, z_2) + V(z_2, z_1))$$

and on the other side we depict the difference in R&D expenditures between the planner's solution and the non-cooperative outcome,

$$\Delta^{c(u^{**}),\, c(u^{*})}(z_1, z_2) = c(u^{**}(z_1, z_2)) - c(u^{*}(z_1, z_2)).$$

One can then deduct that the value of the race to the planner, $W(z_1, z_2)$, indeed exceeds its combined value to firms 1 and 2, $V(z_1, z_2) + V(z_2, z_1)$. However, the reasons for this rent dissipation are quite different in the two cases.

We can illustrate a winner-takes-all situation in which the winning firm is awarded a patent of positive value whereas the losing firm receives nothing, $\overline{P} = 0.0435$ and $\underline{P} = 0$. The reason for this rent dissipation is $c(u^{**}(z_1, z_2)) < c(u^{*}(z_2, z_1))$, i.e. each firm invests excessively in R&D. Since the two firms compete for the same discovery, each additional money invested in R&D brings a firm closer to winning the race and, at the same time, brings its rival closer to losing the race. Hence, its R&D efforts impose a *negative* externality on its rival, and the firm consequently invests excessively in R&D.

The polar case illustrates the situation in which the loser can costlessly and immediately imitate the winner and thus both firms receive the same payoff, $\overline{P} = \underline{P} = 0.0435$. There is again rent dissipation, but the reason is now that firms invest too little in R&D: $c(u^{**}(z_1, z_2)) < c(u^{*}(z_2, z_1))$. In contrast to a winner-takes-all situation, each additional money invested in R&D brings firms closer to the finish line. Hence, a firm's R&D efforts impose a *positive* externality on its rivals, which causes the firm to underinvest in R&D.

To clarify the distinction between the two scenarios, consider the analytically tractable special case of $\gamma = 0$. Let u^* denote firm 1's equilibrium strategy and let u denote an arbitrary strategy for firm 2. Then the Bellman equation (4.3) and the FOC in equation (4.4) can be rewritten as

$$0 = \lambda u^* \overline{P} + \lambda u \underline{P} - (1/\eta)(u^*)^{\eta} - (r + \lambda u^* + \lambda u)V$$
$$0 = \lambda(\overline{P} - V) - (u^*)^{\eta - 1}.$$

Totally differentiating yields

$$dV/du|^{*}_{V,\,u} = \lambda(\underline{P} - V)/(r + \lambda u^* + \lambda u)$$
$$dV/du|^{*}_{V,\,u} = \lambda^2(\underline{P} - V)/(\eta - 1)(u^*)^{\eta - 2}(r + \lambda u^* + \lambda u).$$

Hence, $\underline{P} < V$ if and only if $dV/du|_{V,u} < 0$ if and only if $dV/du|_{V,u} > 0$.

Firms' R&D efforts are therefore strategic complements (strategic substitutes) whenever $\underline{P} < V$ ($\underline{P} > V$), i.e. reaction functions are upward sloping (downward sloping) whenever $\underline{P} < V$ ($\underline{P} > V$).

It follows that a firm's R&D efforts impose a negative externality (positive externality) on its rival whenever $\underline{P} < V$ ($\underline{P} > V$). In other words, depending on whether or not the benefit to imitation $\underline{P}$ is less than the value of continued play V, the character of the R&D race changes from a preemption game into a waiting game. If patent protection is perfect and thus $\underline{P} = 0$, then the R&D race always has the character of a preemption game, whereas if imitation is costless and immediate and thus $\underline{P} = \overline{P}$, then the R&D race always has the character of a waiting game. Finally, if $0 < \underline{P} < \overline{P}$, then the preemption as well as the waiting incentive is operative.

The following proposition formally shows that there is indeed overinvestment (underinvestment) in R&D if and only if a firm's R&D efforts impose a negative (positive) externality on its rival.

Proposition 4.6: *Let* $\gamma = 0$ *and* $Q = \underline{P} + \overline{P}$. *Then* $u^* \gtrless u^{**}$ *if and only if* $V \gtrless \underline{P}$.

Proof. We have $u^* = (\lambda(\overline{P} - V))^{1/\eta - 1} \gtrless \lambda(\overline{P} + \underline{P}) - W))^{1/\eta - 1} = u^{**}$ *if and only if* $W \gtrless V + \underline{P}$. From the proofs of Propositions 4.1 and 4.5 we know that the solution to the non-cooperative game and to the planner's problem are characterized by the zeros of

$$\Delta^V(V) = (\lambda(\overline{P} - V))^{\eta/\eta-1}(1 - 1/\eta) + (\lambda(\overline{P} - V))^{1/\eta-1}(\overline{P} - V) - rV$$

and

$$\Delta^W(W) = 2(\lambda(\overline{P} + \underline{P}) - W))^{\eta/\eta-1}(1 - 1/\eta) - rW,$$

respectively. Let V solve Δ^V (V) $= 0$ and consider $W(V + \underline{P})$ as a candidate to Δ^W $(W) = 0$. Rewriting yields

$$\Delta^W(\mathrm{V} + \underline{P}) = 2\Delta^V(V) + (r + 2\lambda(\lambda(\overline{P} - V))^{1/\eta-1})(V - \underline{P}).$$

Since $\Delta^V(V) = 0$, $\Delta^W(V + \underline{P}) \gtrless 0$ if and only if $V \gtrless \underline{P}$. Since $\Delta^W(W)$ is decreasing, this implies that the actual solution to $\Delta^W(W) = 0$ satisfies $W \gtrless V + \underline{P}$ if and only if $V \gtrless \underline{P}$.

It follows from Proposition 4.6 that $\underline{P} = 0\,(\underline{P} = \overline{P})$ implies $u^* > u^{**}$ $(u^* < u^{**})$. Hence, a sufficient condition for overinvestment (underinvestment) is that patent protection is perfect (imitation is costless and immediate).

Following the literature we consider the case in which firms' R&D efforts are strategic complements and firms overinvest in R&D. In the presence of a tax, the cost incurred to acquire knowledge at a rate u_1 *is* $(1 + \sigma)c(u_1) = (1 + \sigma)/\eta(u_I)^\eta$, where $\sigma \geq 0$ is the tax rate, and equations (4.3) and (4.4) become

$$0 = \lambda u^*(\overline{P} + \underline{P}) - (1 + \sigma)/\eta(u^*)^\eta - (r + 2\lambda u^*)V;$$

where $u^* = (\lambda/(1 + \sigma)(\overline{P} - V))^{1/\eta-1}$.

Generally, governments tend to encourage R&D rather than tax it. Assuming for the moment that governments cannot intentionally make firms worse off, the misallocation of resources in the non-cooperative game can be reduced by reducing the asymmetry in the rewards to winning and losing the R&D race. One way to accomplish this is to partially insure the participating firms against losing the R&D race. Another way is to throw money at all participating firms. In either case, reducing the asymmetry in the rewards reduces the negative externality stemming from firms' R&D efforts and increases the positive externality.

This in turn moves the R&D race away from a preemption game and overinvestment in R&D towards a waiting game and underinvestment in R&D.

4.6 Conclusions

We developed a model of symmetric R&D races that incorporate learning and forgetting. Knowledge plays a productive role in this model that substantiates stages or experience levels in multi-stage races.

The computational strategy employed provides new insights into the strategic interactions among the racing firms. Learning and forgetting shape firms' equilibrium payoffs and strategies. Complementary to the literature on multi-stage races, we find that a firm is able to reduce its R&D investment as its knowledge base increases and that the follower works harder than the leader. In the case of a convex hazard rate, the increasing returns nature of the hazard rate provides the firm with a strong incentive to invest in R&D which gives rise to a counteracting force. Underlying the results is the *pure knowledge effect*. The source of this effect is that a firm's past R&D efforts contribute to its chances of winning the R&D race because the firm's knowledge base enters its hazard rate.

We show that a firm can react either aggressively or submissively to an increase in its rival's knowledge base. In multi-stage models, the follower slows down as he falls further behind whereas the leader may or may not speed up as he gets further ahead. In this model, aggressive or submissive behaviour is not tied to a firm's relative position. Rather, a firm reacts aggressively if it has a sufficiently large knowledge base and/or the value of the patent is sufficiently large. (So here is an explanation for the 'top dog' image described in Chapter 1.) On the other hand, the pure knowledge effect also appears to prevail over strategic considerations. Specifically, the reaction in the firm's current R&D effort due to a change in its own knowledge base dominates the effect of a change in its rival's knowledge base on the firm's current R&D effort.

Lastly, in contrast to the literature on multi-stage races that competition is not necessarily fiercest when firms are neck-and-neck, the model suggests that competition is generally most intense among firms when their knowledge bases are of unequal size and least intense when they are of equal size, although again this need not be the case if the hazard rate is convex.

Empirical studies of the pharmaceutical industry indicate the importance of spillovers (Cockburn and Henderson, 1994, Henderson and Cockburn, 1994, 1996). Spillovers may either directly affect a firm's hazard rate or indirectly operate through the accumulation of knowledge. Moreover, a firm's rivals may either benefit from the firm's current or past R&D efforts. Finally, a firm may be able to absorb spillovers only if it invests in R&D itself. This reflects the view that absorbing spillovers from other firms requires the firm to engage in R&D itself (see e.g. Cohen and Levinthal (1989) and Adams and Jaffe (1996) for empirical evidence).

The view that absorbing spillovers from other firms requires the firm to engage in R&D itself may be extended to 'spillovers' from the firm's own knowledge base. The pure knowledge effect implies that a firm increasingly substitutes past R&D efforts for current R&D effort and, as Proposition 4.2 shows, wins the race for certain and without effort in the limit as the size of its own knowledge base approaches infinity. This is easily avoided by employing a multiplicative hazard rate. In contrast to an additive hazard rate, a multiplicative hazard rate would imply that past R&D efforts are not productive unless they are combined with current R&D effort. I have omitted uncertainty in the accumulation equation.

Empirical studies referred to in Chapters 3 and 5 testify to the important role of product competition, market structure and repeated interactions between competing firms in determining the outcome of an R&D race.

References

Adams, J. and Jaffe, A. (1996) 'Bounding the effects of R&D: an investigation using matched establishment firm data', *Rand Journal of Economics*, **24**(7), 700–721.

Bertsekas, D. P. (1987) *Dynamic Programming: Deterministic and Stochastic Models*, Englewood Cliffs, N.J.: Prentice Hall.

Cockburn, I. and Hendersen, R. (1994) 'Racing to invest? The dynamics of competition in ethical drug discovery', *Journal of Economics and Management Strategy*, **3**(3), 481–519.

Cohen, W. and Levinthal, D. (1989) 'Innovation and learning: the two faces of R&D', *Economic Journal*, **99**, 569–596.

Cohen, W., Nelson, R. and Walsh, J. (2000) 'Protecting their intellectual assets: appropriability conditions and why U.S. manufacturing firms patent or not', Working paper 7552, NBER, Cambridge.

Dorazelski, U. (2003) 'An R&D race with knowledge accumulation', *Rand Journal of Economics*, **34**, 20–42.

Fudenberg, D., Gilbert, R., Stiglitz, J. and Tirole, J. (1983) 'Preemption, leapfrogging and competition in patent races', *European Economic Review*, **22**, 3–31.

Gottinger, H. W. (2004) 'Biotech and pharma industries: application of network economies', *Journal of Policy Studies*, **17**, 27–34.

Grabowski, H. and Baxter, N. (1973) 'Rivalry in industrial research and development', *Journal of Industrial Economics*, **21**, 209–223.

Grabowski, H. and Vernon, J. (1994) 'Returns to R&D on new drug introductions in the 1980s', *Journal of Health Economics*, **13**, 383–406.

Griliches, Z. (1980) 'R&D and the productivity slowdown', *American Economic Review*, **70**(2), 343–348.

Grossman, G. and Shapiro, C. (1987) 'Dynamic R&D competition', *Economic Journal*, **97**(386), 372–387.

Harris, C. and Vickers, J. (1985) 'Perfect equilibrium in a model of a race', *Review of Economic Studies*, **54**(1), 1–21.

Harris, C. and Vickers, J. (1985a) 'Patent races and the persistence of monopoly', *Journal of Industrial Economics*, **33**(4), 461–481.

Harris, C. and Vickers, J. (1985b) 'Perfect equilibrium in a model of a race', *Review of Economic Studies*, **52**(2), 193–209.

Harris, C. and Vickers, J. (1987) 'Racing with uncertainty', *Review of Economic Studies*, **54**(1), 1–21.

Henderson, R. and Cockburn, I. (1994) 'Measuring competence? Exploring firm effects in pharmaceutical research', *Strategic Management Journal*, **15**, 63–74.

Henderson, R. and Cockburn, I. (1996) 'Scale, scope, and spillovers: the determinants of research productivity in drug discovery', *Rand Journal of Economics*, **7**(1), 32–59.

Lee, T. and Wilde, L. (1980) 'Market structure and innovation: a reformulation', *Quarterly Journal of Economics*, **94**, 429–438.

Lerner, J. (1997), 'An empirical exploration of a technology race', *The Rand Journal of Economics*, **28**(2), 228–240.

Lippman, S. and McCardle, K. (1988) 'Preemption in R&D races', *European Economic Review*, **32**, 1661–1669.

Loury, G. (1979) 'Market structure and innovation', *Quarterly Journal of Economics*, **93**, 395–410.

Mansfield, E., Rapoport, J., Romeo, A., Wagner, S. and Beardsley, G. (1977) 'The social and private rates of return from industrial innovations', *Quarterly Journal of Economics*, **91**(2), 221–240.

Mansfield, E., Schwartz, M. and Wagner, S. (1981) 'Imitation costs and patents: an empirical study', *Economic Journal*, **91**, 907–918.

Reinganum, J. (1981) 'Dynamic games of innovation', *Journal of Economic Theory*, **25**, 21–41.

Reinganum, J. (1982) 'A dynamic game of R and D: patent protection and competitive behavior', *Econometrica*, **50**, 671–688.

Reinganum, J. (1989) 'The timing of innovation: research, development and diffusion', Chapter 14 in Schmalensee, R. and Willig, R. D. (Eds), *Handbook of Industrial Organization*, Vol. I, New York: Elsevier Science.

Scherer, F. (1992) *International High-tech Competition*, Cambridge, MA: Harvard University Press.

5 Technological racing in asymmetric industry settings

> Entrepreneurship is rewarded with extraordinary profit. It is the profit for being first. The successful new product or new method of production calls forth imitation that eventually erodes the extraordinary profit.
>
> Morton I. Kamien and Nancy L. Schwartz (1982)

5.1 Introduction

Analyses of 'step-by-step' (sequential) research and development (R&D) that take into account the strategic interaction between firms add a new dimension to technological racing. Grossman and Shapiro (1986) analyse a two-firm, two-stage game, with benefits accruing to the first firm to finish both stages, with the intention of studying how the investment level of a firm depends on its relative position in an R&D race. They show that typically the firm that gets in the lead steps up its effort level, while the firm that falls behind lowers its effort level. Park (1987) carries out a similar analysis. As we shall see below, the analysis of the speed–leadership tradeoff results in quite different resource utilization patterns in similar two-stage games. In their study of multi-stage races, Harris and Vickers (1987) reach the principal conclusion that the race leader invests more than does the follower. This conclusion is also modified in an analysis of multi-stage games. Two-stage strategic interactions have been used in other situations as well, e.g. Fershtman and Kamien (1990) study the strategic interaction between two firms when the product requires development of two complementary technologies. Their focus is on investigating the circumstances under which cross-licensing results at the intermediate stage where each firm has developed only one technology. In a different vein, Vickers (1986) analyses the response of a duopoly to a sequence of process innovations and contrasts situations of 'increasing dominance' where the incumbent firm remains dominant with those of 'action–reaction' where the firm with the current best technology alternates. Beath *et al.* (1987) extend this analysis to consider a sequence of product innovations. A sequential game analysis of firm R&D behaviour has been previously conducted by Iansiti and Khanna (1997).

In this chapter, we model the R&D process as one that is composed of a series of stages, at each of which some set of product characteristics are determined. At each stage, the firm exercises discretion as to the extent to which the particular stage characteristics will be developed. Further, when the firm's R&D resource endowment is fixed and stage-specific, the firm faces a tradeoff between the extent to which it will develop the stage characteristics and the time that it will spend in doing so. We will refer to the degree to which the stage characteristic is developed as the characteristic's (targeted) level of strategic positioning. In this sense, the more 'forward-looking' the product characteristics, the greater the profits the firm can expect to get from the product but the longer the expected time to completion of R&D. We will speak of this as the speed–leadership tradeoff in each stage. When the completion of R&D requires many such tradeoffs to be resolved sequentially, firms will, at the beginning of each stage, want to make their choices in light of the choices made by all firms in previous stages and their stage completion records to that time.

We investigate how firms choose the leadership of their product characteristics when there is a tradeoff between the leadership of product characteristics and the speed with which the stages of R&D are completed, and when strategic interactions between firms are accounted for. While products with more leadership characteristics are 'better' in the sense that they earn the firm greater profits, the required R&D takes more time. This chapter thus incorporates strategic considerations into the analysis of a problem, that of the way in which given resources are used in a sequential R&D context, that is rooted in the framework of Chapter 2. We offer two sets of conclusions to build on the existing literature. First, in the two-stage game (as, for example, in Grossman and Shapiro (1986)), we demonstrate the importance of explicitly accounting for differences in product quality and cost. It is in the assumption of scarce resources for each stage (and the implied tradeoff between product leadership and time to completion of R&D) that this analysis differs from previous analyses. Further, in situations of strategic rivalry, relative R&D resource endowment influences the pattern of resource utilization, as does incumbency (Gilbert, 1989). These analyses are motivated by the empirical work conducted in Cebon *et al.* (2001). Second, this chapter analyses the changing way in which R&D resources are used in a multi-stage game as one firm gets further ahead of the other, and as the firms' proximity to the last stage increases. Throughout, we abstract from 'dropout behavior' (Lippman and McCardle, 1987).

Our purpose is not to analyse how firms vary the amount they spend on R&D as stages in the sequential game evolve (against nature or against a specific opponent). Rather, the issue is: how do firms use given resources in the face of the existence of a very natural tradeoff between the leadership in product characteristics and the speed of execution?

There are two sets of reasons why the resources available for R&D might be limited. The first interpretation is concerned with the assumption that it is the budget available to R&D management that is fixed, and consequently limited use can be made of the services of engineers and scientists to do the R&D (even though there is a well-developed market for such services). The second interpretation is that the resources available to do R&D (in technically complex, knowledge-based industries) are firm-specific and therefore subject to factor market imperfections. As Rubin (1973) points out, '... the value of a resource typically exceeds the market value of the individual parts due to the cohesiveness of the human part of the resource developed through mutual experience within the firm'. In a similar vein, Camerer (1991) suggests that workers might be tied to a firm if they are productive together but cannot collectively agree to move to a better-paying firm. Thus scientists function best (and are worth more in) within the milieu in which much of their work on firm-specific R&D has proceeded. The notion of resources being tied to a firm also implies that a firm cannot augment its resources easily. This 'resource-based' view within the strategy literature (see Connor (1989) for a review) analyses the effects of product market competition taking as given initial differences in firms' resource endowments. Game-theoretic advances (Fudenberg and Tirole, 1989; Tirole, 1988) also suggest foundations for the resource-based view. For example, information asymmetries in the 'new industrial organization' (Jacquemin, 1987; Reinganum, 1989) might preclude the use of an asset outside the boundaries of the firm that owns the asset. Kreps (1990) suggests that in many situations where unforeseen contingencies arise, certain transactions are facilitated by the reputation of a trusted party, i.e. a firm. Thus a firm's resources might not be able to operate independently of the firm. The obvious question that the resource-based view leaves unanswered is that of how the firms came to acquire different endowments in the first place. Here again Rubin (1973) suggests that when resources are firm-specific, then existing resources will have to be used to augment the stock available for future use. Given the opportunity costs of engaging in such augmentation, different firms might opt for different augmentation rates. As such, at any point in time, different firms would have access to different resource endowments.

R&D resources are assumed to be firm-specific. Further, each stage is assumed to have its own resource endowment (non-transferable across stages). This reflects our belief that the nature of work done by, say, other scientists in basic research is considerably different from that done by, say, other scientists and engineers in development. Just as firms would incur significant costs in augmenting their R&D resources, so also is deploying existing resources across the stages assumed to be very costly.

This chapter is organized as follows. Section 5.2 introduces a simple model that will allow us to think about multiple-stage racing and analyses a two-stage interaction between symmetric firms. Section 5.3 elaborates on this analysis to consider some of the asymmetries that empirical work suggests are important (Cohen and Levin, 1989; Lerner, 1997). We show that an incumbent firm will move more aggressively in each stage of the race than will an entrant firm. We analyse factors that cause relatively resource-rich firms to choose differently from resource-poor ones. Also, Section 5.3 is devoted to understanding how getting ahead (and falling behind) alters the choices that a firm makes. We identify the importance of two factors: the magnitude of the lead that one firm has over the other, and the distance between the firms and the finishing line, and suggest that the patterns of resource allocation, in a situation where there is a leadership–speed tradeoff, can be quite different from that suggested in the existing literature. Section 5.4 expands on the theme of multi-stage racing in view of some new complicating issues. Section 5.5 draws some conclusions.

5.2 A model for multi-stage racing

Product development proceeds in a sequence of stages, each responsible for determining a particular characteristic of the product. Each stage involves a number of problem-solving activities, split into two steps, research and development, each of which improves the product characteristic being determined in that stage. The firm exercises discretion as to the extent to which the stage characteristic will be developed. At each stage, the firm has access to an exogeneously determined resource endowment (which cannot be transferred across stages) and has to target the set of activities over which it will allocate these resources. The larger the number of activities in the targeted set, the greater the extent to which the stage characteristic is developed. However, the larger the number of targeted activities, the smaller the resource amount allocated to each activity, and the greater the expected time to completion of the activities in that stage. We will speak of a firm's target being more 'leadership-prone' if its target set contains more activities.

Interactive system–analytic considerations are incorporated here by thinking of each problem-solving activity in a stage as linked to every other problem-solving activity in that stage. We assume that the stage-specific resource is divided equally among the activities in the targeted set and that the problems cannot be solved independently of each other. Thus, at any point in time, the targeted set of problems has either been completely solved or not at all. Let the hazard rate for solving the targeted set of problems be denoted 'h'. Then a higher hazard rate is obtained by allocating more resources to each activity, which implies consideration of fewer activities. Thus a higher hazard rate implies a less-dominant target set. Conversely, the lower the hazard rate of the firm in comparison to its rival in each stage, the higher the payoff assigned to the firm.

We model a two-firm, two-stage interaction. Each firm has two choices for the degree of aggressiveness of its product characteristics in each stage. The choice of the higher rate x in a stage gives the firm a less-dominant stage target, while the choice of the lower hazard rate y corresponds to a more-dominant stage target. While there is uncertainty in the time taken to achieve a particular target, the target itself is chosen at the beginning of each stage. Let p_R, q_R be the choices of hazard rate by firms 1 and 2, respectively, for the first stage (research stage). Similarly, let firms 1 and 2 choose hazard rates p_D, q_D, respectively, for the second stage (development stage). At time 0, firm 1 chooses $p_R = x$ or $p_R = y$, while firm 2 simultaneously chooses $q_R = x$ or $q_R = y$. Assume, without loss of generality, that the first firm to finish the R stage is firm 2. Then firm 2 will choose $q_D = x$ or $q_D = y$ as its hazard rate for the D stage. Finally, when firm 1 is done with stage R (which may or may not be before firm 2 has finished stage D) it will choose $p_D = x$ or $p_D = y$.

Monopoly profits are denoted $M(p_R, p_D)$ or $M(q_R, q_D)$. We abstract from 'dropout behaviour' and assume that both firms will eventually finish the multiple stages. Since we are concerned with the utilization of given R&D resources, flow costs of R&D are fixed regardless of the level of strategic dominance chosen. We do not explicitly include this flow cost term in our expressions. Firms are risk-neutral and maximize discounted expected profits, with r denoting the discount rate. We will look for a subgame-perfect equilibrium in this game (Fudenberg and Tirole, 1989) as indicated in Chapter 2. The solution proceeds by backward induction in three steps after employing notation. Let $V(i)$ be the expected payoff for a firm when it has finished stage D with i points while its rival is still in its D stage (having completed its R stage). Then, for firm 1, we would write

$$rV(i) = M(p_R, p_D) + q_D(D(i) - V(i)) \Rightarrow V(i) = \frac{M(p_D, P_D) + q_D D(i)}{(r + q_D)}$$

where $M(p_R, p_D)$, $D(i)$ denote monopoly, duopoly profits.

Analogously, when $V(i)$ refers to a payoff for firm 2, it is given by $[M(q_R, q_D) + p_D D(i)]/(r + p_D)$. Similarly, let $K(i)$ be the expected payoff for a firm when it is still in stage D, which it will finish with i points, when its rival has already finished its D stage. Then, for firm 1, we would write

$$rK(i) = p_D(D(i) - K(i)) \Rightarrow K(i) = \frac{p_D D(i)}{(r + p_D)}.$$

Analogously, when $K(i)$ refers to a payoff for firm 2, it is given by $[q_D(D(i)]/(r + q_D)$. We assume $V'(i) > 0$, $K'(i) > 0$. Occasionally, we will write the monopoly profit term without its arguments for convenience. The reduced form expressions that we analyse are in terms of $V(i)$ and $K(i)$. However, we will reinterpret the propositions in terms of monopoly and duopoly profit terms. This translation between the $V(i)$, $K(i)$ notation and the monopoly, duopoly notation will be accomplished using the following propositions which are relegated to Appendix A for their technical results.

Turning now to backwards induction, underlying a dynamic programming framework, a firm's strategy consists of its choices of hazard rate (and hence level of leadership) in each of the two stages (R,D). Without loss of generality, we assume that firm 2 is the first firm to finish the R stage. Then the game can be solved in three steps. Step 3, which we analyse first, corresponds to the situation where both firms have had their R stage choices, firm 2 has made its D stage choice, and firm 1 has to make its D stage choice. Step 2 corresponds to the part of the game where firm 1 is still in its R stage, while firm 2 has just finished its R stage and has to make its D stage choice. Step 2 is solved taking as given the results of the Step 3 solution by backwards induction. Finally, Step 1 refers to the part of the game where both firms simultaneously make their R stage choices. We solve Step 1 for specific cases that we identify from the analysis of Steps 2 and 3.

Step 3

Here firm 1 chooses its D stage strategy, p_D, given its own previous choice of p_R, firm 2's choice of q_R and q_D, and information regarding whether firm 2 has completed its D stage or not. (Recall that we assumed that firm 2 was the first to finish the R stage. Consequently, when firm 1 has to choose p_D, we know that firm 2 has already finished its R stage and has chosen q_D.) There are two cases to consider depending on whether firm 2 has completed its D stage or not.

First consider the case where firm 2 has not yet completed its D stage. Then firm 1's expected payoff $W_{1,a}(i)$ is given by

$$rW_{1,a}(i) = p_D(V(i) - W_{1,a}(i)) + q_D(K(i) - W_{1,a}(i))$$

$$\Rightarrow W_{1,a}(i) = \frac{p_D V(i) + q_D K(i)}{(r + p_D + q_D)}.$$

Now, firm 1 chooses $p_D = x$ or $p_D = y$ to maximize this expression for various realizations of p_R, q_R, q_D (i.e. all the previous decisions made by the firms). As an example, consider firm 1's best response to the triple $(p_R, q_R, q_D) = (y, y, x)$. Then for $p_D = y$ to be preferred ($\succ$) to $p_D = x$, we need $W_{1,a}(i+1)$ when $p_D = y$ to be greater than $W_{1,a}(i)$ when $p_D = x$. So

$$y \succ x \Rightarrow \frac{yV(3) + yK(3)}{(r + 2y)} > \frac{xV(2) + yK(2)}{(r + x + y)}.$$

A sufficient condition for this inequality to hold is that more generally,

$$xW(i) \leq yW(i+1) \;\forall\; i \Rightarrow y \succ x(p_R, qR, q_D).$$

Now consider the case where firm 2 has completed its D stage. Then firm 1's expected payoff is given by $W_{1,b}(i)$ where

$$rW_{1,b}(i) = p_D(K(i) - W_{1,b}(i)) \Rightarrow W_{1,b}(i) = \frac{p_D K(i)}{(r + p_D)}.$$

So for $p_D = y$ to be preferred to $p_D = x$, we need

$$y \succ x \Rightarrow \frac{yK(i+1)}{(r+y)} > \frac{xK(i)}{(r+x)}.$$

So $xK(i) \leq yK(i+1) \; \forall \; i \Rightarrow y \succ x \; \forall \; (p_R, q_R)$.

In summary, when $xV(i) \leq yV(i+1)$, $xK(i) \leq y\mathrm{K}(i+1)$ for all i, firm 1 will choose $p_D = y$ regardless of whether firm 2 has completed its D stage.

Step 2

Before firm 1 chooses its D stage strategy, firm 2 is the first firm to finish the R stage and has to choose q_D given (p_R, q_R) and given firm 1's actions as determined by conditions in Step 3 of the backward induction.

Let us define V_2 as firm 2's expected payoff if it finishes stage D before firm 1 finishes stage R.

$$rV_2(i) = p_R(V(i) - V_2(i)) + M(q_R, q_D)$$
$$\Rightarrow V_2(i) = [M(q_R, q_D) + p_R V(i)]/(r + p_R).$$

Similarly, let K_2 be firm 2's expected payoff if it finishes stage D after firm 1 finishes stage R. Then

$$rK_2(i) = q_D(V(i) - K_2(i)) + p_D(K(i) - K_2(i))$$
$$\Rightarrow K_2(i) = \frac{q_D V(i) + p_D K(i)}{(r + p_D + q_D)}.$$

Here $V(i)$ and $K(i)$ are the quantities from Step 3 of the backward induction and refer to payoffs received by firm 2. Recall from Step 3 that $xV(i) \leq yV(i+1)$, $xK(i) \leq yK(i+1)$ for all i implies that firm 1 chooses $p_D = y$ regardless of whether firm 1 chooses its D stage strategy before or after firm 2 finishes its D stage. So firm 1's action is the same in the two cases whose payoffs (for firm 2) are represented by $V_2(i)$ and $K_2(i)$. Thus both $V_2(i)$ and $K_2(i)$ have the same index as argument.

Now, firm 2 chooses q_D to maximize $W_2(i)$ where

$$rW_2(i) = q_D(V_2(i) - W_2(i)) + p_D(K_2(i) - W_2(i))$$
$$\Rightarrow W_2(i) = \frac{q_D V_2(i) + p_R K_2(i)}{(r + q_D + p_R)}.$$

Substituting $V_2(i)$ and $K_2(i)$ into W_2, we have

$$W_2(i) = q_D\left[\frac{M(q_R, q_D) + p_R V(i)}{(r + p_R)}\right] + p_R\left[\frac{q_D V(i) + p_D K(i)}{(r + p_D + q_D)}\right] \Big/ (r + q_D + p_R).$$

Let $k_1 = 2r + p_R + p_D$ and $k_2 = r + p_D$. Then $W_2(i)$ can be rewritten as

$$\frac{p_R k_1 V(i) q_D + p_R V(i) q_D^2 + k_2 M(q_R, q_D) q_D + M(q_R, q_D) q_D^2 + K(i) p_D p_R (r + p_R)}{(r + p_R)(r + p_D + q_D)(r + p_R + q_D)}.$$

Let us compare the magnitudes that the various terms in this expression for $V_2(i)$ take on for firm 2's two possible choices of q_D. We see that the denominator term increases if $q_D = x$ is chosen instead of $q_D = y$. The K term in the numerator takes on the value $p_D p_R (r + p_R) K(i)$ when $q_D = x$ and $p_D p_R (r + p_R) K(i+1)$ when $q_D = y$, and is therefore larger when $q_D = y$. Thus both the K term in the numerator and the reciprocal of the denominator are larger when $q_D = y$ than when $q_D = x$. We need the following assumptions to ensure that the other numerator terms are at least as large when $q_D = y$ than when $q_D = x$:

$$x^2 V(i) \leq y^2 V(i+1)\ \forall\ i. x^2 M(q_R, x) \leq y^2 M(q_R, y) \text{ for } q_R = x, y. \tag{5.1}$$

Recall from the Step 3 analysis above that $xV(i) \leq yV(i+1)$, $xK(i) \leq yK(i+1)$ for all i implied that firm 1 would set $p_D = y$ regardless of firm 2's choice of q_D. When these conditions hold, the inequalities in equation (5.1) above are sufficient to ensure that $q_D = y$ is a dominant strategy for firm 2.

We now want to establish sufficient conditions for firm 2 to choose $q_D = x$. For x to be chosen, W_2 with $q_D = x$ should be greater than W_2 when $q_D = y$. If T_{c1} is a term that includes all non-monopoly terms in the inequality that expresses this comparison, then the relevant condition for x to be chosen is

$$\frac{xM(., x)}{(r + p_R + x)} + T_{c1} > \frac{yM(., y)}{(r + p_R + y)}.$$

This can be given as

$$\theta < \frac{x\,(r + p_R + y)}{y\,(r + p_R + x)} + \frac{T_{c1}}{M(., x)}(1 + \frac{r + p_R}{y}), \text{ where } \theta = \frac{M(., y)}{M(., x)}. \tag{C1}$$

This condition is more likely to hold the closer θ is to 1. Note that T_{c1} may be positive or negative. However, the second term on the right-hand side of the inequality has less of an effect the greater is the monopoly profit term, $M(., x)$, relative to the terms in T_{c1}. Note also that the first term on the right-hand side is greater than 1.

The above analysis of Steps 3 and 2 of the backwards induction can be summarized as follows:

Step 3 (Action when a firm is the last firm to finish R stage and is the last to choose its D stage strategy): Sufficient conditions to choose y:

$$xV(i) \leq yV(i+1), xK(i) \leq yK(i+1) \; \forall \; i.$$

Step 2 (Action when a firm is the first firm to finish the R stage and is the first to choose its stage strategy): Sufficient conditions to choose y:

$$x^2 V(i) \leq y^2 V(i+1) \; \forall \; i, x^2 M(v,x) \leq y^2 M(v,y) v = x, y.$$

Sufficient conditions to choose x: condition (C1) holds.

In what follows, let us distinguish two cases as follows.

Case 1

$$x^2 V(i) \leq y^2 V(i+1), \;\; xK(i) \leq yK(i+1) \; \forall \; i,$$
$$x^2 M(v,x) \leq y^2 M(v,y), \;\; v = x, y$$

In this case, the conditions specified are sufficient to ensure that firms always choose y when they complete the R stage. Proposition 2 suggests that these Case 1 conditions will hold when both duopoly and monopoly profit terms are sufficiently larger under the more-dominant strategy choice (than under the less-dominant choice) to overcome the effect of the longer expected time to completion under the more-dominant choice.

Case 2

$$xV(i) \leq yV(i+1), \;\; xK(i) \leq yK(i+1), \;\; xM(v,x) > yM(v,y), \;\; v = x, y$$

and condition (C1) holds.

In Case 2, the conditions specified are sufficient to ensure that the first firm to finish the R stage chooses x, while the second firm to finish the R stage chooses y. Proposition 5.3 (Appendix) shows that ranges of x and y exist such that these conditions are satisfied.

Step 1

In the game's initial stage, both firms simultaneously choose hazard rates for the R stage. After introducing some preliminaries, we shall solve separately for the two cases identified by the backwards induction thus far.

Let V_1 denote firm 1's expected payoff if it finishes the R stage first and subsequent play is optimal, and let $\overline{V_1}$ denote its expected payoff if it finishes the R stage second and subsequent play is optimal. Then firm 1's overall expected payoff, P_1, is given by

$$rP_1 = p_R(V_1 - P_1) + q_R(\overline{V}_1 - P_1) \Rightarrow P_1 = \frac{p_R V_1 + q_R \overline{V}_1}{(r + p_R + q_R)}.$$

Since we have already derived an expression for firm 2's expected payoff if it finished R first (W_2, in Step 2 of the backwards induction), we can interchange *ps* and *qs* to get an expression for 1's expected payoff if it finishes R first:

$$V_1 = p_D\left[\frac{M(p_R p_D) + q_R W(i)}{(r + q_R)}\right] + q_R\left[\frac{p_D W(j) + q_D L(j)}{(r + p_D + q_D)}\right] \Big/ (r + p_D + q_R).$$

Firm 1 will maximize P_1 with p_D and q_D set by backwards induction. Thus the index i associated with the p_D term in the numerator of V_1 should assume that firm 1 finished the R stage first, while the corresponding index j for the q_R term should assume that firm 1 finished the R stage second. However, for exactly the same reason that V_2 and K_2 in Step 2 of the backwards induction shared the same index as their argument, we have $i = j$ here. In all subsequent references to V_1, the common index will be denoted by i. Also

$$r\bar{V}_1 = p_R(A - \bar{V}_1) + q_D(\zeta - \bar{V}_1)$$

where $A =$ firm 1's payoff when it has finished stage R but its rival has not finished stage D; $\zeta =$ firm 1's payoff when it has not finished stage R but its rival has finished stage D.

Then the terms in the expression can be calculated as follows

$$r\zeta = p_R(B - \zeta) \Rightarrow \zeta = \frac{p_R B}{(r + p_R)}$$

where $B =$ firm 1's payoff when it has finished stage R, and its rival has finished stage D.

$$\text{Also } A = \frac{p_D W + q_D L}{(r + p_D + q_D)}, \quad B = \frac{p_D L}{(r + p_D)}.$$

This leads to the following expression

$$\bar{V}_1 = \frac{p_R A + q_D\left[\frac{p_R B}{(r+p_R)}\right]}{(r + p_R + q_D)}$$

where A, B are as above.

Case 1: Recall that this is the case where the first firm to finish the R stage chooses y, as does the second. We want to look for firm 1's best response to the choices of q_R by firm 2. First suppose that $q_R = y$. When $p_R = x$, P_1 is given by

$$\frac{xV_1(i) + y\bar{V}_1(j)}{(r + x + y)}.$$

Here $V_1(i)$ is what firm 1 gets if it finishes the R stage first. Then $p_R = x$, $q_R = y$, $p_D = y$, $q_D = y$, and the index $i = 1$. Similarly, $V_1(j)$ is what firm 1 gets if it finishes the R stage last. Here also $p_R = x$, $q_R = y$, $q_D = y$, $q_D = y$, and the index $j = 1$. Similarly, when $p_R = y$, P_1 is given by

$$\frac{yV_1(k) + y\bar{V}_1(l)}{(r + 2y)}$$

and the indices are $k = 2$ and $l = 2$. So, when $q_R = y$,

$$\frac{yV_1(2) + y\bar{V}_1(2)}{(r + 2y)} > \frac{xV_1(1) + y\bar{V}_1(l)}{(r + x + y)} \Rightarrow y \succ x.$$

So, firm 1 chooses $p_R = y$ when $q_R = y$ if $yV_1(2) \geq xV_1(1)$.

Now, consider the case where $q_R = x$. By proceeding as above, we have

$$\frac{yV_1(3) + y\bar{V}_1(3)}{(r + 2y)} > \frac{xV_1(2) + y\bar{V}_1(2)}{(r + x + y)} \Rightarrow y \succ x.$$

So firm 1 chooses $p_R = y$ when $q_R = x$ if $yV_I(3) \geq xV_1(2)$.

Combining these results, we can say that $xV_1(i) \leq yV_1(i+1)$ for $i = 1, 2$ is a sufficient condition for y to be a dominant strategy for firm 1 in stage R.

Let us check how this condition relates to the Case 1 conditions. When $p_D = q_D = y$ (as in Case 1), for the case where $q_R = y$, we can rewrite $yV_1(i+1) \geq xV_1(i)$, after some manipulation, as

$$\begin{aligned} &y(r + 2y)[yM(y, y) - xM(x, y)] + y^2(2r + 3y)[yV(i + 1) - xV(i)] \\ &\quad + y^2(r + y)[yK(i + 1) - xK(i)] \geq 0. \end{aligned} \tag{5.2}$$

Now, recall that under Case 1 assumptions, $yV(i+1) \geq xV(i)$, $yK(i+1) \geq xK(i)$ for all i. So the only additional assumption required to make equation (5.2) hold is that $yM(y,y) \geq xM(x,y)$. We can also verify that no additional assumptions are required to make $yV_1(i+1) \geq xV_1(i)$ hold when $q_R = x$. So we have the following proposition.

Theorem 5.1: *When the following conditions hold*

$$x^2 V(i) \leq y^2 V(i+1), \; xK(i) \leq yK(i+1) \; \forall \; i,$$
$$x^2 M(v,x) \leq y^2 M(v,y), \; v = x, y, \; xM(x,y) \leq yM(y,y)$$

both firms choosing dominant strategies (y) in each of the R and D stages constitutes a subgame perfect equilibrium.

The conditions in this theorem can be reinterpreted using Proposition 5.2 (see the Appendix to this chapter). They are the same as saying that both monopoly and duopoly profits are sufficiently larger when the dominant option is chosen to outweigh the cost of achieving these higher profits at a later (expected) date. So this proposition simply establishes that if the advantages to introducing a better product are high enough to outweigh the benefits of introducing the product sooner, then both firms will opt for the dominant targets in both stages.

Case 2: Recall that this is the case where the first firm to finish the R stage chooses its D stage strategy to be x, while the second chooses y. We can now rule out an equilibrium where one firm does x in the R stage, and the other does y in the R stage, when the condition (C1) (under which Case 2 arises) holds. Let us develop a condition as we did condition (C1) in Step 3. Consider firm 1's response to $q_R = x$. We have to compare the payoff P_1 when $p_R = y$ with that when $p_R = x$. Let T_{C2} be the term that includes all non-monopoly terms in the inequality that expresses this comparison. Then the relevant condition for $p_R = y$ to be chosen in response to $q_R = x$ is

$$\frac{yM(y,x)}{(r+x+y)} > \frac{xM(x,x)}{(r+2x)} + T_{C2}.$$

This can be restated as

$$\theta > \frac{x}{y}\left(\frac{r+x+y}{r+2x}\right) + \frac{T_{C2}}{M(x,x)}\left(1 + \frac{r+x}{y}\right), \text{ where } \theta = \frac{M(y,x)}{M(x,x)}. \tag{C2}$$

While T_{C2} may be positive or negative, the T_{C2} term in the condition (C2) inequality has less of an effect the greater is the monopoly profit term, $M(x,x)$, relative to the terms in T_{C2}. Note also that the first term (FT) on the right-hand side is greater than 1. Then the condition (C2) inequality can be approximated by $\theta > FT$. This is less likely to hold the closer θ is to 1. But θ was required to be close to 1 to make condition (C1) hold (which is the condition under which we are in Case 2). So conditions (C1) and (C2) cannot hold simultaneously.

Now define condition (C3) to be condition (C2) with a reversal in the inequality. Then condition (C3) is likely to hold under the same conditions that condition (C1) held. When condition (C3) holds, firm 1 will choose $p_R = x$ in response to firm 2's choice of $q_R = x$. We can now state the following proposition.

Theorem 5.2: *When conditions (C1) and (C3) hold and*

$$xV(i) \leq yV(i+1), \; xK(i) \leq yK(i+1) \; \forall \; i$$

then both firms choosing x *in the R stage, the first firm to finish R choosing* x *in the D stage, and the second to finish R choosing* y *in the D stage, constitute a subgame perfect equilibrium.*

Proposition 5.3 (see the Appendix) establishes that the conditions under which this result holds are indeed met for some parameter values. The intuition behind this proposition is as follows. When the monopoly profits are large enough and do not vary much with the product characteristics, both firms choose the strategy that will enable them to introduce the product quickly. However, following the introduction of the product by one firm (or even the getting ahead of one firm), the rival finds it advantageous to choose the more dominant (and slower) option in the latter stage since the duopoly profits do vary enough with the quality of the product to make the delayed introduction worthwhile. An example of such a situation is as follows. Demand inelasticity implies that aggressive cost reduction (and associated long product development times) is not optimal. Bertrand competition implies that the only way to earn any profits when there is a rival is to have a product that is superior on some stage characteristic. So both firms will start by trying to become the monopolist with the less-dominant strategy, but once one firm has introduced the product, the laggard will prefer the more-dominant strategy.

In these two results, symmetric firms choose identically in the R stage. Depending on whether monopoly profits are large enough, they will choose either the quicker strategy or the more-dominant (slower) strategy. Whereas in the situation analysed in Theorem 5.1, firms are dominant in the D stage regardless of the order in which they finished the R stage, in the Theorem 5.2 situation, whether a firm is dominant or not in the D stage depends on whether it finishes the R stage last or first. The underlying notion is that a firm that gets ahead and closer to the finish line makes different choices from a firm that falls behind. We shall return to develop this notion further in the context of multi-stage races.

5.3 Consideration of asymmetries

Let firm 1 be an incumbent firm that chooses hazard rates p_R, p_D, in the R and D stages, respectively, and let firm 2 be an entrant that chooses hazard rates q_R, q_D similarly. The incumbent earns flow rents I from its existing product, which cease to accrue to it once the new product is introduced by either firm. We will analyse this incumbent–entrant interaction by backwards induction in the same manner as in Section 5.2.

The conditions governing the actions of the entrant firm are the same as those discussed in the symmetric case in Section 5.3. Following the discussion at the beginning of this section, let us assume that the sufficient conditions hold under which the entrant chooses $q_D = y$ if it is the last to finish the R stage and $q_D = x$ if it is the first firm to finish the R stage (Case 2 of Section 5.2). Recall that these were derived in Steps 3 and 2 of the backwards induction in Section 5.2. We need $xV(i) \leq yV(i+1)$, $xK(i) \leq yK(i+1)$ for all i, and condition (C1) should hold. Now we turn to analysing the incumbent's choice of actions.

Step 3

Suppose the incumbent is the last to finish the R stage. If, at the time of its choosing p_D, the entrant has not finished the D stage, then the incumbent earns $W^I_{1,a}(i)$ where

$$W^I_{1,a}(i) = \frac{p_D V(i) + q_D K(i) + I}{(r + p_D + q_D)}.$$

For $p_D = y$ to be preferred to $p_D = x$, the following inequality must hold

$$\frac{yW(i+1) + q_D L(i+1) + I}{(r + y + q_D)} > \frac{xW(i) + q_D L(i) + I}{(r + x + q_D)}.$$

This can be rewritten as

$$\begin{aligned} &I(x - y) + (r + q_D)[yV(i+1) - xV(i)] + xy[V(i+1) - V(i)] \\ &+ q_D(r + q_D)[K(i+1) - K(i)] + q_D[xK(i+1) - yK(i)] > 0. \end{aligned}$$

When $yV(i+1) \geq xV(i)$ for all i, this inequality holds and the incumbent chooses $p_D = y$ regardless of the entrant's choice of q_D.

If, at the time of the incumbent's choice of p_D, the entrant has finished the D stage, then the incumbent earns $S^I_{1,b}(i)$ where

$$S^I_{1,b}(i) = \frac{p_D L(i)}{(r + p_D)}.$$

Here $p_D = y$ is preferred to $p_D = x$ if $yL(i+1) \geq xL(i)$ for all i.

Step 2

If the incumbent is the first firm to finish the R stage, calculations similar to those in the symmetric firm analysis (where we developed an expression for the term $W_2(i)$) show that it gets $W_2^I(i)$ where

$$W_2^I(i) = W_2^I(i) = p_D\left[\frac{M + q_R V(i)}{(r + q_R)}\right] + q_R\left[\frac{p_D V(i) + q_D K(i) + I}{(r + p_D + q_D)}\right] + I/(r + p_D + q_R).$$

We can establish conditions under which the incumbent chooses $p_D = y$ as follows. Let T_{C4} include all non flow-profit (I) terms in the inequality that compares W_2^I when $p_D = y$ with W_2^I when $p_D = x$. Then the relevant condition for $p_D = y$ to be chosen is:

$$E + \frac{T_{c4}}{I} > F$$

where

$$E = \frac{(r + q_R + q_D + y)}{(r + y + q_R)(r + y + q_D)}, \quad F = \frac{(r + q_R + q_D + x)}{(r + x + q_R)(r + x + q_D)}. \tag{C4}$$

Here the q_D term is the same in E and F because the entrant's D stage action, when it is the second to finish the R stage, does not depend on the incumbent's choice of p_D. This follows from the assumptions made at the beginning of the backwards induction.

In condition (C4), the non T_{C4} terms are of the form

$$T^I = \frac{k + p_D}{(\alpha + p_D)(\beta + p_D)}, \quad \text{where } k = r + q_R + q_D, \alpha = r + q_D, \beta = r + q_R.$$

Then

$$\text{sign}\left(\frac{\partial T^I}{\partial p_D}\right) = \text{sign}((\alpha + p_D)(\beta + p_D) - (k + p_D)(\alpha + \beta + 2p_D)) < 0, \Rightarrow E > F.$$

It follows that condition (C4) will hold when I is large enough relative to the terms in T_{C4}, regardless of the sign of T_{C4}. Then the incumbent chooses $p_D = y$ when it is the first to finish the R stage regardless of the entrant's choice of q_D.

Step 1

Now we discuss the incumbent's choice of p_R in the R stage. Let V^I denote the incumbent's payoff if it finishes the R stage before the entrant, and subsequent play is optimal, and let $\bar{V}^I$ denote the incumbent's payoff if it finishes the R stage after the entrant, and subsequent play is optimal. Let P^I denote the incumbent's payoff when it has to choose its strategy p_R for the R stage. Then proceeding as in Step 1 of the backwards induction in Section 5.2, we have

$$P^I(i) = \frac{p_R V^I(i) + q_R \bar{V}^I(i) + I}{(r + p_R + q_R)}.$$

Note that $V^I(i)$ and $\bar{V}^I(i)$ are the analogues of $V_I(i)$ and $\bar{V}_I(i)$ in Step 1 of the earlier backwards induction. We can show that, when the I term in the expressions for $V^I(i)$ and $\bar{V}^I(i)$ is large relative to the other terms, then we can approximate for $V^I(i)$ and $\bar{V}^I(i)$ by the following

$$V^I(i) = I\frac{(r + p_D + q_R + q_D)}{(r + p_D + q_D)(r + p_D + q_R)}\forall i,$$

$$\bar{V}^I(i) = I\frac{(r + p_R + p_D + q_D)}{(r + p_D + q_D)(r + p_R + q_D)}\forall i.$$

Since $V^I(i)$ is what the incumbent gets when it finishes the R stage first, plug in $p_D = y$, $q_D = y$ into its expression. Similarly, since $\bar{V}^I(i)$ is what the incumbent gets when it finishes the R stage second, plug in $p_D = y$, $q_D = x$ into its expression. Then we have

$$V^I(i) = I\frac{(r + 2y + q_R)}{(r + 2y)(r + y + q_R)}\forall i,$$
$$\bar{V}^I(i) = I\frac{(r + x + y + p_R)}{(r + x + y)(r + x + p_R)}\forall i.$$

Substitute these terms into the expression for P^I. Substitute also $q_R = x$ to see what the incumbent's best response is to the entrant choosing the less-aggressive strategy in the R stage. Then we have

$$P^I = I = \frac{\left(\frac{p_R(r + 2y + x)}{(r + 2y)(r + y + x)} + \frac{x(r + x + y + p_R)}{(r + x + y)(r + x + p_R)} + 1\right)}{(r + x + p_R)}.$$

Numerical simulations show that the incumbent's best response to the entrant's choice of $q_R = x$ is 'usually' to choose $p_R = y$. Further, by exactly the same reasoning as in Case 2 of the symmetric firm interaction in Section 5.2, condition (C3) implies that $q_R = x$ is a dominant strategy for the entrant in the R stage. The conditions that we have accumulated in establishing this equilibrium include $xV(i) \leq yV(i+I)$, $xK(i) \leq yK(i+1)$ for all i, and that conditions (C1) and (C3) should hold. But these are exactly the conditions in Proposition 2. So we can state the following proposition.

Theorem 5.3: *When condition (C4) holds and the conditions in Theorem 5. 2 hold, then the following constitutes a subgame perfect equilibrium. The incumbent chooses* y *in each of the R and D stages. The entrant chooses* x *in the R stage,* x *in the D stage if it is the first to finish the R stage, and* y *in the D stage if it is the second to finish the R stage. The entrant behaves exactly as it did in the firm interaction.*

When monopoly profits are large enough and do not vary much with the product characteristics, the entrant chooses the faster, less-aggressive option in each stage as long as it has not fallen behind in the race. The incumbent's behaviour is influenced by what the literature identifies as the 'replacement effect' (see, e.g. Tirole (1988), Chapter 10). The conventional 'replacement effect' says that, in an effort to maximize the discounted value of its existing profit stream, an incumbent monopolist invests less in R&D than an entrant, and thus expects to be replaced by the entrant (in the case where the innovation is drastic enough that the firm with the older technology would not find it profitable to compete with the newer technology). In our model, when the incumbent's flow profits are large enough to make condition (C4) hold, this same replacement effect causes the incumbent to be replaced only temporarily (if the innovation is drastic). Subsequently it regains a dominant position in the market since it has a superior version of the new technology.

5.4 Perspectives on multi-stage races

In Section 5.2, we saw that, under some conditions, identical firms behave differently depending on whether they are the first or the second firm to finish the first stage of a two-stage race. The underlying notion is that getting ahead and closer to the finishing line results in different choices (dominant behaviour or quick behaviour) than does falling behind.

All the existing literature on multi-stage races is in the patent race framework. Harris and Vickers (1987) show that the leader invests more than the follower in a multi-stage patent race scenario. Their result generalizes a similar result due to Grossman and Shapiro (1987) for two-stage games. In contrast, instead of analysing aggregate resource allocation, we discuss how given resources are allocated. As in earlier parts of this chapter, there is a tradeoff between being dominant or being fast in each stage of a multi-stage race. Our focus will be on characterizing the differences in the expected payoff function of firms as they get ahead of their rivals (or fall behind) and closer to the finish line. We will speak of the monopoly (respectively, duopoly) term becoming more important in a payoff expression as the ratio of its coefficient to that of the duopoly (respectively, monopoly) term rises. In particular, we show the following.

Theorem 5.4: *The monopoly term in the expected payoff expression of the leading firm in a two-firm multi-stage race becomes progressively more important as it gets further ahead of its rival, providing the lead meets a minimum threshold. This threshold lead is smaller the closer is the lead firm to the finish line. Conversely, the duopoly term in the expected payoff expression of the lagging firm becomes more important as it falls further behind, subject to the same threshold lead considerations as the leading firm.*

Proof. The detailed analytics are in Appendix B. Here we sketch the method used.

First, we derive an expression for the leading firm's payoff, when it has finished all stages and is reaping monopoly profits, as a function of the lead it has over its rival. Let W_{n-1} be the payoff of the lead firm when it has finished all n stages, and the rival is in stage 1. Let $q_1, \ldots, q_n$. be the lagging firm's choices of hazard rates for the n stages. Then

$$W_{n-1} = \frac{M}{(r+q_1)}[1 + \frac{q_1}{(r+q_{1+1})} + \frac{q_1 q_{1+1}}{(r+q_{1+1})(r+q_{1+2})} + \cdots + \frac{q_1 \cdots q_{n-1}}{(r+q_{1+1})..(r+q_n)}$$

$$+ \frac{q_1 q_{1+1} \cdots q_n}{(r+q_1)(r+q_{1+1})\ldots(r+q_n)} D \ for\, n-1>0$$

$$W_o = \frac{M}{(r+q_n)} + \frac{q_n D}{(r+q_n)}, \quad for\, n-1=0.$$

We show that the coefficient on the monopoly term rises faster than that on the duopoly term as the lead increases.

Then, using this property of the coefficients, we consider the expression for the leading firm's payoff, $V_{n,n-j}$, as a function of its lead j, when it is in the last stage of the n-stage race:

$$V_{n,n-j} = \frac{p_n W_j + q_{n-j} V_{n.n(j-1)}}{r + p_n + q_{n-j}}.$$

Once again, we show that as long as the lead exceeds a threshold lead (which may be 0), the coefficient on the monopoly term rises faster than does that on the duopoly term as the lead increases. This method of recursion, where the relationships on the coefficients when the lead firm is at stage S of the race is used to derive similar relationships when the lead firm is at stage S–1, yields our result.

The procedure is similar for the lagging firm. First we derive an expression for its payoff, as a function of how much it lags the rival, when the rival has finished all stages and the lagging firm is at stage f. Denote this by L_{n-f}. Then

$$L_{n-f} = \frac{p_f \ldots p_n}{(r+p_f)\ldots(r+p_n)} D.$$

So the payoff when the firm is $j+1$ stages behind the leading firm at stage n is given by

$$V_{n-(j+1).n} = \frac{p_{n-(j+1)}V_{n-j} + q_n L_{j+1}}{r + p_{n(j+1)} + q_n}.$$

Then we show that the duopoly coefficients rise faster than the monopoly ones as the lead increases, subject to the threshold lead considerations. The same is shown to be true recursively when the lead firm is at position n–1, n–2, etc.

Hence the result.

This characterization highlights two forces that influence a firm's choices in the various stages: proximity to the finish line and distance between the firms. The probability of reaping monopoly profits is higher the farther ahead a firm is of its rival, and even more so the closer the firm is to the finish line. If the lead firm is far from the finish line, even a sizeable lead may not translate into the dominance of the monopoly profit term, since there is plenty of time for the lead situation to be reversed and failure to finish first remains a probable outcome. In contrast, the probability that the lagging firm will get to be a monopolist becomes smaller as it falls behind the lead firm. This raises the following question. What kinds of actions cause a firm to get ahead? Intuitively, one would expect that a firm that is ahead of its rival at any time t, in the sense of having completed more stages by time t, is likely to have chosen the faster strategy more often. We can construct numerical estimates of the probability that a leading firm is more likely to have chosen this strategy to verify this intuition.

We have shown that the monopoly term is increasingly important to a firm as it gets ahead of its rival, and that the duopoly term is increasingly important to a firm that falls behind. Further simple calculations suggest that the firm that is ahead is likely to have made less-aggressive choices than the firm that is behind in the race.

One question of interest is whether chance leads result in greater likelihood of increasing lead, or in more catch-up behaviour. The existing literature (Grossman and Shapiro, 1987; Harris and Vickers, 1987) has suggested that a firm that surges ahead of its rival increases its investment in R&D and speeds up while a lagging firm reduces its investment in R&D and slows down. Consequently these papers suggest that the lead continues to increase. However, when duopoly competition and system complexity are accounted for, the speeding up of a leading firm occurs only under special circumstances. We suggest that the computer industry is one in which monopoly profits do not change substantially with increased dominance, but duopoly profits do change substantially with increased dominance (Shapiro, 1989). Then a firm getting far enough ahead such that the monopoly term dominates its payoff expression will always choose the fast strategy, while a firm that gets far enough behind will always choose the slow and dominant approach. Then the lead is likely to continue to increase. If, on the other hand, both monopoly and duopoly profits increase substantially with increased aggressiveness then even large leads can vanish with significant probability.

5.5 Further discussion

This chapter sets out to examine and measure racing behaviour on technological positions among firms in high-technology industries. In measuring the patterns of technological evolution in these industries we attempt to answer questions about whether and to what extent their racing patterns differ from those firms in respective industries that do not operate on a global scale. Among the key issues we address is the apparent inability of technology-oriented corporations to maintain leadership in fields that they pioneered. There is a presumption that firms fail to remain competitive because of agency problems or other suboptimal managerial behaviour within these organizations. An alternative hypothesis is that technologically trailing firms, in symmetric competitive situations, will devote greater effort to innovation, so that a failure of technological leaders to maintain their position is an appropriate response to the competitive environment. In asymmetric situations, with entrants challenging incumbents, research could demonstrate whether startup firms show a stronger endeavour to close up to or leapfrog the competitors. Such issues would highlight the dynamics of the race within the given market structure in any of the areas concerned. We observe two different kinds of market asymmetries bearing on racing behaviour: (a) risk-driven and (b) resource-based asymmetries.

When the incumbents' profits are large enough and do not vary much with the product characteristics, the entrant is likely to choose the faster, less-aggressive option in each stage as long as he has not fallen behind in the race. The incumbent's behaviour is influenced by what is known as the 'replacement effect' (Tirole, 1988). The conventional 'replacement' effect says that, in an effort to maximize the discounted value of its existing profit stream, the incumbent (monopolist) invests less in R&D than does an entrant, and thus expects to be replaced by the entrant (in the case where the innovation is drastic enough that the firm with the older technology would not find it profitable to compete with the newer technology). In one of our models, when the incumbent's flow profit is large enough, the same replacement effect causes the incumbent to be replaced only temporarily (if the innovation is drastic). Subsequently, the incumbent is likely to regain a dominant position in the market since she has a superior version of the new technology.

In view of resource-based asymmetries, we observe, as a firm's stage resource endowment increases, it could use the additional resources to either choose more aggressive targets or to attempt to finish the stage quicker, or both. This hypothesis suggests two interpretations, suitable for empirical exploration: (a) if the demand for new products displays different elasticities for different local/regional markets, then we might expect there to be only imperfect correlation between aggressiveness and resource richness when products from different markets are grouped together; (b) if, however, demand for these products is not inelastic enough, then we would expect resource-rich firms to aim for both higher speed in R&D and greater aggressiveness.

A further point of exploration is whether chance leads result in a greater likelihood of increasing the lead, or in more catch-up behaviour. Previous work in this regard (Grossman and Shapiro, 1987; Harris and Vickers, 1987) has suggested that a firm that surges ahead of its rival increases its investment in R&D and speeds up while a lagging firm reduces its investment in R&D and slows down. Consequently, previous work suggests that the lead continues to increase. However, based on related work for the US and Japanese telecommunications industry (Gottinger, 1998) when duopoly and monopolistic competition and product system complexity for new products are accounted for, the speeding up of a leading firm occurs only under rare circumstances. For example, a firm getting far enough ahead such that the (temporary) monopoly term dominates its payoff expression will always choose the fast strategy, while a firm that gets far enough behind will always choose the slow and aggressive approach. Then the lead is likely to continue to increase. If, on the other hand, both monopoly and duopoly profits increase substantially with increased aggressiveness then even large leads can vanish with significant probability.

Overall, this characterization highlights two forces that influence a firm's choices in the various stages: proximity to the finish line and distance between the firms. This probability of reaping monopoly profits is higher the farther ahead a firm is of its rival, and even more so the closer the firm is to the finish line. If the lead firm is far from the finish line, even a sizeable lead may not translate into the dominance of the monopoly profit term, since there is plenty of time for the lead situation to be reversed and failure to finish first remains a probable outcome. In contrast, the probability that the lagging firm will get to be a monopolist becomes smaller as it falls behind the lead firm. This raises the following question. What kind of actions cause a firm to get ahead? Intuitively, one would expect that a firm that is ahead of its rival at any time t, in the sense of having completed more stages by time t, is likely to haven chosen the faster, less-aggressive strategy more often. We will construct numerical estimates of the probability that a leading firm is more likely to have chosen a strategy less-aggressively (faster) to verify this intuition.

Of course, those conditions observed only hold in 'normal' scenarios, in markets that have not been subject to excessive capacity building so that technological racing comes to a standstill because of firms' preoccupation to survive an erosion of market activity (Berman, 2002).

5.6 Conclusion

The main contribution of this chapter has been to formalize the implications of technological racing for firms' R&D strategies. Product development typically proceeds in a sequence of stages, each of which determines the degree to which some subset of product characteristics is developed. In contrast to the existing literature's focus on how firms choose their aggregate level of investment in R&D, we discuss how given R&D resources are utilized in the different stages of the R&D process. Our model makes explicit how scarce R&D resources result in a tradeoff between the dominance of the targeted objectives at each stage and the speed with which that stage is completed. *Ex ante* differences between firms influence the way in which firms exercise their discretion in each stage.

We showed that the fact that higher market share firms appear to choose more aggressive targets is an instance of the 'replacement effect'. Further, we found a weak correlation between a firm's R&D resource endowment and its targeted level of aggressiveness. Theory allows us to suggest some reasons why this might be so. For example, theory suggests that we could expect a strong correlation if the degree of demand elasticity was availability of substitute computing resources.

The second contribution of this chapter was to make explicit the two forces that influence a firm's choices in the various stages of a multi-stage race: proximity to the finish line and distance between the firms. In doing so, we suggested two things. First, the existing literature, by focusing on the issue of aggregate resources devoted to R&D as the sequential game progresses, ignores the factors that influence the way in which these resources are utilized. Second, while conclusions in the existing literature continue to hold when we depart from the conventional patent-race framework, they do so only under special circumstances.

References

Albach, H. (1993) 'Information, Zeit und Wettbewerb' (Information, Time and Competition), in *Jahrestagung des Vereins für Sozialpolitik*, Münster 1993. Berlin: Duncker und Humblot, 113–154.

Albach, H. (1997) 'Global competition among the few', The Ehrenrooth Lectures, Swedish School of Economics and Business Administration, Res. Reports 40. Helsingfors.

Beath, J., Katsoulacos, Y. and Ulph, D. (1987) 'Sequential product innovation and industry evolution', *Economic Journal*, **97** (Conference 87), 32–43.

Berman, D. K. (2002) 'Technology races far ahead of demand and the workplace', *The Wall Street Journal*, 26 September.

Brezis, E., Krugman, P. and Tsiddon, D. (1991) 'Leapfrogging: a theory of cycles in national technological leadership', National Bureau of Economic Research (NBER), Working Paper 3886.

Camerer, C. (1991) 'Does strategy research need game theory?', *Strategic Management Journal*, **12**, 137–152.

Cebon, P. B., Hauptman, O. and Shekar, C. (2001) 'Industries in the making: product modularity, technological innovation and the product lifecycle', Melbourne Business School, mimeo.

Cohen, W. M., and Levin, R. C. (1989) 'Empirical studies of innovation and market structure', in: *Handbook of Industrial Organization, Vol. II*, R. Schmalensee and R. D. Willig (eds.), New York: Elsevier Science Publishers, pp. 1061–1249.

Connor, K. (1989) 'A historical comparison of resource-based theory and five schools of thought within industrial organization economics: do we have a new theory of the firm?', *Journal of Management*, Vol. **17**, 121–154.

Fershtman, C. and Kamien, M. (1990) 'Cross licensing of complementary technologies', Center for Mathematical Studies in Economics & Management Science Discussion Paper 866, Northwestern University.

Fudenberg, D. and Tirole, J. (1989) 'Noncooperative game theory for industrial organization: an introduction and overview', in: R. Schmalensee and R.D. Willig (eds.), *Handbook of Industrial Organization, Vol. I*, New York: Elsevier Science Publ., 260–327.

Gilbert, R. J. (1989) 'Mobility barriers and the value of incumbency', in: *Handbook of Industrial Organization, Vol.I*, R. Schmalensee and R.D. Willig (eds.), New York: Elsevier Science Publishers, pp. 475–535.

Gottinger, H.W. (1998) 'Technological races', *Annual Review of Economics (Japan)*, **38**, 1–9.

Grossman, G. and Shapiro, C. (1986) 'Optimal dynamic R&D programs', *Rand Journal of Economics*, **17**, 581–596.

Grossman, G. and Shapiro, C. (1987) 'Dynamic R&D competition', *Economic Journal*, **97**, 372–387.

Harris, C. and Vickers, J. (1987) 'Racing with uncertainty', *Review of Economic Studies*, **44**, 1–21.

Jacquemin, A. (1987) *The New Industrial Organization, Market Forces and Strategic Behavior*, Cambridge, MA: MIT Press.

Khanna, T. and Iansiti, M. (1997) 'Firm asymmetries and sequential R&D: theory and evidence from the mainframe computer industry', *Management Science*, **43**(4), 405–421.

Kreps, D. (1990) 'Corporate culture and economic theory', in: *Perspectives on Positive Political Economy*, J. Alt & K. Shepsle (eds.) Cambridge, MA: Cambridge University Press.

Lerner, J. (1997) 'An empirical exploration of a technology race', *Rand Journal of Economics*, **28**(2), 228–247.

Lippman, S. and McCardle, K. (1987) 'Dropout behavior in R&D races with learning', *Rand Journal of Economics*, **18**, 287-295.

Park, J. (1987) 'Dynamic patent races with risky choices', *Management Science*, **33**, 1563–1571.

Reinganum, J.F. (1989) 'The timing of innovation: research, development, and diffusion', *Handbook of Industrial Organization, Vol. I*, in: R. Schmalensee and R.D. Willig (eds.), New York: Elsevier Science Publ., 849–908.

Rubin, P. (1973) 'The expansion of firms', *Journal of Political Economy*, **81**, (July–August), 936–949.
Tirole, J. (1988) '*The Theory of Industrial Organization*', Cambridge, MA: MIT Press.
Vickers, J. (1986) 'The evolution of market structure when there is a sequence of innovations', *Journal of Industrial Economics*, September.

Appendix

A. Proposition 5.1: *For any value of points i, if monopoly profits with less aggressive actions in each of the R and D stages exceeds* T(y)D(i), *then (a)* M > V(i) *for all* x, y *combinations that map into* i. *(b)* V(i) > K(i). *Further, the greater is* M(x, x)/D(i), *the greater are the ratios* M(x, x)/V(i) *and* V(i)/K(i).

Proof. For convenience, define the quantity $T(q)=q/(r+q-1)$. The terms can be expressed as follows:

$$rV = M + q(D - V) \Rightarrow V = (M + q_D)/(r + q)$$

where $q=$ rival's D stage strategy

$$M, D = \text{monoply, duopoly profits}$$

$$rK = p(D - K) \Rightarrow K = p\mathrm{D}/(r + p)$$

where $p=$ own D stage strategy

$$M > V(i) \Leftrightarrow M > q_D(i)/(r + q - 1).$$

There are two cases to consider. If $r+q < 1$, then proposition is true trivially. So consider $r+q > 1$. Since $T(q) < 0$, so $T(y) > T(x)$. Also $M(x, x)$ is the lowest value that $M(p_R, p_D)$ can assume. So if $M(x, x) > T(y)D(i)$, then $M > T(q)D(i)$ will hold for all realizations of q, p_R, p_D. Note finally that $M(p, p.) > T(q)D(i)$ suffices to ensure that $M > W(i)$ for all x, y combinations that map into i. Part (a) follows. For part (b),

$$V > K \Leftrightarrow M > r(p - q)D/(r + p).$$

Here p and q are the two firms' D stage strategies. If $p < q$, or $p=q$, then the result follows immediately. If $p=x$, $q=y$ so that $p > q$, then define

$$\gamma = \frac{r(x - y)}{(x + y)}.$$

Then $\gamma > T(y) \Rightarrow r > xy + y^2/[xy - y^2 - (1 - r)(x - y)] > 1$.

So $\gamma < T(y)$. Then $M(x,x) > T(z)D(i) \Rightarrow M(x,x) > \gamma D(i)$.

So part (b) holds.

The fact that M/V and V/K are increasing functions of M/D can be seen by writing the following expressions:

$$M/V = [1 + q/(M/D)]^{-1}, \quad V/K = [(r+p)/(r+q)][1M/pD + q/p].$$

Hence the result.

Proposition 5.2: *The following conditions*

$$\mathrm{yD(i+1)} > \mathrm{xD(i)}, \mathrm{yM(y,y)} > \mathrm{xM(x,y)} \textit{ where } \mathrm{v = x, y}$$

are sufficient to ensure that

$$\mathrm{yK(i+1)} > \mathrm{xK(i)}, \ \mathrm{yV(i+1)} > \mathrm{xV(i)}.$$

Proof: follows directly from the expressions.

Proposition 5.3: *Even when* M *is greater than* V*(*i*)*, K*(*i*)* *in the sense of Proposition 5.1, there exist ranges of values of* x *and* y *such that the following conditions are compatible*:

$$xV(i) < yV(i+1), \quad xK(i) < yk(i+1), \quad xM(.,x) > yM(.,y).$$

Proof. We want to show that the following conditions are consistent.

As in the proof of Proposition 5.1, M is larger than V, K follows if M is larger than D.

As in Proposition 5.2, the condition

$$xD(i) < yD(i+1) \Rightarrow xK(i) < yK(i+1).$$

The only thing that remains to be checked is that it is possible to have $xM(.,x) > yM(.,y)$ and $xV(i) < yV(i+1)$ together. To see this, note that

$$xV(i) < yV(i+1) \Rightarrow q(yD(i+1) - xD(i)) > (xM(.,x) - yM(.,y) - c(x-y).$$

Rewrite this as

$$\frac{q(yD(i+1) - xD(i))}{\left(x\frac{M(.,x)}{M(.,y)} - y\right)} \rangle M(.,y).$$

Let $q = x$ (since we just want to show possibility) and let

$$x = ky, k > 1 \quad \alpha = \frac{D(i+1)}{d(i)}, \quad \beta = \frac{M(.,y)}{D(i)}, \quad \tau = \frac{M(.,y)}{M(.,x)}.$$

Then, the condition can be re-expressed as

$$\frac{(\alpha - k)}{\left(\frac{1}{\tau} - \frac{1}{k}\right)} > \frac{\beta}{y}.$$

This is satisfied for any realization of the RHS if k is small enough.

So, if x and y are sufficiently close, $x > y$, x, y in $(0, 1)$, then the conditions stated in the proposition can hold concurrently.

Hence the result.

B. Proof of Theorem 5.4

We will speak of the monopoly (respectively, duopoly) term becoming more important in a payoff expression as the ratio of its coefficient to that of the duopoly (respectively, monopoly) term rises. We show that the monopoly term in the payoff expression of a leading firm becomes progressively more important than the duopoly term as it gets further ahead of its rival in a multi-stage race (as long as the lead is great enough), while the duopoly term assumes greater significance for the lagging firm as it gets further behind.

Let $p_1, \ldots, p_n$ indicate the choices of hazard rate by the leading firm in each of the n stages, and let $q_1, \ldots, q_n$, indicate the analogous choices by the lagging firm. For convenience, we will denote a firm's flow monopoly profits, which are a function of the its hazard rates, by M and flow duopoly profits, which are a function of the 'points' system introduced in the chapter, by D.

First consider the leading firm. We will derive an expression for its payoff when it has finished all stages while its rival has not and will show that the coefficient of M rises faster than the coefficient of D as the rival gets further behind, as long as the rival is far enough behind. We will then show that this property of the coefficients holds whatever the position of the leading firm (i.e. it need not already have finished all the stages of the race).

Let W_{n-1} be the expected payoff of the leading firm when it has finished all stages, and its rival is still engaged in stage 1. Then

$$rW_{n-1} = M + q_l\left(W_{n-(l+1)} - W_{n-l}\right)$$

where $M = M(p_1, \ldots p_n) =$ leaders of monopoly profits

$q_l =$ laggard's stage l strategy.

$$\text{So } W_{n-1} = \frac{M}{(r+q_l)}(1 + \frac{q_l}{(r+q_{l+1})} + \frac{q_l}{(r+q_l)}\frac{q_{l-1}}{(r+q_{l+1})}W_{n-l-2}.$$

By repeated substitution as above, and by using the fact that

$$W_o = \frac{M}{(r+q_n)} + \frac{q_n D}{(r+q_n)}$$

we can show that

$$W_{n-1} = \frac{M}{(r+q_l)}[1 + \frac{q_l}{(r+q_{l+1})} + \frac{q_{l+1}}{(r+q_{l+1})(r+q_{l+2})}$$
$$+ \dots + \frac{q_l \cdot q_{n-1}}{(r+q_{l+1})..(r+q_n)}$$
$$+ \frac{q_{l+1} \dots q_n}{(r+q)(r+q_{l-1} \dots (r+q_n)} D \quad \textit{for } n-l>0$$
$$W_o = \frac{M}{(r+q_n)} + \frac{q_n D}{(r+q_n)}, \qquad \textit{for } n-l=0.$$

So W_o is the expected payoff of the leading firm when it has finished all stages and the laggard is in the last stage.

Using the following notation

$\eta_j = W_{n-(n-j)}$'s M term coefficient

$\gamma_j = W_{n-(n-j)}$'s D term coefficient

the following recurrence relations hold

$$n_{j+1} = \frac{1}{(r+q_{n-(j+1)})} + \frac{q_{n-(j+1)}}{(r+q_{n-(j+1)})}\eta_j, \quad \gamma_{j+1} = \frac{q_{n-(j+1)}}{(r+q_{n-(j+1)})}\gamma_j.$$

It follows that, when the leading firm has finished all n stages, as the lead widens, the coefficient of M rises relative to that of D since

$$\frac{n_{j+1}}{n_j} > \frac{\gamma_{j+1}}{\gamma_j}, \ \forall j \geq 0.$$

Now consider the situation when the lead firm is in the final stage n. If $V_{n,n-j}$ is its payoff when it is in stage n and the laggard is j stages behind, then

$$V_{n,n-j} = \frac{p_n W_j + q_{n-j} V_{n,n(j-1)}}{r + p_n + q_{n-j}}.$$

If the rival's actions are the same (i.e. q_{n-j} is the same for all j), then we can write this as

$$V_{n,n-j} = \mathrm{A}W_j + \mathrm{B}V_{n,n-(j-1)} - C$$

where A,B,C are constants.

Using the following notation

$$\mu_j = V_{n,n-j}\text{'s } M \text{ term coefficient}$$

$$\lambda_j = V_{n,n-j}\text{'s } D \text{ term coefficient}$$

the following recurrence relations hold

$$\mu_{j+1} = \mathrm{A}\eta_{j+1} + \mathrm{B}\mu_j \tag{i}$$

$$\lambda_{j+1} = \mathrm{A}\gamma_{j+1} + \mathrm{B}\lambda_j$$

where η_j, γ_j are as before.
Then

$$\mu_j = \mathrm{A}\eta_j + \mathrm{B}\mu_{j-1} = \mathrm{A}\eta_j + \mathrm{B}(\mathrm{A}\eta_{j-1} + \mathrm{B}\mu_{j-2}) = \mathrm{A}(\eta_j + \mathrm{B}\eta_{j-1}) + \mathrm{B}^2\mu_{j-1}.$$

Repeating this recursion gives us

$$\mu_j = (\mathrm{A}\eta_j + \mathrm{B}\eta_{j-1} + \mathrm{B}^2\eta_{j-2} + \ldots \mathrm{B}^{j-1}\eta_j) + \mathrm{B}^j\mu_0$$

where $\mu_0 = V_{n,n}$'s M term coefficient.
Similarly,

$$\lambda_j = \mathrm{A}(\gamma_j + \mathrm{B}\gamma_{j-1} + \mathrm{B}^2\gamma_{j-2} + \ldots + \mathrm{B}^{j-1}) + \mathrm{B}^j\lambda_0$$

where $\lambda_0 = V_{n,n}$'s D term coefficient.
From (i) we see that

$$\frac{\mu_{j+1}}{\mu_j} > \frac{\lambda_{j+1}}{\lambda_j} \Leftrightarrow \frac{\mu_j}{\eta_{j+1}} < \frac{\lambda_j}{\gamma_{j+1}}.$$

The inequality on the right-hand side can be expanded to read as

$$\frac{\mathrm{A}(\eta_j + \mathrm{B}\eta_{j-1} + .. + \mathrm{B}^{j-1}\eta_1)}{\eta_{j+1}} + \frac{\mathrm{B}^j\mu_0}{\eta_{j+1}} < \frac{\mathrm{A}(\gamma_j + \mathrm{B}\gamma_{j-1} + .. + \mathrm{B}^{j-1}\gamma_1)}{\gamma_{j+1}} + \frac{\mathrm{B}^j\lambda_0}{\gamma_{j+1}}. \tag{ii}$$

Now, recall from the analysis of the coefficients when the leading firm had finished all the stages that

$$\frac{\eta_j}{\eta_{j+1}} < \frac{\gamma_j}{\gamma_{j+1}}.$$

Further,

$$\frac{\eta_{j-1}}{\eta_{j+1}} = \frac{\eta_{j-1}}{\eta_j}\frac{\eta_j}{\eta_{j+1}} < \frac{\gamma_{j-1}}{\gamma_j}\frac{\gamma_j}{\gamma_{j+1}} = \frac{\gamma_{j-1}}{\gamma_{j+1}}.$$

Similarly,

$$\frac{\eta_{j-i}}{\eta_{j+1}} < \frac{\gamma_{j-i}}{\gamma_{j+1}}, \forall\, i \in (0,j).$$

Inequality (ii) will be satisfied for j large enough regardless of the relative magnitude of μ_0, and λ_0 (since $B < 1$). It follows that the coefficients of the monopoly term rise faster than those of the duopoly term once the lead (j) gets large enough and the leader is in the final stage.

Thus far we have used our knowledge of the fact that the monopoly coefficients rise faster than the duopoly ones when the lead firm has finished all stages to derive a similar result for the corresponding coefficients when the lead firm is in the final stage and the lead is large enough. A similar recursive analysis (i.e. using the pattern of coefficients when the leader is at stage n–m to derive the pattern of coefficients when the leader is at stage n–m–1) establishes that the monopoly coefficients will rise faster than the duopoly ones regardless of the lead firm's position as long as the lead is large enough.

Further, it is possible to show that the further the lead firm gets from the finishing line, the greater is the lead needed before the property about this monopoly and duopoly coefficients holds. This can be seen if we reinterpret the notation from above as follows. Let the η and γ terms refer to the monopoly and duopoly coefficients when the leader is at stage n–m, and let λ and μ refer to the coefficients when the leader is at n–m–1. Let j refer to the lead as before. Call the following inequalities *condition (iii)*

$$j = 0\text{: For } \frac{\mu_1}{\mu_o} > \frac{\lambda_1}{\lambda_0},\ \textit{thus}\ \frac{\mu_0}{\eta_1} < \frac{\lambda_0}{\gamma_1}$$

$$j = 1\text{: For } \frac{\mu_2}{\mu_1} > \frac{\lambda_2}{\lambda_1}\ \textit{thus}\ \frac{\mu_1}{\eta_2} < \frac{\lambda_1}{\gamma_2} \Rightarrow \frac{A\eta_1 + B\mu_0}{\eta_2} < \frac{A\gamma_1 + B\lambda_0}{\gamma_2} \qquad \text{(iii)}$$

$$j = 2\text{: For } \frac{\mu_3}{\mu_2} > \frac{\lambda_3}{\lambda_2},\ \textit{thus}\ \frac{\mu_2}{\eta_3} < \frac{\lambda_2}{\gamma_3}$$

$$\Rightarrow \frac{A(\eta_2 + B\eta_1) + B^2\mu_0}{\eta_3} < \frac{A(\gamma_2 + B\gamma_1) + B^2\lambda_0}{\gamma_3}.$$

Similarly for $j = 3, 4, \ldots$

For any m, the condition (iii) inequality for a particular lead j must hold true for the monopoly coefficient to rise faster than the duopoly coefficient when the leader is at stage n–m–1.

Now consider the conditions (iii) when $m=S$. If μ_0 is larger than λ_0, this is more likely to cause a problem for the conditions for low j than for higher j. Suppose the conditions above do not hold for $j=0$ and $j=1$, but do hold for $j>1$ (in stage $m=S$). Then, in the next recursion when $m=S+1$, we have

$$\frac{\eta_1}{\eta_0}<\frac{\gamma_1}{\gamma_0},\ \frac{\eta_2}{\eta_1}<\frac{\gamma_2}{\gamma_1},\ \frac{\eta_{j+1}}{\eta_j}>\frac{\gamma_{j+1}}{\gamma_j}\ \forall j\geq 2.$$

Then we see that in conditions (iii) for $m=S+1$, it is even more likely that the low j inequalities will not hold, and we will need j to be higher than for the $m=S$ situation for the inequalities to hold. Thus the higher is m, i.e. the greater is the distance of the lead firm from the end of the race, the greater has to be the lead before the monopoly coefficient starts rising faster than the duopoly coefficient.

Now we turn our attention to the lagging firm. The analysis procceds in a manner similar to that for the leading firm. Let L denote the lagging firm's payoff when the leader has finished all n stages. Let $p_1..,p_n$ denote the choices of hazard rate for each of the n stages by the lagging firm. Then

$$L_{n-f}=\frac{p_f..p_n}{(r+p_f)\dots(r+p_n)}D.$$

As above, we are interested in understanding how the coefficients of the M and D terms behave. Consider the lagging firm's payoff when it is at stage $n-(j+1)$ and the leading firm is at stage n. Then

$$V_{n-(j+1)n}=\frac{p_{n(j+1)}V_{n-j}+q_nL_{j+1}}{r+p_{n-(j+1)}+q_n}.$$

Adopting notation similar to that used above, we have

$$\gamma_j = L_{n-j}\text{'s } D \text{ term coefficient}$$
$$\mu_j = V_{n-j,n}\text{'s } M \text{ term coefficient}$$
$$\lambda_j = V_{n-j,j}\text{'s } D \text{ term coefficient.}$$

Then

$$\mu_{j+1}=\frac{p_{n-(j+1)}}{r+p_{n-(j+1)}+q_n}\mu_{j+1},\ \lambda_{j+1}=\frac{p_{n-(j+1)}}{r+p_{n-(j+1)}+q_n}\lambda_j+\frac{q_n}{r+p_{n-(j+1)}+q_n}\gamma_{j+1}$$
$$\Rightarrow\frac{\lambda_{j+1}}{\lambda_j}>\frac{\mu_{j+1}}{\mu_j}\forall_j$$

i.e. duopoly term coefficients rise faster than the monopoly term coefficients as the lagging firm gets further behind the leading firm when the latter is at stage n.

Then we carry out the recursion. So when the leader is at stage n–1, and the laggard is $j+1$ stages behind, the latter's payoff is

$$V_{n-1(j+1)n-1} = \frac{p_{n-j-2}V_{n-1-jn-1} + q_{n-1}V_{n-1(j+1)n} - c}{r + p_{n-j-2} + q_{n-1}}.$$

Then we let η_j and γ_j be the M and D term coefficients when the leader is at stage n, and μ_j, and λ_j be those when the leader is at stage n–1, and we derive recurrence relations for the μ and λ coefficients in much the same way as the analysis for the leading firm. The conclusions are similar. The coefficients of the duopoly term in the lagging firm's payoff expression rise faster than those of the monopoly term when the lead is large enough.

6 Stochastic racing and competition in network markets

> If there is a chance that today's products will be replaced by a major innovation, a leader's survival depends on bringing this innovation to market and thereby replacing itself before others do.
>
> Evans and Schmalensee (2001)

6.1 Introduction

Many high-technology industries are characterized by positive network externalities. Firms essentially compete against each other for goods that share a network.

Our model of competition contains two special features that apply equally well to network markets. One is the uncertainty in technological development or uncertainty in the realization of a firm's R&D effort. The other is the dynamic nature of price competition between firms in the presence of network effects.

Firms compete with each other over an extended period of time and must therefore choose prices strategically as the market shares of the firms evolve.

Most existing models of network markets focus on only one of the two features. Further, almost all the models have the commonality that one firm captures the market instantaneously and sells to all consumers from then on (the 'winner-takes-all' market). However, the history of high-technology industry abounds with instances where rival firms (and technologies) have had extended battles for providing the industry standard (Uttenback, 1994).

In network markets evolution of market share is a very interesting phenomenon, especially in the face of uncertainty about future product quality. This chapter also attempts to provide a framework that could capture the richness of market share evolution in the presence of network externalities.

We rely on the stochastic racing model in Chapter 2, and also take some features from Schumpeterian racing in the seminal contribution by Futia (1980).

In order to motivate the setup we start from some unconventional ideas on antitrust analysis in dynamically competitive industries (Evans and Schmalensee, 2001):

> Firms that are not leaders in network industries generally have little hope of reaching that status unless they come up with a major innovation – one that can defeat the natural advantage that network effects bestow on the industry leaders. Incremental innovation – making slight improvements in the leaders' products – will not enable a small firm to overtake a leader that enjoys the benefits of network economies. Similarly, the possibility of being displaced by a major innovation will shape leaders' research agendas. If there is a chance that today's products will be replaced by a major innovation, a leader's survival depends on bringing this innovation to market and thereby replacing itself before others do. As a result, competition in network industries often involves intense R&D efforts aimed at capturing or retaining market leadership. It is not atypical for a fringe firm that invests heavily to displace the leader by leapfrogging the leader's technology . . .
>
> (Evans and Schmalensee, 2001, pp. 10–12)

This situation described may also be encountered in 'dynamic oligopoly', where exogenously emerging new technologies are rapidly eroding costs, or where market structure responds endogenously to intense racing behaviour (Shapiro, 1989, Section 5).

We are exploring this problem of tradeoff between 'network dominance' and 'radical innovation' that could tip the market the other way, with a significant caveat added that breakthrough R&D is highly uncertain. From a strategic perspective, in this environment, for any two firms of asymmetric size, both compete dynamically over prices to win market share. In this dynamic process there are two ways to achieve (temporary) monopolistic status. The 'smaller' firm can use dynamic pricing competition to delay the time in which the 'larger' firm wins a critical market share in the hope to hit the innovation first and displace it. If the probability of innovation and the discount factor are sufficiently high, there is an equilibrium in which duopoly persists (no firm achieves a critical market share) until one of the two firms wins the race for innovation.

We briefly summarize a historical example from markets for IT products. In the personal computer (PC) operating system market, Microsoft products (MS DOS and Windows) have been dominating the market since the mid-1980s. An operating system is the fundamental program that controls the allocation and use of computer resources. Thus, the utility that operating systems provide to consumers depends on the number of compatible applications. As a general rule, an application that relies on a specific operating system will not function on another operating system unless it is ported to that specific operating system. Therefore, because of its dominance, the majority of applications have been written to run on Microsoft operating systems (MS DOS). The domination of MS DOS has become even stronger since the arrival of Windows 95 in the PC operating system market. This, in turn, has provided a great indirect positive network externality to PC owners who adopted Windows 95 as an operating system. Many other firms, such as IBM and BEA Systems, Inc., introduced their own operating systems and tried to compete with Windows. These products, however, lacked sufficient compatible applications to compete efficiently with Microsoft products. The lack of compatible applications prevented enough application developers and consumers from regarding OS/2 Warp or BeOS as a viable alternative to the dominant incumbent, Windows. This obstacle prevented these potential entrants from obtaining a sizeable market share. Their failure to enter the market successfully, however, was not due to the inferior quality of their operating systems. In fact, OS/2 Warp was reported to be at least as good as Windows, and BeOS offers superior support for multimedia applications. If consumers who use multimedia applications frequently adopt Windows at the expense of BeOS, they have to give up the convenience that is provided by BeOS. Thus, for multimedia-specific users, adopting BeOS as their operating system at the expense of another operating system might provide the highest utility. Nevertheless, the lack of compatible applications, which in turn implies the lack of positive network externality, has prevented consumers from adopting OS/2 Warp or BeOS. As a conclusion, the positive network externality for the dominant incumbent (Windows) has worked as an entry barrier against entrants (OS/2 Warp or BeOS) which do not have network externality. Such an entry barrier could only be overcome by a radical innovation, virtually leapfrogging the dominant incumbent.

In Section 6.2 we provide paradigmatic cases from ICT industries that indicate restrictions to overcome technological barriers. Section 6.3 presents the context of technological races in network industries. Section 6.4 outlines the equilibrium conditions attainable for multi-state Markov games resulting in Markov equilibria. Section 6.5 addresses some extensions of the framework and draws conclusions.

6.2 Market histories

We substantiate our claim on a natural tradeoff between network size and quality change (through significant technological change) by looking independently at four market histories for the IT industry.

(a) Internet Explorer

The first version of Internet Explorer was introduced for the web browser market in July 1995. This version of Internet Explorer was reportedly inferior to Netscape's Navigator. Microsoft made many improvements in its next version, Internet Explorer 3.0, in August 1996. Internet Explorer 3.0 matched nearly all the features in Navigator. However, Microsoft was not able to draw significant market share at the expense of Navigator with this enhanced version. Internet Explorer 4.0, released in late 1997, finally managed to obtain roughly the same number of reviewers who preferred it to Navigator. In 1996, most consumers were much more familiar with Navigator, the dominant web browser, than with Internet Explorer. Thus although the quality of Internet Explorer was much improved since its release into the market, more than 80% of consumers were still using Navigator until late 1996. It was evident that Internet Explorer could not draw a significant percentage of consumers away from Navigator by simply improving the quality of its web browser. Senior executives within Microsoft concluded that Internet Explorer could not catch up with Navigator's market share by itself. J. Allchin, senior executive within Microsoft, wrote to P. Martiz, in late December 1996:

> I don't understand how IE (Internet Explorer) is going to win. The current path is simply to copy everything that Netscape does, packaging and product wise. Let's [suppose] IE is as good as Navigator/Communicator. Who wins? The one with 80% market share. Maybe being free helps us, but once people are used to a product it is hard to change them. Consider Office. We are most expensive today, but we're still winning . . .

From 1996 to 1998, Internet Explorer's market share increased dramatically. The estimated market share of Navigator fell from 80% in January 1996 to 55% in November 1997, while that of Internet Explorer increased from 5% to 36% over the same period. These changes in market shares, however, were mainly due to the fact that Microsoft leveraged Windows to increase the usage of Internet Explorer, rather than constantly improving the quality of its browser.

(b) Server Operating System

The environment of the server operating system market is different from that of the PC operating system in the sense that a server is maintained by a group of computer experts (not by novices as in the case of a PC) and is intended for specific purposes, rather than for general uses. Moreover, a choice of server operating systems requires many technical features, such as functionality, system management, performance, reliability (stability) and security. Functionality and performance are important technical features for both PCs and servers, but other aspects are not very important for PCs. This difference comes from the fact that while a PC is designed for use by one person at a time, a server is designed to provide data, services, and functionality through a digital network to multiple users. Most individual PC owners do not have an incentive to buy a server operating system, since server operating systems generally cannot function efficiently on PC hardware. In other words, the consumers in the two markets are basically separated. According to Data Quest, the Unix System is currently the most widely used operating system platform. It expanded its market share from 36% in 1996 to 43% in 1998. The newcomer in the server operating system market, Windows NT, also achieved great success during that period; its market share rose from 10% in 1996 to 16% in 1998. Linux licence shipments in this year grew at a rate of 212.5%, accounting for more than 17% of all server operating system units shipped. Other than Unix servers and Windows NT, Open VMS, Novel NetWare and OS/2 are the major competitors in the server operating system market. A small product differentiation in a server operating system enables it to create and to maintain its own demand. In sum, a dominant product cannot eliminate other products with its prevailing positive network externalities when substitutability for products is low.

(c) OS/2 Warp

In late 1994, IBM introduced its Intel-only-compatible operating system, OS/2 Warp, as an alternative to Windows. OS/2 was originally designed to be optimized for Intel microprocessors. This resulted in a lack of portability. Because the Intel-compatible PC dominated the PC market, the lack of portability did not prevent consumers from adopting OS/2 as an operating system. Furthermore, OS/2 was not reported to be inferior to Windows. The greatest disadvantage of adopting OS/2 as an operating sytem was the lack of compatible software, approximately 2500 applications were available. In 1995, IBM still attempted to compete with Windows 95 in the PC operating system market with its release of OS/2 2.0. However, IBM figured that it would lose between 70 and 90% of its sales volume if it installed OS/2 to the exclusion of Windows 95. In 1996, IBM withdrew from the PC operating system market and, instead, has been repositioning OS/2 for the server operating system market. The failure of OS/2 Warp to gain enough market share to survive is due mainly to its lack of applications rather than inferiority to Windows. By the time OS/2 Warp was released, Windows was already the predominant operating system and ran an overwhelming number of applications. Most consumers used Windows not because it was superior in quality to its competitors, but rather because it provided the greatest positive network externality among the available PC operating systems. In other words, Microsoft successfully deterred entry into the PC operating system market by using its predominating network externality.

(d) BeOS

BEAS Systems, Inc. (BEAS), founded in 1990 by a former executive of Apple's product division, released its cross-platform operating system, BeOS. BeOS functions virtually identically on both Intel-compatible PCs and Macintosh systems. In addition, BeOS is optimized for multiprocessor media-based systems. These two features provide great convenience for multimedia-specific users. BeOS has stated from the beginning that it intends to coexist within the Windows and Apple World as a specialized operating system. Even taking this specific purpose into account, the market share of BeOS is too small compared to the number of Windows users. The most outstanding disadvantage is the lack of compatible applications (approximately 1000). BeOS, Inc. tried to overcome this problem by maintaining a website which provided a list of freeware and shareware applications. BeOS, Inc. failed to gain sizeable market share, and was eventually sold to Palm. However, Palm continues to produce BeOS and to maintain its BeWare website. BeOS still exists as a niche operating system, in spite of its superior support for multimedia applications. The failure of BeOS to obtain a significant market share is evidence in support of the proposition that the incumbent can deter entry by using its network externality.

6.3 Asymmetry of market evolution

For network economies, Katz and Shapiro (1992) look at situations where there is too much or too little technological innovation in a market with network externalities, compared to the social optimum. In particular, they explore whether a new product which embodies technological progress is introduced too early or too late compared to the social optimum. In their model, technological progress is deterministic. Only the entrant gets the benefit of improving technology, but once it enters the market its product is fixed. Therefore, in any equilibrium of their model, either the new firm captures the market as soon as it enters or it does not enter at all. Their model is thus not able to capture the richness of pricing strategies and market evolution as a function of market shares. For instance, it is unable to explain the phenomenon of two competing technologies fighting for superiority in a market when both are making sales until one firm finally leaves the market.

Earlier, Harris and Vickers (1987) presented two models of races in which there is both technological uncertainty and strategic interaction between competitors as the race unfolds.

Their aim is to see how the equilibrium effort levels of competitors vary with the intensity of rivalry between them. They show that under certain conditions the leader in the race makes greater efforts than the follower, and efforts increase as the gap between competitors in their progress decreases. The model presented here can be seen along similar lines to that of Harris and Vickers, but more realistically, with the network advantage as the state variable instead of progress, and prices providing strategic interaction between firms instead of the amount of effort put into R&D. While there is one ultimate goal for each firm in the multi-stage race of Harris and Vickers to reach the finishing line of the race first, we leave the choice open for firms which network they want to build. It is this feature that can lend profit maximization by firms to a complex network evolution pattern. Also the amount of effort put into R&D stochastically moves firms to a new stage in the race, whereas in our model prices – the variables of strategic interaction – deterministically move firms to a new market share.

While a lot of work has been done in the area of R&D, very little of it addresses markets with network externalities specifically. Katz and Shapiro look at the choice of product compatibility in a market with technological progress. They look at both the cases of deterministic and uncertain technological progress. However, technological progress in their model is exogenous and they look at a two-period game which does not provide a very suitable framework for looking at the dynamic pricing strategy of firms and the resulting evolution of market shares. Choi (1994) looks at a two-period model of a monopoly in a market with network externalities. He studies the incentive of the monopolist to introduce an incompatible improved product in the presence of network externalities. Kristiansen (1998) studies the consequences of network externalities on the riskiness of R&D projects chosen by an entrant and an incumbent. He shows that the incumbent chooses a too-risky project that too often lets a new firm with an incompatible technology enter. In addition, the entrant has an incentive to choose more certain projects than are socially optimal and these strengthen the possibility of adoption of an incompatible technology.

This chapter develops and analyses a model of technological uncertainty where firms compete strategically in prices in the presence of network externalities. Technological uncertainty is modelled as an improvement in the quality of the firm's product.

Whether the firm gets the innovation, however, is uncertain. In the simplest model, once a firm gets the innovation, the race ends and there is no further innovation.

Once the innovation occurs, when there is no uncertainty, the model looks like that of Katz and Shapiro – the firm with the quality plus network advantage sells to all the consumers subsequently. We attempt to analyse how firms share the market by strategically choosing their prices before a firm succeeds in getting an innovation.

The advantage of a firm is measured by the sum of its network size and the quality of its product as a 'proxy' of its innovation intensity. If the firms are in a technological race but the probability of innovation is small then again the firm with the bigger advantage sells to all the consumers in the absence of innovation. If, however, those firms expect innovation with a high probability then so long as neither firm has a very big advantage both firms take turns in selling to the consumers in the absence of innovation. Thus if the firms are on an equal base neither firm gets pushed out unless one of the firms gets an insurmountable advantage in the process. That is, because with firms that are sufficiently close in terms of their advantage, neither firm finds it worthwhile to push the other firm out of the market, since the 'losing' firm prices more aggressively than the 'leader' as it has more to gain by staying in the market.

We provide some anecdotal observations across various industries that show a strong rivalry on the technological front to gain market share between a leader in the field and its approaching rival(s). Clearly those interactive patterns reveal technological races. We can even extend those observations to megamergers and alliances as products of network economies that did not protect those networks from relentless technological attacks by upstarts, new entrants or Schumpeterian entrepreneurs hailing 'creative destruction'. This asymmetric view reflects the match between the 'Old Guard vs the Vanguard' (Hamel and Switzer, 2004) in the sense that an incumbent's position endows her with a natural network advantage which could only be overcome by a breakthrough innovation of a new entrant depending on some probability of innovation success. We have manifestations of intense rivalry and 'neck-and-neck' competition across industries as revealed and detailed in Chapter 1.

6.4 Modelling technology racing in a network

We start out with a simple model in which two firms, Y and Z, are producing a good using incompatible technologies, and the good is infinitely durable. The firms have products of identical qualities which we normalize to zero. For simplicity, the marginal cost of production is assumed to be zero for both firms.

We define: the network base of firm i at time t, b_{it}, is the total number of consumers it has sold to until time t is $i \in \{Y, Z\}$.

The network advantage of firm i at time t, $n_{it} \equiv b_{it} - b_{jt}$, $i, j \in \{Y, Z\}$.

A new consumer arrives in the market every period. Each consumer has an inelastic unit demand for the good and must decide at which firm to buy as soon as she enters the market. Consumers care about how many other consumers buy the same product as they do because of network externalities. That is, they are likely to choose products of the larger network over those of the smaller. They also care about the quality of the product and its price. For simplicity, all 'representative' consumers have identical preferences. We may state their utility functions as $u(n_t, q_t, p_t)$ where q_t is the quality of the product as a perfect substitute for the network advantage and p_t is its price. Throughout the phase of adding uncertainty to the innovation process the consumers are expected to be risk-neutral. We assume that there is a threshold difference in network sizes of the two firms, D, such that if neither firm's network is D (critically) bigger than the other's then consumers get the same network benefits from the two firms. More realistically, one could think of D not just as a number but as an 'order of magnitude'. Therefore, if neither firm has a network advantage of D then consumers buy from the firm which charges a lower price if the quality is the same, and are indifferent between the two firms if they charge the same price. Once a firm establishes a network advantage of D over the other firm, consumers get a much bigger benefit from it. Thus, if both firms charge the same price then consumers prefer the firm which has the network advantage of D (or more). It follows that once a firm gets a network advantage of D it becomes a (temporary) monopolist of the market.

If both firms did not have identical qualities to start out with, then we could think of the network advantage as comprising the network difference plus the quality difference, i.e. treat network and quality as perfect substitutes in the consumer's utility.

The proposed rule for consumer decisions implies that consumers are myopic, i.e. the current base of consumers with the firms is what matters to them. While in practice we would expect consumers to form expectations about the future networks of firms, and thus introduce the possibility of multiple fulfilled-expectations equilibria, our assumption of myopic consumers would help select the most efficient equilibrium outcome among them. Both assumptions – of myopic consumers and consumers with fully rational expectations – are extreme, and what is more realistic is something in between the two. Having myopic consumers has the added advantage of making the analysis more tractable.

Let n_{Yt} represent the state of the system (we could equivalently choose n_{Zt} to be the state variable). Thus state 1 denotes the situation where firm Y has one (order of magnitude) more consumer than firm Z, state -1 denotes the situation where firm Z has one (order of magnitude) more consumer than firm Y. We define 3 to be the absorbing state for firm Y and -3 the absorbing state for firm Z, i.e. we let D equal 3. The net discounted payoff of a firm in its absorbing state is M, for the other firm likewise. The absorbing state can be thought of as the threshold advantage in quality or network which is big enough to deter the other firm from ever trying to catch up. Thus the firm that has this advantage becomes a monopolist in the market. It is the sole seller in the market, making monopoly profits of M, whereas the other firm leaves the market and thus gets zero profits.

The product is subject to technological innovation. We assume that an innovation pushes a firm to its absorbing state. Thus we assume that the innovation is equivalent to a big enough network advantage that gives a firm an insurmountable lead over the other firm, in many given cases of asymmetrical network advantage the innovations need to be 'radical'. The innovation, however, is uncertain – either a firm gets the required quality improvement or there is no improvement at all. The probability that a firm gets the innovation in any time period is $\alpha \in [0,1/2]$, which is the same for both firms. We could think of $\alpha > 0$ as representing the technological environment in which the firms operate. We assume that once one firm succeeds, there are no further innovations in the market. We also assume for notational simplicity that at most one firm gets the innovation. Thus the probability that neither firm gets the innovation in any one time is $1–2\alpha$. In this situation, we look at the innovation process as 'exogenous' (although we could make it 'endogenous' by letting the probability of innovation success depend on the network size as a proxy for the degree of accumulated knowledge (learning) in determining innovation success).

We could think of the technological innovation as a new feature that the firms try to introduce in their respective products. Once one firm gets its innovation it can get a patent even though the firms produce incompatible products. We assume that the new improved product that a firm might introduce is compatible with its older product. Compatibility, in this model, translates into network-sharing – compatible products share a network.

Firms compete in prices and thus must decide what price to charge at each of the states in $S=\{-2,-1, 0, 1, 2\}$ in order to maximize their respective profits given by

$$\pi_Y(s_0,p) = \sum_t \delta^t(1-2\alpha)^{t-1}[(1-2\alpha)p_{Yt}x_{Yt} + \alpha M] + \delta^{\tau+1}I(s_{\tau+1}=3)M,$$

$$t = 0, 1, 2, \ldots, \tau$$

$$\pi_Z(s_0,p) = \sum_t \delta^t(1-2\alpha)^{t-1}[(1-2\alpha)p_{Zt}x_{Zt} + \alpha M] + \delta^{\tau+1}I(s_{\tau+1}=3)M,$$

$$t = 0, 1, 2, \ldots, \tau$$

$$s_{t+1} = s_t + x_{Yt}, \text{ with}$$

$$x_{it} = \begin{cases} 1 & \text{if } p_{it} < p_{jt} \\ 0 & \text{if } p_{it} > p_{jt} \\ 0 \text{ or } 1 & \text{if } p_{it} = p_{jt} \end{cases}$$

where p_{it} is the price charged by firm i in period t, $x_{it} \in \{0, 1\}$ is the number of period t consumers that firm i sells to, $\tau+1$ is the time period in which a firm reaches its absorbing state even without getting the innovation, $s_0 \in S$ is the initial state and $p=(p_1,p_2)$ where $p_i=\{p_{it}\}$, $t=0, 1, 2, \ldots, \tau$.

We thus have a game where the firms are identical except in the sizes of their networks. There is uncertain technological innovation but the size of the innovation and the probability of innovation are again identical for both firms. Therefore, by comparing the solution of a game with no uncertainty with the solution of our game we should be able to gain valuable insights into the effects that uncertainty has on competition between two firms in the presence of network externalities. The pertinent technical results are relegated to the Appendix of this chapter.

The major results can be summarized as follows.

Propositions 2 and 3 illustrate the effect of uncertainty on equilibrium behaviour in a market with network externalities. In Proposition 1 we see that when the only difference between two firms is in the sizes of their networks of consumers, the firm with the bigger network sells to all consumers that subsequently enter the market.

However, once we introduce uncertainty about the quality of a firm's product this is no longer true. Even if the uncertainty faced by the firms is fully symmetric and the only difference between the firms is in the size of their respective networks, the firm with the bigger network does not always sell in equilibrium. For a sufficiently high success rate in innovation, it is the firm with the smaller network that sells provided the network distance is not too large.

The reason for this strikingly different outcome compared to the situation without uncertainty is as follows. When there is no uncertainty, the only way either firm can realize positive profits is by reaching its absorbing state. Outside of their respective absorbing states both firms make the same profits, i.e. zero. Thus the way for either firm to make positive profits is to sell to all the consumers and reach its absorbing state. However, since the firm with the larger network needs a smaller addition to its network to reach its absorbing state, it needs a shorter time span to reach its absorbing state, hence its discounted profits from selling to all consumers is larger. Not selling to any consumers, on the other hand, gives both firms zero profits. Thus since the firm with the bigger network has more to gain from selling to all subsequent consumers while not selling gives it the same profits as the firm that is behind, it is willing to price more aggressively in order to lure the consumers. Thus in equilibrium the firm that is ahead always sells.

The introduction of uncertainty, even when it is uniform across the firms, alters the incentives of the firms to fight for the consumers. When the difference in network (size) between the two firms is sufficiently small and the probability of success is high, the presence of uncertainty reduces the incentive of the firm that is ahead to fight for the current consumer relative to that of the firm with the network disadvantage. That is because with a high rate of innovation the leading firm has a high probability of getting its monopoly profits whether it sells to the current consumer or not. The firm with the network disadvantage, on the other hand, has a bigger stake in winning the current consumer since by not letting the other firm get close to its absorbing state, it keeps its own chances of getting the innovation and thus reaping the high monopoly profits alive. However, when the leading firm has a big network advantage, i.e. is close to its absorbing state, its incentive to sell is much higher since it can then guarantee itself the monopoly profits by selling to the current customer. (Thus as $\delta \to 1$ and $\alpha \to 1/2$, we see the firm with a network advantage of 2 selling and also the firm with a network disadvantage of 1.)

Until now all the discussion has been in terms of network advantage or disadvantage. We could equally well have had the discussion in terms of quality advantage or disadvantage since network and quality are perfect substitutes in a consumer's utility. Thus without uncertainty, in the presence of network externalities, if the only difference between two firms was in the quality of their products then the firm with the better quality product would sell to all the consumers. With uncertain quality improvement, however, if the probability of quality improvement were high enough then the firm with the lower quality would sell to the consumer. Both firms would remain in the market until one of them got an innovation that gave it an insurmountable lead over the other.

6.5 Extensions and conclusions

We have observed competitive situations in network markets where there is uncertain technological development in product/process technologies. Firms 'price' compete in those markets to gain market share before any of them succeed in getting an innovation to move ahead of its rival(s). If the firms are in a technology race but the probability of innovation success is small, then a firm with a bigger network advantage is likely to attract more customers in the absence of innovation. If, however, any of those firms expect innovation with a high probability, and none of them have a big advantage, then they continue to share the market until the innovation occurs.

We have looked at a simple model of a market with network externalities where there is technological innovation. It would be interesting to explore what kind of modifications to our basic model might change some of the results. The model is based on the assumption of symmetry in terms of the innovation that the firms expect.

If we introduce asymmetric innovation this would permit us to endogenize innovation to the extent that we could look at the firms' decisions of how much to improve quality. If we look at a more dynamic model of having sequential innovation and multiple races, as in Chapter 5, we might be able to see whether there is a clustering of innovations, and by examing the size of innovations, whether firms would rather tend toward one big risky innovation or rather for a series of small ones which are less risky. Under welfare theoretic considerations there may be related questions on whether there is too much or too little innovation in view of a social optimum (Gottinger, 2003).

Our simple model shows how competing firms behave in a market with network externalities before the resolution of technological uncertainty. If our firms are in waiting (for the breakthrough innovation to come true) but the probability of innovation is small, then the firm that has a network advantage makes all the sales. However, if the probability of innovation is large, then it is not always the firm with the bigger network that sells. While a firm with a network advantage of 2 would make the sale, it is the firm with a network disadvantage of 1 that makes the sale if she is lucky to catch this innovation.

Alternatively, it could also serve as an explanation for why a firm that is ultimately driven out of the market may just hang on there for a while, with a network disadavantage, before she finally gives up and exits (or is exited from) the market.

Furthermore, as some sort of by-product of this analysis, we could pursue the interaction between network effects and innovation for market structures. If network effects are 'competitive barriers' that could only be overcome by radical innovation then, by implication, the synergy of network effects and radical innovation through the incumbent would constitute a formidable powerhouse to consolidate monopolistic positions in dynamic product markets where 'winner-takes-all markets' are created.

References

Choi, J. P. (1994) 'Irreversible choice of uncertain technologies with network externalities', *Rand Journal of Economics*, **25**, 382–401.

Evans, D. S. and Schmalensee, R. (2001) 'Some economic aspects of antitrust analysis in dynamically competitive industries', National Bureau of Economic Research, Cambridge, MA. Working Paper 8268, http://www.nber.org/papers/w8268.

Fudenberg, D. and Tirole, J. (1993) *Game Theory*, Cambridge, MA MIT Press.

Futia, C. (1980) 'Schumpeterian competition', *Quarterly Journal of Economics*, **94**, 675–695.

Gottinger, H. W. (2002) 'Modeling stochastic innovation races', *Technological Forecasting and Social Change*, **69**, 607–624.

Gottinger, H. W. (2003) *Economies of Network Industries*, London: Routledge.

Hamel, G. and Switzer, L. (2004) 'The Old Guard vs. the Vanguard', *The Wall Street Journal* (Online Edition), 23 Febraury.

Harris, C. and Vickers, J. (1987) 'Racing with uncertainty', *Review of Economic Studies*, **54**, 1–21.

Katz, M. and Shapiro, C. (1992) 'Product introduction with network externalities', *The Journal of Industrial Economics*, **40**, 55–83.

Kristiansen, E. G. (1998) 'R&D in the presence of network externalities: timing and compatibility', *Rand Journal of Economics*, **29**, 531–547.

Shapiro, C. (1989) 'Theories of oligopoly behavior', in *Handbook of Industrial Organization, Vol. 1*, R. Schmalensee and R.D. Willig (eds.), Amsterdam: North Holland, 329–414.

Uttenback, J. M. (1994) *Mastering the Dynamics of Innovation*, Boston: Harvard Business School Press.

Appendix

We look for Markov equilibria (Fudenberg and Tirole (1993)), which are non-collusive and symmetric. In such an equilibrium we look for a firm that if it does not sell to the consumer it is indifferent between selling or not selling. Thus if firm Y is the firm not selling at state s

$$p_Y(s) + \delta\,\Pi_Y(s+1) = \delta\,\Pi_Y(s-1)$$

where $\Pi(s)$ represents the continuation payoff starting from state s. If Z is the firm not selling at state s then

$$p_Z(s) + \delta\,\Pi_Z(s-1) = \delta\,\Pi_Z(s+1).$$

In such an equilibrium the non-selling firm ensures that even if against all expectations the consumer buys its product then it will not make negative profits in equilibrium.

Definition 1: *An equilibrium is non-collusive if neither firm charges a positive price at a non-absorbing state. Alternatively, if either firm charges a positive price at a non-absorbing state, equilibrium is collusive.*

Definition 2: *The equilibrium is symmetric if* $p_Y(s) = p_Z(-s)$ *for all permissible* s *and* $x_Y(s) = x_Z(-s)$ *for all permissible* $s \neq 0$.

If instead of having an infinite horizon game we look at a game with a long but finite horizon, the non-collusive equilibria are the ones we expect would most likely be sustainable.

We first consider the case in absence of uncertainty, i.e. $\alpha = 0$.

Proposition 1: *For* $\alpha = 0$*, the symmetric non-collusive Markov equilibrium prices are unique: the firm with a positive network advantage makes all the sales in equilibrium.*

Proof. Suppose firm Z sells at $s = 2$. If firm Y were to sell to the consumer at $s = 2$, then in the next period it would get a profit of M. Therefore, if in a non-collusive equilibrium firm Y does not sell at $s = 2$ as we have assumed, it must be getting a continuation payoff at least as big as M next period. There are two possibilities for $s = 1$ – either firm Y sells or firm Z does.

Suppose firm Y sells at $s = 1$. Then once the state reaches 1 it must alternate between 1 and 2 forever more from then on. The continuation payoff to firm Y at state 1 then is just the discounted sum of the price it charges to the consumer at $s = 1$ every other period. For this to be at least M, the price at $s = 1$ must be positive and equilibrium is then collusive. Suppose firm Z sells at $s = 1$ also. Then once $s = 1$ is reached it must remain within the states –1, 0, 1 forever more. But then again for firm *Y*'s continuation payoff at $s=1$ to be at least M, it must get a positive payoff at $s=0$ or -1, a contradiction. Thus *Y* must sell at $s=2$. Suppose firm *Z* sells at $s=1$ and *Y* sells at $s=2$. Then once state 1 is reached the state must remain within the states –1, 0, 1. In a non-collusive equilibrium where neither firm charges a positive price, *Y*'s continuation payoff at $s=0$ cannot be positive. *Z*'s continuation payoff at $s=2$ is zero. Similarly, the smallest price *Z* is willing to charge at $s=1$ cannot be less than zero. However, firm *Y*'s continuation payoff at $s=2$ is then positive and at most zero at $s=0$, it is therefore willing to charge a negative price at $s=1$ in order to lure the consumer into buying its product. But that means the consumer should buy from firm *Y* at $s=1$ since its price is lower than *Z*'s, a contradiction. Thus firm *Y* must sell at $s=1$ also.

We have so far shown that if a symmetric, non-collusive Markov equilibrium exists then it must have the firm with the network advantage selling. We now show that such an equilibrium does indeed exist and that it is (almost) unique.

If in equilibrium the firm with the network advantage wins always then the continuation payoffs of both firms at $s=0$ must be zero in a symmetric equilibrium since both firms are willing to give up exactly the same surplus in order to win the consumer. This surplus is equal to the continuation payoff from winning the current which has to be at least zero.

At $s=1$, the lowest firm *Z* is willing to charge is zero, since whether it sells to the current consumer or not it is going to get a payoff of zero from then on. But that means it must charge a price of zero at $s=2$ also. The continuation payoff at $s=0$ from winning is then $\delta^2 M$.

Thus both firms charge $-\delta^3 M$ at $s=0$ and get zero discounted payoffs. It can be checked that these are equilibrium prices.

Thus the unique equilibrium prices are as follows:

$$p(2) = p(-2) = 0$$
$$p(1) = p(-1) = 0$$
$$p(0) = -\delta^3 M.$$

Both firms charge the same prices. Firm *Y* sells to the consumer at $s=1,2,\ldots,$ firm *Z* sells to the consumer at $s=-1,-2$, either firm could sell at $s=0$.

Thus we see that in the absence of uncertainty the firm that is already ahead in terms of network size is the 'winning' firm. The other firm that has a network disadvantage is unable to overtake the firm that is ahead. This result is similar to that obtained by Katz and Shapiro (1992). Once a firm acquires a bigger network the other firm is unable to overcome its disadvantage. Since in our model quality and network are perfect substitutes in the consumer's utility, we could replace network by quality to get the same result. The firm with the higher quality sells to all consumers, the lower-quality firm is unable to push out the better-quality firm from the market given that both firms start out with the same network size.

We now introduce uncertainty into the game and see how firms behave. As in the previous case, the only difference between the firms is in their networks. The uncertainty faced by firms is symmetric: both firms have the same probability of getting an innovation and the effect of an innovation is symmetric for both firms.

Proposition 2: *There exist δ_* and α_* such that for $\delta_* < \delta < 1$ and $0 < \alpha < \alpha^*$ the non-collusive symmetric equilibrium prices are unique. The firm with the bigger network sells to the consumer, either firm could sell when $s = 0$.*

Proof. The proof proceeds along a sequence of lemmas.

Since we are looking for symmetric equilibria, at $s = 0$ the lowest price both firms are willing to charge must be the same. Thus in equilibrium both firms must be indifferent between selling or not at $s = 0$, i.e. at $s = 0$ both firms have the expected payoffs. Again, since we are looking at symmetric equilibria we need only focus on behaviour at $s = 0$, 1, 2. The different possibilities at $s = 1$ and $s = 2$ determine the different candidates for equilibrium outcomes. There are four possible candidates for equilibrium network advantage evolution given that no innovation occurs in the meantime.

(i) $x_Y(1) = 1$, $x_Y(2) = 1$,
(ii) $x_Z(1) = 1$, $x_Y(2) = 1$,
(iii) $x_Z(1) = 1$, $x_Z(2) = 1$,
(iv) $x_Y(1) = 1$, $x_Z(2) = 1$,

where $x_i(s) = 1$ means that at state s firm i sells to one consumer (and thus $j \neq i$ sells to no one since there is only one consumer in the market at any time period) in the absence of innovation having taken place.

Define $p_i{}^*(s)$ be the lowest price firm i is willing to charge and $p(s)$ to be the equilibrium price at s. The lowest a firm is willing to charge at any state is just the difference between its profits if it lost the current consumer and if it won it. In other words, the maximum a firm is willing to give up in order to attract a consumer is exactly the amount of extra profit the firm can expect to make by selling to that consumer. Define $\pi_i{}^W(s)$, $\pi_i{}^L(s)$ to be the expected profits of firm i next period onwards from, respectively, winning and losing the current consumer at state s, let $\pi_i(s)$ be the *ex ante* expected payoff of i at state s.

We prove the proposition using the following lemmas.

Lemma 1: *Case (iii) cannot occur in equilibrium.*

Proof. Suppose (iii) described the equilibrium network evolution. Then

$$p_Y^*(1) = p(1) = \delta(1 - 2\alpha)p(0). \tag{1}$$

This is just the difference in payoffs to firm Y from losing versus winning the current consumer:

$$\pi_Y^W(1) = 0 + \delta\alpha M + \delta(1 - 2\alpha)\pi_Y(1)$$

$$\pi_Y^L(1) = (1 - 2\alpha)p(0) + \delta\alpha M + \delta(1 - 2\alpha)\pi_Y(1)$$

$$p_Y^*(1) = \pi_Y^L(1) - \pi_Y^W(1).$$

Similarly,

$$p_Z^*(0) = p(0) = \delta(1 - 2\alpha)p(1). \tag{2}$$

Equations (1) and (2) together imply that

$$p(1) = \delta^2(1 - 2\alpha)^2\, p(1).$$

But that in turn implies that either $p(1) = 0$ or $\alpha\delta = 1$ and $\delta = 0$. Since we are looking at the case where $\alpha\delta \neq 0$, $p(1)$ must be zero, which means from equation (2), $p(0) = 0$. Thus

$$\pi_Y(0) = \pi_Z(0) = \frac{\alpha M}{1 - \delta(1 - 2\alpha)} \equiv \pi(0).$$

We can calculate $p_Z^*(2)$ and $p_Y^*(2)$ (=$p(2)$ since Y is the losing firm at $s = 2$)

$$p_Z^*(2) = -\delta\alpha M - \delta^2(1 - 2\alpha)\pi(0)$$

$$p_Y^*(2) = \delta\alpha M + \delta^2(1 - 2\alpha)\pi(0) - \delta M.$$

But for $\delta < 1$, $p_Y^*(2) < p_Z^*(2)$ which means firm Y must be the one selling to the consumer at $s = 2$, a contradiction.

Lemma 2: *Case (iv) describes the network evolution in equilibrium only when* $\delta^2(1 - 2\alpha)^2 > 1/2$.

Proof.

$$\begin{aligned} &\pi_Z^W(1) = 0 + \delta\alpha M + \delta(1 - 2\alpha)\pi_Z(1) \\ &\pi_Z^L(1) = (1 - 2\alpha)p(2) + \delta\alpha M + \delta(1 - 2\alpha)\pi_Z(1) \\ &p_Z^*(1) = p(1) = \pi_Z^L(1) - \pi_Z^W(1) \\ &\Rightarrow p(1) = \delta(1 - 2\alpha)p(2). \end{aligned} \tag{3}$$

We can calculate $p(2) = p^*_Y(2)$ using the same method:

$$p(2) = \frac{\delta\alpha M}{1 - \delta(1 - 2\alpha)} + \frac{\delta(1 - 2\alpha)}{1 - \delta^2(1 - 2\alpha)^2} p(1) - \delta M$$

$$p^*_Z(2) = \frac{\delta\alpha M}{1 - \delta(1 - 2\alpha)} - \frac{\delta^2(1 - 2\alpha)^2}{1 - \delta^2(1 - 2\alpha)^2} p(2). \qquad (4)$$

Substituting the expression for $p(2)$ in equation (3) we get

$$p(1) = \frac{\delta^2(1 - 2\alpha)^2\alpha M}{1 - \delta(1 - 2\alpha)} + \frac{\delta^2(1 - 2\alpha)^2}{1 - \delta^2(1 - 2\alpha)^2} p(1) - \delta^2(1 - 2\alpha)M. \qquad (5)$$

We similarly calculate $p(0)$ using the relation of (3) to get $p(0) = 0$ since

$$p(0) = \frac{\delta^2(1 - 2\alpha)^2}{1 - \delta^2(1 - 2\alpha)^2} p(2) - \frac{\delta(1 - 2\alpha)}{1 - \delta^2(1 - 2\alpha)^2} p(1).$$

This in turn implies that $p^*_Y(1) = 0$.

For case (iv) to occur in equilibrium therefore $p(1)$ must be non-negative. That leads to $p^*_Y(1) \leq p(1)$; it also implies that $p(2)$ is non-negative, therefore $p^*_z(2) \leq 0 \leq p(2)$ from equation (4). From (5) we see that $p(1)$ is non-negative only if $\delta^2(1 - 2\alpha)^2 \geq 1/2$.

Lemma 3: *As $\delta \to 1$ and $\alpha \to 0$, the pattern of network evolution described in (iv) can only occur in a collusive equilibrium.*

Proof. As $\delta \to 1$, $\alpha \to 0$, $\delta^2(1 - 2\alpha)^2 \geq 1/2$. Thus from lemma 2 case (iv) can occur in an equilibrium. But when $\delta^2(1 - 2\alpha)^2 \geq 1/2$, we see from (5) that $p(1) > 0$, i.e. the equilibrium is collusive.

Lemma 4: *Case (i) describes the pre-innovation network evolution in a non-collusive equilibrium only when $3\alpha\,[1 + \delta(1 - 2\alpha)] \leq 1$.*

Proof. We calculate the lowest prices firms are willing to charge as described earlier on

$$\begin{aligned}
P^*_Y(2) &= \delta\alpha M + \delta^2(1 - 2\alpha)\alpha M + \delta^3(1 - 2\alpha)^2 M - \delta M \\
&\quad + \delta(1 - 2\alpha)p(1) + \delta^2(1 - 2\alpha)^2 p(2) \\
p^*_Z(2) &= p(2) = -\delta\alpha M - \delta^2(1 - 2\alpha)\alpha M \\
p^*_Y(1) &= \delta^2(1 - 2\alpha)\alpha M + \delta^3(1 - 2\alpha)^2\alpha M - \delta^2(1 - 2\alpha)M - \delta(1 - 2\alpha)p(2) \\
p^*_Z(1) &= p(1) = -\delta^2(1 - 2\alpha)\alpha M - \delta^3(1 - 2\alpha)^2\alpha M \\
p(0) &= -\delta(1 - 2\alpha)p(1) - \delta^2(1 - 2\alpha)^2 p(2) - \delta^3(1 - 2\alpha)^2 M \\
&= \delta^3(1 - 2\alpha)^2 M[2\alpha + 2\delta\alpha(1 - 2\alpha) - 1].
\end{aligned}$$

We can use the above equations to solve for $p(0)$, $p(1)$, and $p(2)$ simultaneously. It can be checked that $p^*_Y(2) \leq p(2)$ always holds, but that $p^*_Y(1) \leq p(1)$ holds only if

$$3\alpha[1 + \delta(1 - 2\alpha)] \leq 1. \tag{6}$$

It can be checked that when inequality (6) is satisfied then $p(0)$, $p(1)$, and $p(2)$ are all negative. Note that (6) holds when $\delta \to 1$ and $\alpha \to 0$. Thus when inequality (6) is satisfied the network evolution described in case (i) occurs as a result of the non-collusive equilibrium described by the prices given above. Both firms charge $p(s)$, $s \in \{-2, -1, 0, 1, 2\}$, where $p(0)$, $p(1)$, and $p(2)$ are as given above and $p(-2) = p(2)$, $p(-1) = p(1)$. At $s = 1,2$ firm Y sells to the consumer in the market, at $s = -1, -2$ firm Z sells to the consumer, and at $s = 0$ either firm could sell.

We see that the equilibrium when there is no uncertainty is just the limit of the above equilibrium as $\alpha \to 0$.

Lemma 5: *Case (ii) describes the pre-innovation network evolution in a non-collusive equilibrium only when* $1 - 3\delta^2(1 - 2\alpha)^2 \leq 0$ *or when* $\delta^4(1 - 2\alpha)^4 + 4\,\delta^2(1 - 2\alpha)^2 - 1 \leq 0$ *and* $\alpha - 3\delta^2\,(1 - 2\alpha)^2 - (1 - 2\alpha)\,(1 - \delta) \geq 0$.

Proof. Suppose (ii) describes the pattern of equilibrium network evolution in the absence of innovation. Then, $p^*_Z(2) = p(2)$ and $p^*_Y(1) = p(1)$:

$$p^*_Z(2) = \frac{\delta\alpha M}{1 - \delta(1 - 2\alpha)} - \frac{\delta(1 - 2\alpha)}{1 - \delta^2(1 - 2\alpha)^2} p(1) \tag{7}$$

$$p^*_Y(2) = \frac{\delta\alpha M}{1 - \delta(1 - 2\alpha)} + \frac{\delta^2(1 - 2\alpha)^2}{1 - \delta^2(1 - 2\alpha)^2} p(0) - \delta M \tag{8}$$

$$\begin{aligned} p^*_Y(1) = {} & \frac{\delta\alpha M}{1 - \delta(1 - 2\alpha)} + \frac{\delta(1 - 2\alpha)^2}{1 - \delta^2(1 - 2\alpha)^2} p(0) \\ & - \delta\alpha M - \delta(1 - 2\alpha) p(2) \delta^2(1 - 2\alpha) M \end{aligned} \tag{9}$$

$$p^*_Z(1) = \frac{\delta\alpha M}{1 - \delta(1 - 2\alpha)} - \frac{\delta^2(1 - 2\alpha)^2}{1 - \delta^2(1 - 2\alpha)^2} p(1) + \delta\alpha M \tag{10}$$

$$p(0) = \delta(1 - 2\alpha) p(1). \tag{11}$$

Using equations (7) and (11) to substitute in the values of $p(2)$and $p(0)$ in equation (9) we get

$$p(1) = -\frac{[1 - \delta^2(1 - 2\alpha)^2]\delta^2(1 - 2\alpha)^2(1 - \delta)M}{[1 - \delta(1 - 2\alpha)][1 - 3\delta^2(1 - 2\alpha)^2]}. \tag{12}$$

For (ii) to occur in equilibrium it must be true that $p^*_Z(1) \leq p(1)$ and $p^*_Y(2) \leq p(2)$. It can be checked that $p^*_Y(2) \leq p(2)$ iff $1 - 3\delta^2\,(1 - 2\alpha)^2 \leq 0$ or

$\delta^4(1-2\alpha)^4+4\delta^2(1-2\alpha)^2-1\leq 0$. It can also be checked that $p^*_Z(1)\leq p(1)$ iff $1-3\delta^2(1-2\alpha)^2\leq 0$ or $\alpha-3\delta^2\alpha(1-2\alpha)^2-(1-2\alpha)(1-\delta)\geq 0$, thus proving the lemma.

Lemma 6: *For $\delta\to 1$ and $\alpha\to 0$ the equilibrium described in lemma 5 is collusive.*

Proof. For $\delta\to 1$ and $\alpha\to 0$, $1-3\delta^2(1-2\delta)^2<0$, thus case (ii) can occur in equilibrium from lemma 5. But it can be checked from (12) that for $1-3\delta^2(1-2\alpha)^2\leq 0$ and $\delta<1$, $p(1)$ is positive.

Thus from lemmas 1 to 6 we can infer that as $\delta\to 1$ and $\alpha\to 0$, the (almost) unique non-collusive equilibrium shows the following properties. In the absence of innovation both firms charge the prices as below: As $\delta\to 1$ and $\alpha\to 0$, the equilibrium prices in the absence of innovation are:

$$p(2)=p(-2)=-\delta\alpha M-\delta^2(1-2\alpha)\alpha M$$
$$p(1)=p(-1)=-\delta^2(1-2\alpha)\alpha M-\delta^3(1-2\alpha)^2\alpha M$$
$$p(0)=\delta^3(1-2\alpha)^2\,M[2\alpha+2\delta\alpha(1-2\alpha)-1].$$

As $\alpha\to 0$, the equilibrium prices in the absence of innovation are:

$$p(2)=p(-2)\to 0$$
$$p(1)=p(-1)\to 0$$
$$p(0)\to -\delta^3\,M.$$

Firm Y sells to the consumer in the market at $s=1$, 2 and firm Z sells when $s=-1,-2$. When $s=0$, either firm could sell but both earn the same expected profits when $s=0$, i.e. $x_Y(1)=x_Y(2)=x_Z(-1)=x_Z(-2)=1$ and $x_Y(0)=1$ or $x_Z(0)=1$.

Thus in the limit as the probability of innovation approaches zero, the equilibrium approaches the non-collusive equilibrium without uncertainty.

Proposition 3: *There exist δ_* and α_* such that for $\delta_*<\delta<1$ and $\alpha_*<\alpha<1/2$, the non-collusive symmetric equilibrium prices are unique. The firm with network advantage of 2 and the firm with network advantage of 1 sell.*

Proof. For lemmas 1, 2, 4, and 5, we infer that the only possible outcome in equilibrium has the firm with the advantage of 2 and the firm with the disadvantage of 1 sell prior to an innovation occurring. From lemma 5 we see that for δ close to 1 and α close to 1/2, the equilibrium prices are negative. The equilibrium is therefore non collusive. As $\delta\to 1$ and $\alpha\to 1/2$, the equilibrium prices are as follows:

$$p(2)=p(-2)\to -M/2$$
$$p(1)=p(-1)\to 0$$
$$p(0)\to 0.$$

7 High-speed technology competition

Asked about his everyday thoughts (former) Sony President Idei said: "Samsung, Samsung, Samsung".

7.1 Introduction

In highly competitive technological industries new challenges and opportunities are arising in the new product development arena. Driven by global markets, global competition, the global dispersion of scientific/engineering talent, and the advent of new information and communication technologies (ICT), a new vision of product development is that of a highly disaggregated, distributive process with people and organizations spread throughout the world. At the same time, products are becoming increasingly complex, requiring numerous engineering decisions to bring them to market. Competitive pressures mean that 'time to market' has become a key to new product success. However, at the same time, it is important to maintain the innovation and quality dimensions of the new product at their optimal level.

A central question to address is how firms should invest in innovation and the implications of such investments for competitive advantage. Understanding why a firm benefits from investments in innovation and quality illuminates issues of competitive strategy and industrial organization. In the field of competitive strategy, much attention has been devoted to the concept of core capabilities (Teece *et al.*, 1997). Understanding how firms make optimal investments in the face of competition reveals the nature of competition and provides theoretical and managerial implications for developing core competence and dynamic capabilities.

In major parts of competitive analysis involving R&D decisions the focus is on breakthrough innovations which could create entirely new markets, for example, in studies featuring patent racing between competing firms (Chapters 1 and 6). In more common competitive situations we observe firms, however, competing by investing in incremental improvements of products. It is an important aspect when innovation is considered to be manifested in product quality, process improvements and in the overall quality culture of an organization. For example, after product launch, incremental improvement of different aspects of product quality, improvements in various business processes and an incremental adoption of a quality culture are quite real-world phenomena. Some firms operate in a simultaneous product launch situation while others compete sequentially by adopting the role of leader or follower. The strategic implications in these diverse circumstances can be treated within a unified framework of dynamic stochastic differential games.

In recent years with the emergence of e-business and a supply chain view for product development processes, multiple firms with varying and at times conflicting objectives enter into collaborative arrangements. In such situations, a competitive strategy based on quality and innovation could permeate in those collaborative setups. A recent example is the collaborative venture of Sony and Samsung to build a cutting-edge plant for LCD flat screen TVs, although displaying ongoing fierce rivalry in new product launches in exactly the same product categories. When innovation and quality levels form the core of a firm's capabilities, each member in the supply chain would have an incentive to invest and improve their dynamic capabilities. This leads to tacit competition among collaborative product development partners by means of active investment in innovation and quality.

Although the forces of innovation are central to competition in young, technically dynamic industries, they also affect mature industries where, historically; life cycles were relatively strong, technologies mature, and demands stable.

A strategy for technology must confront primarily what the focus of technical development will be. The question is what technologies are critical to the firm's competitive advantage. In this context, technology must include the know-how the firm needs to create, produce and market its products and deliver them to customers. As a major step in creating a technology strategy it has to define those capabilities where the firm seeks to achieve a distinctive advantage relative to competitors. For most firms, there are a large number of important areas of technological know-how but only a handful where the firm will seek to create truly superior capability.

Having determined the focus of technical development and the source of capability, the firm must establish the timing and frequency for innovation efforts. Part of the timing issue involves developing technical capabilities, and the rest involves introducing technology into the market. The frequency of implementation and associated risks will depend in part on the nature of the technology and the markets involved (e.g. disk drive vs automotive technology), but in part on strategic choice. At the extreme, a firm may adopt a rapid incremental strategy, that is, frequent, small changes in technology that cumulatively lead to continuous performance improvement. The polar opposite might be termed the great leap forward strategy. In this approach, a firm chooses to make infrequent but large-scale changes in technology that substantially advance the state of the art.

As an example of the importance of innovation strategy in product development we notice that IBM created and continues to dominate the mainframe segment, but it missed the emergence of the minicomputer architecture and market by many years. The minicomputer was developed and its market applications exploited by firms such as the Digital Equipment Corporation (DEC) and Data General (DG) (Chapter 1).

To represent the features of leader–follower-type interactions in a sequence of technological racing we consider a class of differential games in which some firms have priority of moves over others.

The firm that has the right to move first is called the leader and the other competing firm is called the follower. A well-known example of this type of sequential move game is the Stackelberg model of duopoly. In this type of interaction the open-loop Nash equilibrium conditions in a sequential move game can be derived. It would lead to a comparison of the strategies of leader and the follower.

Section 7.2 presents the essence of leader–follower-type interactions through the format of a differential game which is accommodated to a competitive market situation in Section 7.3. The properties of their interactions are derived in Section 7.4. Section 7.5 handles various market asymmetries in the leader–follower-type situation. Finally, Section 7.6 summarizes and draws the conclusions.

7.2 A differential game formulation

For our modelling effort we follow the notations and symbols as explained below.

N: number of firms, *T*: finite time horizon for the strategies, *i,j*: superscripts to denote competing firms in a duopoly, *t*: an instant of time in the dynamic game setup, $u(t)$: investments in innovation effort (expenditure per unit time), $R(t)$: net revenue rate for the firm at *t*, R_0 product category net revenue rate for the existing product, R_1: product category net revenue rate for the new product, x_0: quality level of the existing product. Several examples illustrate the importance of innovation and quality as competitive weapons. In the computer mainframe arena IBM replaced Remington Rand and Sperry Univac as market leader in the 1950s, and its subsequent growth outperformed its competitors so rapidly that in 1963 its data process revenues were four times larger than the combined revenues of its eight main rivals in the US market (Hofmann, 1976).

In the model we assume that enhancements of quality are achieved by climbing a performance ladder that may squarely embrace technology but could expand to other criteria uniquely identified with quality. Let the quality of the product at time *t* be $x(t)$. In the context of total quality management, as quality levels increase it becomes even more difficult to climb the performance ladder. The hypothesis behind the formulation is that a firm needs to make higher innovation investments targeted towards improvement of quality. To capture this dynamics a negative feedback effect of the present state quality on the rate of change of product quality ($\dot{x}(t)$) is considered. The state dynamics is

$$\dot{x}(t) = K[u(t)]^{\alpha} - \mathrm{L}x(t) \tag{7.1}$$

where K is proportional to the level of capital investment in development technology and L is the proportionality constant for the influence of present quality on the speed of further quality improvements, α is the innovation resource productivity parameter.

Based on (7.1) the quality of the product at time t is

$$x(t) = x_0 + \int_0^t [K[u(s)]^{\alpha} - \mathrm{L}x(s)]\, ds. \tag{7.2}$$

The product quality provides a means for evaluating the product's attractiveness in the market in the presence of other competing products. The firm's market share is a function of both its own product quality and the product quality of rivals. The difference lies in the planning horizon. In this competitive set-up, the planning horizon extends beyond the date of launch and the competing firms continue investing in product–process innovation till the end of growth phase of a product:

$$R(t) = \begin{cases} R_0 \cdot \dfrac{x_0}{x_0^i + x_0^j}, & 0 \le t < T_p \\[2ex] R_1 \cdot \dfrac{x^i(T)}{x^i(T) + x^j(T)}, & T_p \le t < T. \end{cases} \tag{7.3}$$

The cumulative development cost of the new product at time t is given as:

$$TC(t) = \int_0^T \{u(s)\}\, ds. \tag{7.4}$$

The firm's cumulative profit at time t is determined as follows:

$$T\Pi(t) = TR(t) - TC(t) \tag{7.5}$$

where $TR(t)$ and $TC(T)$ are total revenues and costs at time t, respectively. The total revenue function is given by:

$$TR(t) = \int_0^T R(s)\, ds \tag{7.6}$$

where $R(\cdot)$ is given in (7.3). The firm's decision set is $\Delta = \{u(t)\}$. Notice that the firm pre-commits on the date of product launch T_p. The cumulative profit function, $T\,\Pi(\delta)$, is defined as the total profit by end of the window of opportunity with decision $\delta\in\Delta$. The firm's decision problem can be stated as

$$\max_{\delta\in\Delta} T\,\Pi(\delta) = T\,R(\delta^*) - T\,C(\delta^*) = T\,\Pi^*(\delta^*). \tag{7.7}$$

The combination of equations (7.1) to (7.6) generates an explicit representation of firm *i*'s cumulative profit by the end of time horizon. This substitution yields:

$$\begin{aligned} T\,\Pi^*(\delta^*) = \max_{\delta\in\Delta} \Bigg[& \mathrm{R}_0 \cdot \frac{x_0}{x_0^i + x_0^j} \cdot T_p \\ & + R_1 \cdot \frac{x_0 + \int_0^T K_1[u(s)]^\alpha - \mathrm{L}x(s)\,ds}{\mathrm{x}_0 + \int_0^T K[u(d)]^\alpha - \mathrm{L}x(s)\,ds + x^j(T)} \cdot (T - T_p) \\ & - \int_0^T \{u(s)\}\,ds \Bigg]. \end{aligned} \tag{7.8}$$

where x_0^j and $x^j(T)$, respectively, are the quality of existing and new products of the competitor. Considering a duopoly, the optimization problem written above can be reformulated as a differential game problem with state variable for the firm *i* (the competing firm is represented by superscript *j*) given as $x^i\,(t)$; the control variable $u^i(t)$. In the terminology used in optimal control and differential games, the salvage term for firm *i*, $\Phi^i\,(T,\,x(T))$ is defined as follows:

$$\begin{aligned} \Phi^i(T,\,x(T)) \underline{\underline{\Delta}}\, R_0 \cdot & \frac{x_0^i}{x_0^i + x_0^j} \cdot \mathrm{T}_p^i \\ + R_1 \cdot & \frac{\left\{x_0^i + \int_0^T K_1\left[u^i(s)\right]^{\alpha^i} - \mathrm{L}_1 x^i(s)\,ds\right\}(T - T_p)}{x_0^i + \int_0^T K_1[u^i(s)]^{\alpha^i} - \mathrm{L}_1 x^i(s)\,ds + x^j(T)} \end{aligned} \tag{7.9}$$

where

$$x^j(T) = x_0^j + \int_0^T \left[K_2\left[u^j(s)\right]^{\alpha^i} - \mathrm{L}_2 x^j(s)\right] ds. \tag{7.10}$$

7.3 A sequential differential game between two competing firms

We extend the conceptualization of a hypercompetitive scenario by considering a sequential differential game as being representative of a leader–follower or incumbent–entrant competitive situation. The game is characterized by information asymmetry where the follower is aware of the innovation and quality levels of the leader's products. The motivation for considering this scenario is its close correspondence with many real life competitive cases as described in Chapter 1.

Knowing the investment strategy of the leader, the rival firms can formulate their own strategies. Therefore, the firm acting as a leader chooses a decision path that maximizes the objective for all conceivable responses that can be taken by the follower(s). In the case of sequential games a hierarchical play differential game approach is used to model the competitive situation and to obtain the open-loop Stackelberg Nash equilibrium. The issue of subgame perfectness and commitment is extremely important in these solutions. Adding to the previous notation we have T_p, date of product launch by leader, $T_p + \tau$, date of product launch by follower.

The leader is represented by the superscript i and the follower is represented by the superscript j. Since it is a sequential game the time of product launch for the two players is such that the leader launches the product at time T_p. Later, the follower launches the product after time τ at $T_p + \tau$. A continuous improvement in the product is considered and therefore the entire time interval $t \in 0,T$ for the leader, and $t \in [T_p T]$ for the follower needs to be optimized. The state dynamics of the leader is

$$\dot{x}^i(t) = K_1[u^i(t)]^\alpha - L_1 x^i(t). \tag{7.11}$$

The rate of quality improvement $\dot{x}^i(t)$ increases with investments $u^i(t)$. The factor $L_1 x^i(t)$ suggests a natural decay in quality in the absence of any investments. The differential game formulation for leader is:

$$\max T\Pi^i(u^i(t), T_p) = -\int_0^T \{(u^i(s))\}\, ds + \Phi^i(T, x^i(T))\} \tag{7.12}$$

where $\Phi^i(T, x^i(T))$ is defined as

$$\begin{aligned}\Phi^i(T, x^i(T)) &\cong R_0 \cdot x_0^i / \left(x_0^i + x_0^j\right) \cdot T_p^i + R_1 \cdot x^i(T)(\tau) / \left(x^i(T) + x_0^j\right) \\ &+ R_1 \cdot x^i(T)(T - T_p - \tau) / \left(x^i(T) + x^j(T)\right)\end{aligned} \tag{7.13}$$

subject to

$$\dot{x}(t) = K_1\left[u^i(t)\right]^\alpha - L_1 x^i(t), x^i(0) = x_0^i, T \text{ fixed}, \; x^i(T) \text{ free}. \tag{7.14}$$

Next, the follower's problem formulation is discussed. Owing to the sequential nature of the game, the information about the leader's quality and about the leader's investments is known to the follower.

Specifically, it is assumed that the knowledge gained by the leader's investments 'spills over' to the follower's quality improvement dynamics. It is plausible to assume that in the context of a leader–follower competition, the level of quality improvements of the follower depends not only on its own innovation efforts but also on the knowledge pool available because of the leader's investments. The fact that the leader often cannot wholly conceal his efforts nor can he credibly announce the commitment he has made make the situation quite complicated.

To address these issues fully would require a subtle and rich analysis of games with incomplete information.

A formal treatment of 'spillover effects' would enrich this model.

The spillover effect is modelled by considering a linear additive term $M_2u^i(t)$. The follower's state dynamics is

$$\dot{x}^j(t) = K_2[u^j(t)]^\alpha - L_2x^j(t) + M_2u^i(t). \tag{7.15}$$

It is assumed that the spillover is a function of the investments made by the leader $u^i(t)$. In this term M_2 is a very small number which quantifies the amount of spillover from the leader to the follower. The term $M_2u^i(t)$ would restrict the analysis to the case where the follower emulates the innovation and product quality of the leader, as opposed to setting his own technology and quality standards. Now with all considerations wrapped up, a leader–follower dynamics can be stated.

The follower's differential game formulation is given as:

$$\begin{aligned} \max T\Pi^j(u^j(t), T_p) &= -\int_{T_p}^{T} \{u^j(s)\}\, ds + \Phi^j(T, x^j(T)) \\ &= -\int_0^{T-T_p} \{u^j(s+T_p)\}\, ds + \Phi^j(T, x^j(T)) \end{aligned} \tag{7.16}$$

where $\Phi^i\ (T,\ x(T))$ is defined as

$$\begin{aligned} \Phi^j(T, x(T)) \triangleq R_0 \cdot \frac{x_0^j}{x_0^i + x_0^j} \cdot T_p + R_0 \cdot \frac{x_0^j}{x^i(T) + x_0^j} \cdot \tau \\ + R_1 \cdot \frac{\{x^j(T)(T - T_p - \tau)\}}{x^i(T) + x^j(T)} \end{aligned} \tag{7.17}$$

subject to

$$x^j(t) = K_2\left[u^j(t)\right]^\alpha - L_2x^j(t) + M_2u^i(t) \tag{7.18}$$

where $M_2u^i(t)$ represent the spillover of knowledge, assumed to be a function of investments by the leader, M_2 is a small constant such that $0 < M_2 \ll 1$.

$$x^j(0) = x_0^j, \; T \text{ fixed}, \tag{7.19}$$

$$x^j(T) \text{ free} \tag{7.20}$$

where

$$x^i(T) = x_0^i + \int_0^t \left[K_1 \left[u^i(s) \right]^{\alpha^i} - L_1 x^i(s) \right] ds \tag{7.21}$$

$$x^j(T) = x_0^j + \int_{T_p}^t \left[K_2 \left[u^j(s) \right]^{\alpha^j} - L_2 x^j(s) + M_2 u^i(s) \right] ds. \tag{7.22}$$

7.4 Analysis of the model and discussion

A conventional tool for solving the problem is the application of Pontryagin's maximum principle for open-loop Stackelberg equilibrium conditions. Interested readers can refer to Dockner *et al.* (2000) for details regarding Pontryagin's approach to solving sequential differential game problems. The equilibrium results are expressed in the form of properties as given below (see also the Appendix).

Equilibrium results

Property 1: The maximized costate variables for the follower are a function of time and are given by

$$\lambda_1^* = \lambda_{01} e^{L_2 t} \tag{7.23}$$

$$\lambda_2^* = \lambda_{02} e^{L_1 t} \tag{7.24}$$

where λ_1^* and λ_2^* are the costate variables reflecting the marginal price for a unit increase in the follower firm's own state and the state of the leader i, $\lambda_1(0) = \lambda_{01}$, $\lambda_2(0) = \lambda_{02}$ are known constants.

Property 2: The Stackelberg equilibrium investment by the follower in product development is given as:

$$u^j(t)^* = \left[K_2 \alpha^j \lambda_{01} e^{L_2 t} \right]^{\frac{1}{1-\alpha^j}}. \tag{7.25}$$

Next, the leader's problem is investigated. The leader knows the follower's best response to each control path $u^i(\cdot)$.

Property 3: The maximized costate variables for the leader are given by

$$\psi_1^* = \psi_{01} e^{L_1 t} \tag{7.26}$$

$$\psi_2^* = \psi_{02} e^{L_2 t} \tag{7.27}$$

$$\psi_3^* = -\psi_{03} e^{-L_2 t} \tag{7.28}$$

where ψ_1^*, ψ_2^* and ψ_3^* are the costate variables reflecting the marginal price for a unit increase in the leader's own state, the state of the follower and the costate of the follower. ψ_{01}, ψ_{02} and ψ_{03} are constants.

Property 4: The Stackelberg equilibrium investment by the leader in product development is given as:

$$u^i(t)^* = \left[\frac{K_1 \alpha^i \psi_{01} e^{L_1 t}}{1 - M_2 \psi_{02} e^{L_2 t}}\right]^{\frac{1}{1-\alpha^i}}. \tag{7.29}$$

Property 5: The equilibrium state trajectory of performance improvement of the follower is given as:

$$\begin{aligned} x^j(t) = \frac{e^{-L_2 t}}{L_2} \Bigg[& L_2 x_0^j + (-1 + e^{L_2 t}) M_2 \left[\frac{K_1 \alpha^i \psi_{01} e^{L_1 t}}{1 - M_2 \psi_{02} e^{L_2 t}}\right]^{\frac{1}{1-\alpha^i}} \\ & + (-1 + e^{L_2 t}) K_2 (K_2 \alpha^j \lambda_{01} e^{L_2 t})^{\frac{\alpha^j}{(1-\alpha^j)^2}} \Bigg]. \end{aligned} \tag{7.30}$$

Property 6: The equilibrium state trajectory of performance improvement of the leader is given as:

$$x^i(t) = \frac{e^{-L_1 t}}{L_1} \left[L_1 x_0^i + (-1 + e^{L_1 t}) K_1 \left[\frac{K_1 \alpha^i \psi_{01} e^{L_1 t}}{1 - M_2 \psi_{02} e^{L_2 t}}\right]^{\frac{\alpha^i}{1-\alpha^i}} \right]. \tag{7.31}$$

For the analysis, the leader and the follower are assumed to be symmetric and therefore the corresponding values for the follower j are also assumed to be the same.

The first observation is that the costate variable of both the leader and the follower increases more rapidly with time.

The marginal utility of a unit increase in quality exhibits a convex increasing trajectory. With comparable values of parameters the plot of the costate trajectory of the follower has an upward exponential drift with time. The plot begins at time $t = T_p$, since the follower initiates product development activities only after the leader has already launched the product. Regarding the costate trajectory of the leader we see that ψ_1 represents marginal utility from a unit increase in quality of the leader's product. The costate variable for both the leader and the follower firm increase for the entire planning horizon. This follows since the leader and follower are competing on the basis of their quality levels. Firms increase their market shares based on their relative quality to that of the competitor. After launch the product faces its introduction and growth stage.

In these phases, maintaining higher quality becomes even more important since the product sales are directly influenced by product quality. This leads to a convex increasing trajectory for the costate variable. Thus, as a second observation:

> *The investment in innovation made by the follower increases more rapidly with time.*

The investment strategies of the follower are purely a function of its own costate variable. The follower's costate variable increases in a convex fashion. This in turn results in a convex increase in the follower's control trajectory. With identical parameter values we can derive that the follower compensates for a delayed entry into the market by increasing its investment intensity. Such an increase in investment results in increased quality and therefore high total revenue for the follower. Thus we observe:

> *The investment in innovation made by the leader initially increases rapidly with time but later increases at a decreasing rate.*

The leader's investment trajectory is sigmoidal or S-shaped. Because of the nature of the game, the leader enters the market before the follower. While formulating his equilibrium investment strategy, the leader takes into account the evolution of his own costate and also the costate of the follower.

Being early in the market, the leader takes into account all possible courses of action that a follower may choose.

We note that the leader's rate of investment increases in the initial phase and then starts to decrease. That is, the leader *capitalizes* on the advantage of early market entry by increasing the intensity of investments and gaining a high market share. Subsequently, after the follower's entry, the leader reduces the intensity of investments and thereby reduces the amount of spillover that is potentially possible.

> *When the parameter values of the leader and follower are identical, the follower invests higher than the leader.*

In the course of time, the follower initially maps its investment strategy to that of the leader. However, subsequently the leader starts reducing the rate of investments, while the follower continues with the high investment rate-based strategy.

> *As a further observation, the rate of increase of the follower's product quality increases with time.*

The quality of the follower's product increases for the entire time-horizon. Moreover, this increase is a convex function. As can be inferred from the problem formulation, the quality level of the follower is a function of investments made by the leader and the follower. Specifically, the quality trajectory is influenced by both the convex profile of investments made by the follower and also by the spillover effect of the investments made by the leader. This results in a convex increase in quality improvements. The state trajectory is therefore upward sloping in convex form.

The follower starts accruing revenue only after $T_p + \tau$, where T_p is the date of launch of the leader. In the game context of quality-based competition, the follower compensates for the delayed entry by increasing the quality levels at a fast rate.

> *The rate of increase of the leader's product quality decreases with time.*

The leader's investment trajectory follows an S-shaped trajectory. In the problem formulation the leader's quality improvement is only a function of its own investments. Moreover, it is assumed that the leader does not obtain the advantages of spillover of knowledge gained by the follower's investments. Furthermore, the leader needs to deliberately avoid maintaining very high investments to ensure that the follower does not achieve huge gains from spillovers. Under such a setup it must be noted that the leader chooses an S-shaped trajectory for investments. In such an investment profile the investment increases rapidly in the initial phase but then eventually tapers off. Corresponding to this investment profile, the rate of quality improvement is high in the initial phase but then eventually starts decreasing. Thus, the quality trajectory of the leader is concave. This leads to another observation.

> *When the parameter values of the leader and follower are identical, the quality of the follower's product is higher than that of the leader's product.*

With identical parameter values the follower and leader are perfectly symmetric in their capabilities. Moreover, the leader has the advantage of earlier market entry, whereas the follower obtains the gains of information asymmetry and the associated knowledge spillovers from the leader.

It is reasonable to assume that the follower employs his inherent competence and the benefit of spillover of knowledge from the leader's investments to increase quality at a much faster rate. Under such circumstances the overall gain and loss in the competitive game is dictated by the relative revenues achieved and the costs incurred by the two players. With higher investments than the leader the follower incurs higher costs. At the same time these investments also lead to higher quality levels and therefore revenues for the follower. In the event of the value of R_1 being very high the follower wins the game while the results favour the leader if R_1 is low.

7.5 Firm asymmetries

We now turn to firm asymmetries and their implications. It is likely that firms take up the role of leader and follower based on inherent strengths and weaknesses. Therefore, an explicit consideration of firm asymmetries is very important and could lend more insights in a sequential game setup, as first outlined in Chapter 5. Asymmetries may be addressed in at least four ways.

First, it is possible that organizational approaches and techniques (such as total quality management) can be used to make the product development process more cost-efficient and effective. This would influence the value of the resource productivity parameter α.

Second, it is possible to have an advantage in product development by making capital investment in technology development. This is equivalent to considering asymmetries in the value of K.

Third, it is possible that the obsolescence parameter of firms denoted by L is different.

Finally, different values of M suggest different levels of spillover of knowledge from the leader's investments to the follower. The firm with the higher value of L has a higher obsolescence or decay in quality. We restrict assessment of asymmetries by considering different values for the parameters: (α^i and α^j), (K_1 and K_2), (L_1 and L_2) and (M_1 and M_2).

Innovation resource productivity

Parameters α^i and α^j denote the innovation resource productivity parameter of the two firms. An asymmetry could result if the competing firms have differing capabilities in making a productive use of investments. The skill set of employees, training and development activities, the quality culture are some of the reasons for such an asymmetry. For example, $(\alpha^i = 0.2) > (\alpha^j = 0.1)$ suggest that firm i has a higher innovation productivity over j. An investigation of different values of α suggest that the firm with the higher innovation productivity also invests a higher amount in product development.

From the results we observe that the follower has a relatively higher level of investments in innovation as compared to the leader. Therefore, an increase in the value of α of the follower will only result in making these investments still higher. Instead, an increase in the leader's α would provide some interesting insights in that it will show that the leader sails ahead with the follower having no chance of catching up. It clearly shows that the leader has a higher investment than does the follower for almost the entire planning horizon.

As can also be noted, the quality of the leader's product is always higher than that of the follower. Hence an asymmetry in terms of resource productivity parameters provides a totally different result than that observed when the competing firms were symmetric. It suggests that the leader with an advantage in terms of early market entry as well as a higher level of resource productivity continues having higher revenues. Under this situation the leader does not worry much about the spillover to the follower since its own resource productivity enables attainment of higher revenues by increasing quality levels relative to that of the follower. With a high value of R_1 it can be conjectured that the result favours the leader.

Capital investment parameters

Firm asymmetries can be analysed by evaluating different values for K and the implications on control and state trajectories. K_1 and K_2 are proportional to the level of capital investment in development technology by the two competing firms. If $(K_1 = 20) > (K_2 = 10)$, it suggests that the leader i has higher levels of capital investment in development technology as compared to the follower j. The implications of a higher value of K for the leader can easily be derived. The parameter K exerts a positive effect on innovation investments. Similar to the conjecture regarding different values of α, the intuition behind this effect is that with a higher level of capital investments in development technology, a firm targets the investments towards increasing the product quality. Thus with a higher value of K the investment by the leader is relatively higher than that of the follower for most of the planning horizon. However, in the latter part of the planning horizon, the follower's investment in fact shoots up while that of the leader tapers off. Such an investment profile would have an impact on the state trajectory.

Furthermore, the product quality of the leader is higher than that of the follower for most part of the planning horizon. Eventually, at the end of the planning horizon, the high growth in the follower's investment results in higher quality than that of the leader. In the model, the relative difference in resources due to strategic investments is captured in asymmetries in the value of K. Given adequately high values of R_1, the leader i would have relatively higher profits than the follower j.

The obsolescence parameter

L_1 and L_2 characterize the decay or in other words the obsolescence effect of quality. Different values of L_1 and L_2 help evaluate the difference among the two firms regarding this effect. $(L_1 = 1) > (L_2 = 0.7)$ suggests that the obsolescence effect for leader *i* is higher than that for follower *j*. An investigation into the dynamics of innovation investments reveals that with higher values of the parameter L, a firm would make higher levels of investments.

Interestingly, with a difference in L_1 and L_2 the shape of the leader's investment trajectory is now convex. That is, when the leader faces a high obsolescence effect the investments are increased at a faster rate to ensure an increasing state trajectory. However, as opposed to parameter α and K such an increased control trajectory does not translate into higher quality. In fact, in this situation, the leader has lower product quality in spite of higher investments. A high decay effect puts the leader into a disadvantageous position. In such a situation, under equilibrium control the leader always loses the competitive game.

Spillover factor M_2

M_2 characterizes the amount of spillover from the leader to the follower. In this section the impact of reduction of the spillover factor M_2 is considered. Evidently, in this case the leader would be less worried about the amount of advantage the follower obtains from the leader's own investments.

If the spillover effect is low, the leader pursues investments more aggressively. Under such circumstances, the control trajectory of the leader is also convexly increasing and is only marginally lower than that of the follower.

Regarding the impact of such investments on the state trajectory, the quality of the follower is higher than that of the leader and both obtain a convex increasing state trajectory. But unlike α, K and L, a different value of M_2 does not correspond to any advantage related to resources, core competence or dynamic capabilities. This is purely an exogenously defined variable based on technology and industry types. Under changed circumstances with respect to M_2 the leader does not really gain much in the competitive game.

The results in fact point to the follower winning the game due to much higher levels of quality and only marginally higher levels of investments.

7.6 Conclusions

There is apparently a paradox in trying to assess, both empirically and theoretically, the impact of competitive pressure on innovation and growth. On the one hand, according to the tradition originating in Schumpeter (1942) the prospective reward provided by monopoly rent to a successful innovator is required to stimulate sufficient R&D investment and technological progress. On the other hand, the incentives to innovate are weaker for an incumbent monopolist than for a firm in a competitive industry (Arrow, 1962). When competition is intense in the product market, innovation may even be seen as the only way for a firm to survive. In neo-Schumpeterian models of endogenous growth (Grossman and Helpman, 1991; Aghion and Howitt, 1992), innovation allows a firm in an industry to take the lead and gain profit. But the monopoly rent enjoyed by the winner is only temporary, and a new innovator, capitalizing on accumulated knowledge, is always able to 'leapfrog' the leader unless the leader is endowed with advantages of firm asymmetries. In the recent research literature, Aghion *et al.* (1997, 2001), supposing a duopoly in each sector, both, at the research and production levels, have introduced what they call 'step-by-step innovation', according to which technological progress allows a firm to take the lead, but with the lagging firm remaining active and eventually capable of catching up. This model can be extended allowing for the possibility that the lagging firm leapfrogs the leader, without driving it out of the market.

Here we adopt a simplified approach to consider spillover effects. Griliches (1979) lays out the conceptual framework and provides an early discussion of the importance of spillover effects of R&D. Later in 1992, Griliches reviewed the recent empirical evidence on spillovers, and tentatively concluded that spillover effects may be substantial.

The conventional approach used in innovation and R&D research viewed the process as one with constant returns, competitive output and factor markets and no externalities. However, such a framework does not offer a full explanation of productivity growth. For a better understanding it therefore becomes very important to consider increasing returns to scale, R&D spillovers and other externalities and disequilibria.

References

Aghion, P., Harris, C. and Vickers, J. (1997) 'Competition and growth with step-by-step innovation: an example', *Review of Economic Studies*, **41**, 771–782.

Aghion, P., Harris, C., Howitt, P. and Vickers, J. (2001) 'Competition, imitation and growth with step-by-step innovation', *Review of Economic Studies*, **68**, 467–492.

Aghion, P. and Howitt, P. (1992) 'A model of growth through creative destruction', *Econometrica*, **60**, 323–352.

Arrow, K. (1962) 'The economic implications of learning by doing', *Review of Economic Studies*, **29**, 155–173.

Dockner, E., Jorgensen, S., VanLong, N. and Sorger, G. (2000) *Differential Games in Economics and Management Science*, Cambridge: Cambridge University Press.

Griliches, Z. (1979) 'Issues in assessing the contribution of research and development to productivity growth', *The Bell Journal of Economics*, **10**(1), 92–116.

Griliches, Z. (1992) 'The search of R&D spillovers', *The Scandinavian Journal of Economics*, **94**, S29–S47.

Grossman, G. and Helpman, E. (1991) *Innovation and Growth in the Global Economy*, Cambridge, MA: MIT Press.

Hoffman, W. D. (1976) 'Market structure and strategies of R&D behavior in the data processing market – theoretical thoughts and empirical findings', *Research Policy*, **5**, 334–353.
Schumpeter, J. A. (1942) *Capitalism, Socialism and Democracy*, New York: Harper.
Teece, D. J., Pisano, G. and Shuen, A. (1997) 'Dynamic capabilities and strategic management', *Strategic Management Journal*, **18**(7), 509–533.

Appendix

Mathematical preliminaries, definitions and theorems

This section covers some of the salient analytical aspects specific to a sequential play game.

For a finite horizon T let L and F denote the leader and follower, respectively. Let x denote the vector of state variables, u^L the vector of control variables of the leader, and u^F the vector of control variables of the follower. Assume $x \in R^n$, $u^L \in R^{mL}$ and $u^F \in R^{mF}$.

Definition 1: *The initial value of the, follower's costate variable λ_i is said to be noncontrollable if $\lambda_i(0)$ is independent of the leader's control path $u^L(t)$. Otherwise, it is said to be controllable.*

The definition suggests that if the costate variable is controllable, the follower's control variable $u^F(t)$ at time t depends also on future values of $u^L(.)$; that is, on values $u^L(s)$ with $s > t$.

Theorems and proofs.

Theorem 1: *The maximized costate variables for the follower are a function of time and is given by*

$$\lambda_1^* = \lambda_{01}e^{L_2 t} \tag{A.1}$$
$$\lambda_2^* = \lambda_{02}e^{L_1 t} \tag{A.2}$$

where λ_1^* and λ_2^* are the costate variables reflecting the marginal price for a unit increase in the follower firm's own state and the state of the leader i; $\lambda_1(0) = \lambda_{01}$, $\lambda_2(0) = \lambda_{02}$ are constants.

Proof. Stackelberg equilibrium conditions are derived by constructing the Hamiltonians. The analytical solution is derived for the follower firm j by writing the Hamiltonian as:

$$\begin{aligned} H^j = {} & -\left[u^j(t)\right] + \lambda_1\left[K_2 u^j(t)^{\alpha^j} - L_2 x^j(t) + M_2 u^i(t)\right] \\ & + \lambda_2\left[K_1 u^i(t)^{\alpha^i} - L_1 x^i(t)\right]. \end{aligned} \tag{A.3}$$

The necessary conditions for optimality are:

$$H^j_{u^j} = 0 \tag{A.4}$$

$$\lambda_1^* = -H^j_{x^j}; \quad \lambda_2^* = -H^j_{x^i} \tag{A.5}$$

$$x^j(0) = x^j_0. \tag{A.6}$$

From (A.5)

$$\lambda_1^* = \mathrm{L}_2\lambda_1; \quad \lambda_1^* = \lambda_{01}e^{\mathrm{L}_2 t} \tag{A.7}$$

$$\lambda_1^* = \mathrm{L}_1\lambda_2; \quad \lambda_2^* = \lambda_{02}e^{\mathrm{L}_2 t} \tag{A.8}$$

where $\lambda_1(0) = \lambda_{01}$ is the known positive constant denoting the initial value of the costate λ_1 and $\lambda_2(0) = \lambda_{02}$ is a known negative constant denoting the initial value of the costate λ_2.

Theorem 2: *The Stackelberg equilibrium investment by the follower in product development is given as*

$$u^j(t)^* = \left[K_2\alpha^j\lambda_{01}e^{\mathrm{L}_2 t}\right]^{\frac{1}{1-\alpha^j}}. \tag{A.9}$$

Proof. Differentiating (A.3) with respect to $u^j(t)$, we obtain:

$$H_{uj} = -1 + \lambda_1\alpha^j K_2\left[u^j(t)\right]^{\alpha^j - 1}. \tag{A.10}$$

By equating (A.10) to zero and some algebraic manipulations, the following expression is obtained for optimal effort in product development:

$$u^j(t)^* = \left[K_2\alpha^j\lambda_1^*\right]^{\frac{1}{1-\alpha^j}}. \tag{A.11}$$

Substituting the expression for λ_i in (A.11):

$$u^j(t)^* = \left[K_2\alpha^j\lambda_{01}e^{\mathrm{L}_2 t}\right]^{\frac{1}{1-\alpha^j}}. \tag{A.12}$$

Theorem 3: *The maximized costate variables for the leader are given by*

$$\psi_1^* = \psi_{01}e^{\mathrm{L}_1 t} \tag{A.13}$$

$$\psi_2^* = \psi_{02}e^{\mathrm{L}_2 t} \tag{A.14}$$

$$\psi_3^* = -\psi_{03}e^{-\mathrm{L}_2 t} \tag{A.15}$$

where ψ_1^*, ψ_2^* and ψ_3^* are the costate variables reflecting the marginal price for a unit increase in the leader's own state, the state of the follower and the costate of the follower; ψ_{01}, ψ_{02} and ψ_{03} are constants.

Proof. The Stackelberg equilibrium conditions are derived by constructing the Hamiltonian. An analytical solution is derived for the leader firm i by writing the Hamiltonian as:

$$H^i = -\left[u^i(t)\right] + \psi\left[K_1 u^i(t)^{\alpha^i} - \mathrm{L}_1 x^i(t)\right] + \psi_2\left[K_2 u^j(t)^{\alpha^j} - \mathrm{L}_2 x^j(t) + M_2 u^i(t)\right] + \psi_3[\lambda_1 \mathrm{L}_2]. \tag{A.16}$$

The necessary conditions for optimality are

$$H^i_{u^i} = 0; \tag{A.17}$$

$$\psi_1^* = -H^i_{x^i} \quad \psi_2^* = -H^i_{x^j} \quad \psi_3^* = -H^i_{\lambda_1} \tag{A.18}$$

$$x^i(0) = x^i_0. \tag{A.19}$$

From (A.18)

$$\psi_1^* = \mathrm{L}_1\psi_1; \quad \psi_1^* = \psi_{01}e^{\mathrm{L}_1 t} \tag{A.20}$$

$$\psi_2^* = \mathrm{L}_2\psi_2; \quad \psi_2^* = \psi_{02}e^{\mathrm{L}_2 t} \tag{A.21}$$

$$\psi_3^* = -\psi_3\mathrm{L}_2; \quad \psi_3^* = -\psi_{03}e^{-\mathrm{L}_2 t} \tag{A.22}$$

where ψ_1^*, ψ_2^* and ψ_3^* are the costate variables reflecting the marginal price for a unit increase in the leader's own state, the state of the follower and the costate of the follower; ψ_{01}, ψ_{02} and ψ_{03} are constants.

Theorem 4: *The Stackelberg equilibrium investment by the leader in product development is given as:*

$$u^i(t)^* = \left[\frac{K_1\alpha^i\psi_{01}e^{\mathrm{L}_1 t}}{1 - M_2\psi_{02}e^{\mathrm{L}_2 t}}\right]^{\frac{1}{1-\alpha^i}}. \tag{A.23}$$

Proof. Differentiating (A.3) with respect to $u^i(t)$, we obtain:

$$H_{u^i} = -1 + \psi_1\alpha^i K_1\left[u^i(t)\right]^{\alpha^i - 1} + \psi_2 M_2. \tag{A.24}$$

By equating (A.24) to zero and some algebraic manipulations, the following expression is obtained for optimal effort in product development:

$$u^i(t)^* = \left[\frac{K_1\alpha^i\psi_1^*}{1 - M_2\psi_2^*}\right]^{\frac{1}{1-\alpha^i}}. \tag{A.25}$$

Substituting the expression for ψ_1^* and ψ_2^* in (A.25) yields

$$u^i(t)^* = \left[\frac{K_1\alpha^i\psi_{01}e^{L_1 t}}{1 - M_2\psi_{02}e^{L_2 t}}\right]^{\frac{1}{1-\alpha^i}}. \tag{A.26}$$

Theorem 5: *The equilibrium state trajectory of performance improvement of the follower is given as*

$$\begin{aligned} x^j(t) = \frac{e^{-L_2 t}}{L_2}\Bigg[& L_2 x_0^j + (-1 + e^{L_2 t})M_2\left[\frac{K_1\alpha^i\psi_{01}e^{L_1 t}}{1 - M_2\psi_{02}e^{L_2 t}}\right]^{\frac{1}{1-\alpha^i}} \\ & + (-1 + e^{L_2 t})K_2(K_2\alpha^j\lambda_{01}e^{L_2 t})^{\frac{\alpha^j}{(1-\alpha^j)^2}}\Bigg]. \end{aligned} \tag{A.27}$$

Proof. The expression for the optimal state trajectory can be obtained by considering the state dynamics given in equation (7.11). Substituting $u^i(t) = u^i(t)^*$ in (7.11) the following first-order differential equation can be obtained:

$$x^j(t) = K_2\left[K_2\alpha^j\lambda_1^*\right]^{\frac{\alpha^j}{1-\alpha^j}} - L_2 x^j(t)\,(T_p + \gamma - t) + M_2\left[\frac{K_1\alpha^i\psi_1^*}{1 - M_2\psi_2^*}\right]^{\frac{1}{1-\alpha^i}}. \tag{A.28}$$

The first-order differential equation can be solved with the initial condition $x^j(0) = x_0^j$. The resulting expression for $x^j(t)^*$ is:

$$\begin{aligned} x^j(t) = \frac{e^{-L_2 t}}{L_2}\Bigg[& L_2 x_0^j + (-1 + e^{L_2 t})M_2\left[\frac{K_1\alpha^i\psi_{01}e^{L_1 t}}{1 - M_2\psi_{02}e^{L_2 t}}\right]^{\frac{1}{1-\alpha^i}} \\ & + \Big(-1 + e^{L_2 t})K_2(K_2\alpha^j\lambda_{01}e^{L_2 t})^{\frac{\alpha^j}{(1-\alpha^j)^2}}\Bigg]. \end{aligned} \tag{A.29}$$

Theorem 6: *The equilibrium state trajectory of performance improvement of the leader is given as:*

$$x^i(t) = \frac{e^{-L_1 t}}{L_1}\left[L_1 x_0^i + (-1 + e^{L_1 t})K_1\left[\frac{K_1\alpha^i\psi_{01}e^{L_1 t}}{1 - M_2\psi_{02}e^{L_2 t}}\right]^{\frac{\alpha^i}{1-\alpha^i}}\right]. \tag{A.30}$$

Proof. The expression for the optimal state trajectory can be obtained by considering the state dynamics given in equation (7.14). Substituting $u^i(t) = u^i(t)^*$ in (7.14), the following first-order differential equation can be obtained:

$$x^j(t) = K_1\left[\frac{K_1\alpha^i\psi_1^*}{1 - M_2\psi_2^*}\right]^{\frac{\alpha^i}{1-\alpha^i}} - \mathrm{L}_1 x^i(t). \tag{A.31}$$

The first-order differential equation can be solved with the initial condition $x^i(0) = x_0^i$. The resulting expression for $x^i(t)^*$ is:

$$x^i(t) = \frac{e^{-\mathrm{L}_1 t}}{\mathrm{L}_1}\left[\mathrm{L}_1 x_0^i + (-1 + e^{\mathrm{L}_1 t})K_1\left[\frac{K_1\alpha^i\psi_{01}e^{\mathrm{L}_1 t}}{1 - M_2\psi_{02}e^{\mathrm{L}_2 t}}\right]^{\frac{\alpha^i}{1-\alpha^i}}\right]. \tag{A.32}$$

8 Dynamic rivalry in technology platforms

> Today the CEOs of the world's top 1,000 companies ... lie awake at night worrying that a faster, more innovative, and lower-cost competitor could spring up any day, eager to take big bites from their cash cows and star performers.
>
> *Teaching Elephants to Dance*, MIT Technology Review (2004)

8.1 Introduction

Research and development (R&D) is a crucial phenomenon both from the point of view of the individual firm and the economy as a whole. Since innovation could be regarded as a public good, society as a whole benefits from innovation. However, the private benefits to a firm from innovating are likely to be different from the social benefits. In the absence of any mechanism preventing it, the benefits to an innovating firm are likely to be quickly dissipated by the entry of other imitating firms. In such a scenario, firms are unlikely to innovate. Thus according to conventional thinking, firms need to have some sort of reward for innovating. Intellectual property rights such as patents and copyrights provide this compensation. A big portion of the R&D literature has focused on the optimal patents' duration and breadth and the incentive of firms to innovate.

However, a different trend has emerged these days, especially with the increasing proliferation of hi-tech (network) industries. Instead of trying to get exclusive ownership rights, an increasing number of firms are making their technology freely available, i.e. their technology is no longer proprietary or 'open-source' (Cane, 2004).

In the so-called browser war in the 1990s, we witnessed intense competition in the market for internet browsers between Microsoft and Netscape. Netscape had a major head start on Microsoft, controlling 90% of the browser market by 1996 before Microsoft started aggressively selling in the market. With the entry of Microsoft, both firms engaged in a race to have the best available product.

Given the intense competition between the two firms, by the end of 1997 Microsoft was pricing the Internet Explorer free. In contrast, Netscape was charging corporations licensing fees for using their browser. By the end of 1997, Microsoft had stolen a large chunk of Netscape's market share. Netscape eventually followed suit and started giving away its browser for free.

The extended battle between Microsoft and Netscape had its toll on the profits of both companies. In 1998 Netscape came up with a new strategy and decided to release its source code, the actual line of programming language, for the Netscape Communicator. This allowed users and developers to look inside the workings of the browser, to modify the software and even to redistribute the new version under their own brand name, provided that the modified source code was also freely available. The whole idea is to turn the entire internet community into a vast research division for Netscape's browser.

The term 'open-source software' has been widely used in the popular and professional literature. Instead of keeping their technology proprietary, the firms distribute it freely. It is this phenomenon that this chapter attempts to explore. We wish to study the decisions of firms about whether to keep their technologies proprietary or not.

Even though the whole unorthodox open approach may seem counter-intuitive, Netscape was not the only firm to employ it. Apache, a program for serving world-wide web sites, and Sendmail, a program that routes and delivers internet electronic mail, are examples of free open-source programs that dominate the market. Open-source approaches have been expanded to the biotechnology and health care industries (The Economist, 2004). Linux, an increasingly popular operating system created in 1991, is another classic example of successful open-source software (OSS). Many of the programmers and software designers advocating OSS may share a utopian vision of software development, or they may simply want to prove themselves to be better than software giant Microsoft. However, the whole idea of OSS may not be so anti-capitalistic as it seems. It is hard to believe that profit-aiming firms will employ the OSS strategy without considering more pragmatic matters. (As recently being pointed out, open-source software such as Linux refers to freedom not prices – Cane, 2004.) The emergence of OSS as an observable phenomenon may be because the markets under consideration are no longer conventional markets. These markets exhibit 'network externalities' – a market has network externalities when buyers of a good exert positive benefits on the other users of the same good. For instance, consumers are likely to value computer hardware higher the more users of the hardware there are. This could be because there is likely to be a better support system with a larger network of consumers buying the product. Similarly, it is more likely for improved software to get written for the computer hardware the bigger the network of consumers buying it. But network externalities can be working both ways, positive or negative; for example, incompatibility with network systems on a rival operating system (such as MS Office) is a major obstacle in the OSS pursuit of the desktop, although low prices for OSS products would be a strong incentive to switch and for new customers to enter.

In such a market time is of utmost importance in the race for product improvement. Firms cannot afford to let their competitors get ahead in the race for technological innovation since that would give them the added advantage of a bigger network. Also, consumers in these markets tend to exhibit a very high level of loyalty. That is because learning to use the product involves a cost. Once a consumer becomes familiar with particular software, she is unlikely to switch to a completely different brand performing the same tasks. Instead, she would rather purchase new releases of the same brand even though there can be various close substitutes with similar qualities available in the market. This enhances the effect of network externalities in the long run. Further, OSS *can* feasibly translate into better quality in markets such as those for computers.

The effect of OSS on product improvement is two-pronged. Making the technology freely available means that there can be more people directly working on improving the product. For example, ever since Linux went fully OSS, thousands of programmers have volunteered elaborate improvements of their own design for no more reward than the respect of the geek subculture. It is like expanding the R&D department, so larger improvements in quality can be realized. Second, there is likely to be a better supply of complementary goods. For instance, giving out the source code for an operating system is likely to lead to more software being developed for it, which is in essence equivalent to having a better quality operating system, i.e. consumers now find this OS more attractive.

On the other hand, making technology freely available means a loss in licence fees. There is also the fear of technology being stolen. But in a market with network externalities, if the firm giving away its technology already has a sufficiently large network then it is more difficult for other firms just entering the market to steal the technology and get ahead, since they would also have to overcome the network advantage of the existing firm (Gottinger, 2003). Besides in this digital era, the relative ease of creating software with similar functionalities using different programming codes has made the whole idea of keeping technology proprietary less relevant.

We can thus think of OSS as increasing the rate of product improvement or increasing the success rate of R&D. We model OSS via licence fees and assume that OSS increases product development deterministically.

A lower licence fee represents a less-proprietary technology. A zero or negative licence fee means that the technology is totally non-proprietary. Positive licence fees represent a proprietary technology – the firm is not willing to freely distribute its technology. The more 'open' a firm is the higher is its rate of R&D – in our model R&D translates directly into the quality of the product. The greater the R&D, the higher the quality of the product. OSS improves the quality of the product in our model by increasing the supply of people or firms working on improving the product.

We look at a market with network externalities with an incumbent and an entrant. The incumbent, unlike the entrant, already has an installed base of consumers. Our objective is to explore the decisions of the firms regarding how proprietary they want to make their technology, i.e. how copyright or 'open' to make their product. The decision of copyright or open-source mentioned above is modelled via licence fees. We wish to see whether the incumbent could use 'open' as an entry deterring strategy. We also compare the incumbent's decision with that of a monopoly.

We explicitly model the direct and indirect effects of network externalities. As in most models, we have the network term showing up in the consumers' utility; the bigger the network, the better off the consumers are. Consumers prefer to use a popular word processor because they know the format of their work can be easily transported to other users' computers. This is the direct effect of network on consumers' utilities. In our model there is also an indirect effect of network externalities – a larger network translates into better quality. For example, more software companies are willing to produce programs for an operating system if it has a larger consumer base. That is because the downstream software companies thus can tap into this larger network of customers. This improves the quality of the OSS. This is the indirect effect of network on consumers' utilities.

This chapter is organized as follows: the next section (8.2) looks at the literature related to this work. In Section 8.3 we present the model. In Section 8.4, we look at the equilibrium results and give interpretations. Section 8.5 provides extensions to our model. Section 8.6 draws conclusions on further use.

8.2 Patenting, licensing and open-source technologies

The literature related to this paper includes the works on network externalities, R&D, entry deterrence and licensing.

R&D is an extensively researched area in industrial organization starting with Arrow's pioneering article where he asked: 'What is the gain from innovation to a firm that is the only one to undertake R&D, given that its innovation is protected by a patent of unlimited duration?' (Arrow, 1962.) Since then, there has been a spate of research on R&D covering issues such as the incentives to innovate, patent races, welfare implications of R&D, choice of technology and the adoption of technologies. However, most of this work has focused on conventional markets rather than on markets with network externalities in which dynamic or 'Schumpeterian' competition evolves (Evans and Schmalensee, 2001). We explicitly look at the effect of network externalities on R&D competition and introduce the possibility of firms not wanting their technology to be proprietary which is not recognized by the traditional R&D literature.

Within the topic of R&D, there has been some work devoted to licensing. This literature takes as the starting point one or more firms having a patent. Licensing is then a means of disseminating an innovation. Among the incentives for licensing are product market competition which creates incentives for managers who would otherwise exploit their monopoly positions, cost savings to the licensees which could be appropriated by the licensing firm, and lowering rivals' incentives to invent around the innovation. Katz and Shapiro (1985b) look at the incentives to engage in licensing once an innovation is developed in a world of perfect patents. They also look at the incentives to innovate, given the feasibility of licensing. Katz and Shapiro (1986a) examine the optimal licensing strategy of a research lab selling to firms who are product market competitors. They show that the seller's incentives to develop an innovation may be excessive and the incentives to disseminate information may be too low. Kende (1998) explores the conditions under which a monopolist selling a system consisting of a main component and differentiated secondary components can increase profits by allowing competition in the market for the secondary components. Opening the system in this fashion can increase profits by giving consumers an added incentive to incur the setup cost of purchasing the main component.

The results show that an open system is likely to be more profitable than a closed one when it is more elastic, when secondary-component variety is more valued, and when the share of the main component in the total system budget of the consumer is high.

Licensing plays an important role in our model. However, the relation between licensing and innovation has been reversed compared to the traditional licensing literature described above. Instead of licensing being used as a means of diffusing an innovation *after* it has occurred, in our model licensing actually *leads to* higher innovation.

Farrell and Saloner (1985, 1986) and Katz and Shapiro (1985a,b, 1986a,b) are the pioneering works exploring the implications of network externalities in industrial organizations. Farrell and Saloner concentrate on the demand side and show that the existence of network externalities leads to coordination problems and thus to a multiplicity of equilibria. In a model where two conflicting technologies compete, they show that 'excess inertia' might exist in equilibrium, i.e. the adoption of a new standard might be too slow compared to the social optimum.

Katz and Shapiro extend the scope of Farrell and Saloner's works by including the supply side. They look at the issues of compatibility and pricing in the presence of network externalities. The model of Katz and Shapiro (1985a) will serve as a building block for our model.

The issues we address, however, are different. They use their model to reinforce the importance of consumer expectations in markets with network externalities. They also show that the private decisions of firms regarding compatibility is greatly affected by whether firms can act unilaterally (or if a consensus is required) and whether side payments are feasible. Katz and Shapiro (1986b) show that in the presence of network externalities the private and social incentives to achieve compatibility may diverge.

They show the conditions under which firms may use compatibility as a medium for reducing competition.

While a lot of work has been done in the area of network externalities, little of it specifically addresses R&D. Katz and Shapiro (1992) look at whether there is too much or too little technological innovation in a market with network externalities compared to the social optimum. In particular, they look at whether a new product which embodies technological progress is introduced too early or too late compared to the social optimum. Their conclusion is that contrary to what was earlier believed to be true, there is *excess momentum* in equilibrium, i.e. a new product is introduced too soon compared to the social optimum. That happens because the sooner the product is introduced, the sooner the firm can start building up a network and reaping its benefits.

Choi (1994) looks at a two-period model of a monopoly in a market with network externalities. He studies the incentive of the monopolist to introduce an incompatible improved product in the presence of network externalities. Kristiansen (1996) studies the consequences of network externalities on the riskiness of R&D projects chosen by an entrant and an incumbent. He shows that the incumbent chooses a too-risky project that too often lets a new firm with an incompatible technology enter as compared to the social optimum. In addition, the entrant has an incentive to choose more certain projects than are socially optimal and these strengthen the possibility of adoption of an incompatible technology. Regibeau and Rockett (1996) analyse the compatibility choices of two firms which must also decide when to introduce their goods in a market characterized by network externalities. They show that the firms' incentives to achieve compatibility depend crucially on the time at which the degree of compatibility must be chosen. The current chapter is also related to the literature on investment as a means of entry deterrence. Firms compete not only with existing firms, as is transparent, but also with potential entrants. In an extension of the Spence model, Dixit (1980) shows that an incumbent firm may make irrevocable commitment of investment in order to alter the initial conditions of the post-entry game to its own advantage. One of the questions we are interested in is whether the incumbent chooses its licence fees so as to discourage entry. That is, the incumbent could be thought of as choosing investment in *quality* as a means of deterring entry.

8.3 Network competition under licensing

There are two firms, an incumbent, I, and an entrant, E. The incumbent has an installed base of consumers of $x_I > 0$ unlike the entrant ($x_E = 0$). We consider a two-stage game in a market with network externalities. The products of the firms are incompatible (i.e. they have separate networks).

In the first stage, firms charge a licence fee (f) to other downstream firms for use of their product (for example, Microsoft licenses its Windows operating system to other companies to develop applications software). The number of downstream firms willing to work on a firm's product (m_i) depends on the licence fee charged and also the initial network of the firm (the consumer base x_i for firm i):

$$m_i = k + x_i - f_i, \quad k > 0 \tag{8.1}$$

k *is* a measure of the potential market size of downstream firms independent of the licence fee and initial network.

The smaller the licence fee charged, the more freely the firm distributes its technology and in our terminology, the more 'open' the firm is. We assume no competition in getting downstream firms to buy the firms' licences. m_i decreases as f_i increases because as the licences become more expensive fewer downstream firms buy them. This is just the standard argument for a downward-sloping demand in price. As for the relationship between m_i and x_i, since the downstream firms are developing products that could be used in conjunction with the firm's product (like complements), the larger the installed base of customers that the firm has, the larger the potential demand for the downstream firms' product. Thus the more the downstream firms are willing to pay for the licence.

Definition 8.1: *If $f_i < f_j$ then firm i chooses to be more open than firm j. (Alternatively, firm j chooses to be less open than firm i.)*

In the second stage firms engage in Cournot competition to sell their products directly to the consumers. The inverse demands for the products of the two firms are given by:

$$p_i = A + q_i + \gamma x_i - (x_I + x_E) \tag{8.2}$$

where $i \in \{I.E\}$. p_i is the price charged by firm i. $x_i \geq 0$ is the amount of output firm i sells, x is the initial network of firm i. q_i is the quality of firm i's product.

Note that ideally we would want to include the *expected* network size in the demand functions instead of just the initial network size. However, the implications of forming expectations for equilibria in games similar to ours have been extensively studied (see Katz and Shapiro and Farrell and

Saloner). Since we are not interested in analysing the role of expectations and would like to keep our model simple, we just include the initial network size in the consumer's utility function.

The more downstream firms (m_i) a firm licenses out to, the better the quality q_i of its product will be. We assume the following functional form for q_i:

$$q_i = m_i, \quad i \in \{I,E\}. \tag{8.3}$$

Firm *i's* profit is:

$$\Pi_i = x_i p_i + f_i m_i, \quad i \in \{I,E\}.$$

The first term represents the profits from direct sales and the second term represents the revenues from licensing. Both firms are assumed to have identical costs, which we have normalized to zero. By staying out of the market the entrant makes zero profit.

As the profit function above shows, the licence fees (and hence the decision about how closed or open a firm wants to be) affect a firm's profits through two avenues. The first is the direct effect on profits through the licensing revenues. Then there is also an indirect effect through the quality of the firm's product which affects the profits made from direct sales.

The timing of the game is as follows.

1. Incumbent I sets its licence fee f_I. This determines quality q_I.
2. Entrant E decides whether to enter.
3. If E enters, it chooses a licence fee f_E from its set of acceptable licence fees. This determines quality q_E.
4. Firm(s) play output game (Cournot).

Firms choose their licence fees and output levels to maximize their profits. However, we do restrict the strategy set of the entrant. We do this in order to rule out situations where the entrant enters in order to make money solely out of licensing knowing that at the fee it charges or close to it, it could not sell any output in the second stage. Since the downstream firms could be thought of as producing complementary goods, it would be unreasonable to expect them to buy licences knowing that the entrant has no potential to sell output. Hence we restrict the entrant to choosing a licence fee from its set of 'acceptable licence fees'. An acceptable licence fee for the entrant, given that the incumbent charges a licence fee f_I, is f_E such that for every $\varepsilon > 0$, $x_E\ (f_I, f_E - \varepsilon) > 0$ where $x_E\ (f_I, f_E)$ is the optimal output for the entrant in the ensuing Cournot game. Thus, the 'downstream firms' are not willing to buy licences unless the entrant has the potential to sell output by making an arbitrarily small reduction in its licence fees.

As in other two-stage games, we will now proceed to solve the second-stage output game first in order to determine the subgame perfect equilibria.

8.4 Equilibrium licensing

Given the results from the second-stage output game, we will move on to solving the first stage. Note that if q_i were independent of m_i, then the only component of the firm's profit that depends on f_i is the revenue from licensing which is maximized when firm i charges a licence fee of $[k+x_i]/2$. This is our full 'closed' benchmark.

Before we go on to see how the two firms strategically choose their licence fees we look at the decision of the incumbent if it were a monopolist in this market.

Case 1. Monopoly: No potential entrant

Proposition 8.1: *The monopolist charges a licence fee* $f_I^M = [\mathrm{k} + (1-\gamma)\times \mathrm{x_I} - \mathrm{A}]/3$.

Proof. See the Appendix to this chapter.

We see in Proposition 8.1 that if $\gamma < 1$, the optimal licence fee is *increasing* in x_I; if $\gamma > 1$; the optimal licence fee is *decreasing* in x_I. To see why, note that the firm's profit has two components – the revenue from selling licences and the revenue from direct sales. As we saw above, the revenue from licences is maximized when the licence fee is $[k+x_I]/2$. However, by lowering its licence fee the firm can improve its product's quality and thus make more profits from its direct sales. So the optimal licence fee is less than $[k+x_I]/2$. Thus increasing the licence fee by a small amount beyond the optimal fee increases the revenue from licences and decreases that from direct sales, and the two exactly balance at the optimal licence fee. When $\gamma < 1$, the contribution of network size to consumer utility is small compared to its contribution to m. The result is that starting from an optimal f for a given x_I, a higher x_I means that increasing f_I will increase the revenues from licences by dx_I but decrease the revenue from direct sales only by $(1+\gamma)dx_I/2 < dx_I$. Therefore at the higher x_I profits have to be increasing in f at the previously optimal f. Thus, given the concavity of the profit function in f, the optimal licence fee must be *increasing* in x_I. Exactly the opposite argument holds when $\gamma > 1$. Thus the optimal licence fee is *decreasing* in x_I when $\gamma > 1$.

Case 2. Duopoly: With incumbent and potential entrant

We now look at a market that has an incumbent firm (*I*) with an installed consumer base of x_i^0 and a potential entrant (*E*).

Once the firms have decided their licence fees, they play a Cournot output game (Shapiro, 1989). The result of the Cournot game is described by the intersection of the two reaction functions that gives the unique interior Cournot outputs, i.e.:

$x_I = (A + 2q_I + 2\gamma x_I - q_E)/3$, $x_E = (A + 2q_E - \gamma x_I - q_I)/3$. Call these optimal output levels $x_i(f_I, f_E)$, $i \in \{I,E\}$.

First we make the following definitions.

If the incumbent charges its monopoly licence fee f_I^M, *and the entrant stays out of the market, then we say that entry is blocked.*

If the incumbent charges a fee different from f_I^M *in equilibrium, and the entrant stays out of the market, then we say that entry is deterred.*

If the entrant enters but sells nothing in equilibrium then we say that entry is restricted.

If the entrant sells a positive amount in equilibrium, we say that entry is unrestricted.

We can now go on and characterize the entry decision of firm E and the equilibrium licence fees for the two firms. But before we do that we briefly state the optimal response of the entrant, assuming that the firms produce optimally in the Cournot subgame, to the different licence fees that the incumbent can charge. For proofs please refer to the appendix. We can show that there exists a licence fee f_I^D for the incumbent such that for all $f_I < f_I^D$ the optimal response of the entrant is to stay out of the market and for $f_I < f_I^D$ it is to enter. There exists $f_I^R > f_I^D$ such that in the interval $[f_I^D, f_I^R]$ the best response of the entrant is to charge the largest acceptable licence fee that makes optimal x_E exactly 0, and for $f_I > f_I^R$ it is optimal for the entrant to enter and charge its unconstrained optimal licence fee resulting in a positive x_E.

Proposition 8.2: *The levels of* x_I^0, A *and* k *determine the entry decision and licence fees in equilibrium as follows:*

1. *Entry is blocked if* $(1+2\gamma)\, x_I^0 \geq A + 2k$.
2. *For* $A + 2k \geq (1+2\gamma)\, x_I^0 \geq A + k/2$, *entry is restricted with the incumbent charging* f_I^R *and earning its monopoly profits.*
3. *For* $A + k/2 > (1+2\gamma)\, x_I^0 > 2A/5 + k/5$. *Entry is restricted with the incumbent charging* f_I^R *and earning less than its monopoly profits.*
4. *For* $2A/5 + k/5 > (1+2\gamma)\, x_I^0$, *entry is unrestricted and the firms just charge the unconstrained Stackelberg licence fees.*

Proof. See Appendix.

Call the equilibrium duopoly licence fees f_i^*. Figure 8.1 represents the duopoly equilibrium more visually.

A is a measure of the size of the potential market for direct sales, k is a measure of the potential market for licences. x_I^0 measures the base of consumers already with firm I. A and k are terms common to both firms whereas x_I^0 is specific to firm 1. Therefore if firm I's network x_I^0 or its network advantage represented by $(1+\gamma)\, x_I^0$ is much larger than A

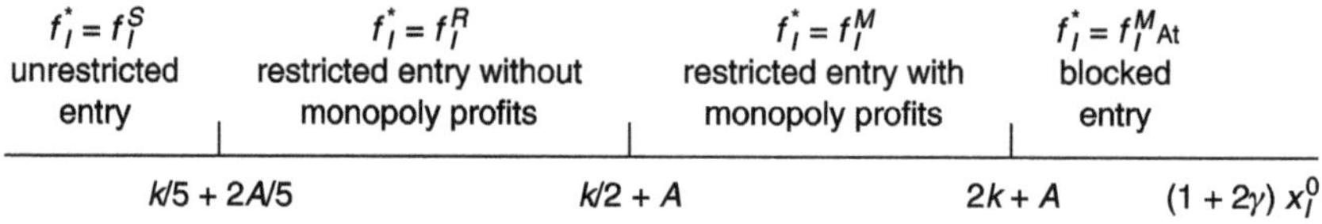

Figure 8.1 Duopoly licence fees.

or k, then the incumbent already has an insurmountable advantage in the market.

The entrant never finds it profitable to enter this market. The incumbent is the sole seller in the market in such a scenario and behaves as if there were no potential entrant.

When the network advantage of the incumbent is not large enough to block entry, the entrant enters the market. However, if the advantage is still sufficiently large then the incumbent is able to charge its monopoly fees and make its monopoly profits while restricting the entrant to zero output in the Cournot output game.

Given that the network advantage of the incumbent is not large enough to block entry, we saw in Statement 2 of Proposition 8.2 that if it is still sufficiently high then the incumbent can restrict entry and make monopoly profits. If its advantage is, however, not large enough to do that, even then a sufficiently high network advantage allows it to restrict entry. It then charges f_I^R which is different from its monopoly licence fee and earns less than its monopoly profits. Thus in this case the incumbent is able to act as a monopolist in the output market but not in licence fees.

If the initial network advantage of the incumbent is not very big compared to A and k then not only is it unable to keep the entrant out of the market but it is unprofitable to restrict its entry too. In that case the entrant enters the market and both firms sell positive amounts of output.

As corollaries to Proposition 8.2 we get the result that there is no entry deterrence and we also get a comparison between the monopoly licence fees of the incumbent and its licence fees when a potential entrant exists.

Corollary 8.1: *There is never entry deterrence.*

At f_I^D the entrant produces nothing and the incumbent just makes the profits it would as a monopolist in the second-stage output game. Since f_I^D is the largest licence fee for the incumbent that deters entry, if $f_I^M \leq f_I^D$ then there is blocked entry. If $f_I^M > f_I^D$, however, then entry is not blocked. But it also means that the incumbent's profits must be increasing at f_I^D (since the monopoly profits are concave in f_I and from A2 (Appendix) we know that in the interval $[f_I^D, f_I^R]$, the incumbent makes monopoly profits). Hence the incumbent will not find it profitable to charge f_I^D or less. Thus entry is not deterred.

As a corollary to Proposition 8.2 we get a comparison between the monopoly licence fees of the incumbent and its licence fees when a potential entrant exists.

Corollary 8.2: *$f_I^* \leq f_I^M$ always and sometimes $f_I^* < f_I^M$.*

Proof. See Appendix.

Thus we see that the presence of another firm makes the incumbent choose at most the licence fee it would if it were a monopolist. In other words, the incumbent chooses its technology to be at most as closed as that of the monopolist.

If its initial network is not very big then the presence of the other firm makes it want to be more open. The firm is then willing to give up more of the licensing revenue in order to improve quality since that will help it in the second-stage output competition. If its network advantage is already very big then it does not need to raise quality to give it a competitive edge in the second stage and it continues charging its monopoly licence fees.

Proposition 8.3: *For $(1+2\gamma)\ x_I^0 < A + 2k$ (i.e. when the entrant enters), there exists x_I^{0*}, such that for $x_I^0 > x_I^{0*}$, $f_I^* > f_E^*$ and for $x_I^0 < x_I^{0*}$, $f_I^* < f_E^*$, i.e., if the incumbent's network is smaller than x_I^{0*}, then the incumbent chooses to be more open and if the network is bigger than x_I^{0*}, then the entrant chooses to be more open.*

Proof. See Appendix.

Consider first the scenario where the incumbent has a small network advantage, namely the case of unrestricted entry and restricted entry without monopoly profits. Because of the installed base of consumers that firm *I* has, firm *E* is at a disadvantage in the output market so cannot expect to make as large a profit from direct sales as can firm *I*. Firm *I*, on the other hand, when it has a small network advantage, would want to compound its advantage in the output market by increasing its quality more than firm *E*. Given this initial imbalance in network advantage, firm *E* is less willing to afford a loss in licence revenues in order to improve quality. Thus the incumbent charges a lower licence fee than the entrant; in other words, the incumbent is more open than the entrant.

To examine this phenomenon in more detail, consider first the case of unrestricted entry. In this case we can isolate two effects that make the incumbent choose to be more 'open' than the entrant, namely the 'Stackelberg' effect and the 'network' effect. The 'Stackelberg' effect can be discerned by comparing the Cournot licence fee (when the two firms choose their licence fees simultaneously) with the Stackelberg licence fee in the absence of network externalities. In the Cournot case with $x_I^0 = 0$, both firms charge exactly the same fees, whereas in the Stackelberg case the incumbent charges a lower fee. Thus part of the reason that the incumbent chooses to be more open is just the 'first mover' or Stackelberg effect. Now if we reintroduce the network term, we see that the incumbent chooses to be even more open while the entrant chooses to be less open. From the Appendix (equation A10) we see that starting from an optimal f_I increasing x_I^0 by a small amount increases the marginal effect of f_I on licence revenues by less than it decreases the marginal effect of f_I on direct sales. Thus the introduction of x_I^0 requires a reduction in f_I in order to bring the marginal effects of f_I on direct sales revenues and licensing revenues into balance again. For the entrant, we see from the reaction function (equation A5) that both a larger x_I^0 and a smaller f_I, lead E to charge a higher f_E.

In the case where there is restricted entry without monopoly profits, the entrant is held down to zero output but the incumbent has to lower its licence fee below its monopoly level to do so. The entrant, on the other hand, is able to charge its full copyright licence fee. Thus again, the incumbent chooses to be more open than the entrant. If, however, the incumbent has a large network advantage so that it can still charge its monopoly licence fee and make its optimal monopoly profits, then the entrant has to fight to even stay alive in the market. The bigger the incumbent's network, the more difficult it is for the entrant to stay alive, i.e. the smaller the licence fee it charges. Thus there exists a sufficiently large network advantage x_I^{0*} for the incumbent beyond which the entrant charges a lower licence fee, i.e. chooses to be more open.

8.5 Conclusions and extensions

This chapter has shown that being 'open' could be a very sensible, in fact, optimal equilibrium strategy for profit-maximizing firms in a market with network externalities, for example, the internet and computer-related markets. As in most other theoretical models, we could not possibly include all the factors relevant to this recently observed phenomenon. Instead, we construct one way of modelling 'open', trying to capture the central idea of why 'open' makes sense. We devote this section to discussing what can be done or added on to our model in future research efforts on this topic.

In our model we have excluded the traditional 'technology stealing' effect. One might argue that that is precisely why we need copyrights and patents – to prevent the influx of copycats into the market driving profits to zero. However, as we mentioned in the introduction, in a market with network externalities, there is a natural barrier against imitators getting ahead of an innovator with an installed base of consumers, namely that consumers will perceive their product not to be as attractive as the incumbent's, even if they are of the same quality. It would still be interesting to see in future research the result of this 'business stealing' effect being put back into the model. We would expect it there to still be 'open' in equilibrium, although to a lesser extent. The equilibrium now needs to strike a balance between the two opposing effects of going 'open', namely, the indirect network effect of quality improvement, and the 'technology stealing' effect. We could also make the probability of technology being stolen vary inversely with network size, in which case we would expect the result that the incumbent goes more open to be reinforced.

Another possibility is to introduce competition in the first stage of the game. As of now, the incumbent and the entrant do not compete for downstream firms; each faces an identical and independent demand for its licences. This could be justified if there are numerous downstream firms looking for prospective technologies to develop. A more realistic setting may be one where both the incumbent and the entrant are competing in licence fees for downstream firms to buy their licences. Each downstream firm buys at most one unit of the licence.

However, this should not change our results drastically and we should still see 'open' as an equilibrium strategy. In fact, with competition, both the incumbent and the entrant are likely to go more open because now they need to compete to sell their licences.

We could introduce costs into the technology of the firms. In our current model, we have normalized both firms' costs to zero. Differences in the technologies of the firms may sway our results either way depending on who has the better (less-costly) technology. It would also be interesting to see how the coupling of the 'technology stealing' effect and positive costs may work. For example, if the incumbent freely distributes its technology then the costs for the entrant may fall because it can now 'steal' the technology of the incumbent.

In our model, we do not have entry deterrence but we have restricted entry in which the entrant sells licences in the first stage, but produces nothing in the second stage. This may seem a bit unrealistic. With positive fixed costs, we should be able to have the standard entry deterrence result. We would still expect restricted entry but one where the entrant is 'restricted' to some positive output level.

Another possible extension is making innovation stochastic, i.e. replacing the link of $q_i = q(m_i)$ with $q_i = q(m_i, \varepsilon_i)$, where ε_i is stochastic. We argue that this again should not change our results qualitatively so long as the resulting distributions for qualities exhibit second-order stochastic dominance – the larger the m, the more likely it is for firm i to have a bigger innovation.

Making the game more dynamic is also another interesting extension. As of now, we have two stages in the game. If we add in more stages, the network effect may get compounded since it now lasts longer, a large early network carries all the way into later stages too. It is even more essential for firms to build up a network fast to lure in more customers, the 'open' result thus may be even more pronounced.

We can bring back expectations in consumers utility. Instead of only caring about the initial size of the network, consumers now form expectations on how big the future network size for a firm will be before they make their purchase decisions. We then need to deal with much more complicated algebra and to adopt a more refined equilibrium concept in the second stage to handle this problem. For example, we can use the Fulfilled Expectations Cournot Equilibrium (FECE) as in Katz and Shapiro (1986). We suspect that doing so will not buy us anything new in our results.

References

Arrow, K.J. (1962) 'The economics of learning by doing', *Review of Economic Studies*, **29**, 155–173.

Cane, A. (2004) 'No free lunch, no free software', *Financial Times*, 31 March.

Choi, Jay Pil (1994) 'Network externality, compatibility choice, and planned obsolescence', *The Journal of Industrial Economics*, **30**, 85–105.

Dixit, A.K. (1980) 'The role of investment in entry-deterrence', *Economic Journal*, **90**, 95–106.

Economides, N. (1996) 'The economics of networks', *International Journal of Industrial Organization*, **14**, 673–699.

The Economist (2004) 'An open-source shot in the arm', *Technology Quarterly*, 12 June.

Evans, D.K. and Schmalensee, R. (2001) 'Some economic aspects of antitrust analysis in dynamically competitive industries', National Bureau of Economic Research, Cambridge, MA, Working Paper 8268, www.nber.org/papers/w8268.

Farrell, J. and Saloner, G. (1985) 'Economic issues in standardization', in James Miller (ed.), *Telecommunications and Equity: Policy Research Issues*, North Holland: New York, 165–178.

Farrell, J. and Saloner, G. (1986) 'Installed base and compatibility: innovation, product preannouncements, and predation', *American Economic Review*, **76**, 940–955.

Fudenberg, D. and Tirole, J. (1991) *Game Theory*, MIT Press: Cambridge, MA.

Gottinger, H.W. (2003) *Economies of Network Industries*, Routledge: London.

Katz, M.L. and Shapiro, C. (1985a) 'Network externalities, competition, and compatibility', *American Economic Review*, **75**, 424–440.

Katz, M.L. and Shapiro, C. (1985b) 'On the licensing of innovations', *Rand Journal of Economics*, **16**, 504–520.

Katz, M.L. and Shapiro, C. (1986a) 'Product compatibility choice in a market with technological progress', *Oxford Economic Papers*, **38**, 146–165.

Katz, M.L. and Shapiro, C. (1986b) 'How to license intangible property', *Quarterly Journal of Economics*, **101**, 567–590.

Katz, M.L. and Shapiro, C. (1992) 'Product introduction with network externalities', *The Journal of Industrial Economics*, **40**, 55–83.

Kende, M. (1998) 'Profitability under an open versus a closed system', *Journal of Economics and Management Strategy*, **20**, 72–85.
Kristiansen, E.G. (1996) 'R&D in markets with network externalities', *International Journal of Industrial Organization*', **29**, 531–547.
Regibeau, P. and Rockett, K. (1996) 'The timing of product introduction and the credibility of compatibility decisions', *International Journal of Industrial Organization*, **14**, 801–823.
Shapiro, C. (1989) 'Theories of oligopoly behavior', in R. Schmalensee and R.D. Willig (eds), *Handbook of Industrial Organization, Vol. I*, North Holland: Amsterdam, 370–414.

Appendix

Proof of Proposition 8.1

When the monopolist has a product of quality q_I and charges a price of p_I the only consumers who buy from it are the ones for whom $r+q_I+\gamma x_I^0-p_I \geq 0$. Therefore demand for the monopolist's product equals $A+q_I+\gamma x_I^0-p_I$. Its revenue (R) equals $p_I(A+q_I+\gamma x_I^0-p_I)$. The optimal price for selling its output is thus given by:

$$\partial R/\partial p_I = A + q_I + \gamma x_I^0 - 2p_I = 0$$

or

$$p_I^M = x_I^M = (A + q_I + \gamma x_I^0)/2. \tag{A.1}$$

The monopolist's profit is therefore

$$\Pi_I^M = [(A + q_I + \gamma x_I^0)/2]^2 + f_1(k_1 + x_I^0 - f_I).$$

Π_I^M is concave in f_I.
Therefore

$$\partial \Pi_I^M/\partial f_I = (A + (1+\gamma)x_I^0 + k - f_I)/2 + k + x_I^0 - 2f_I = 0$$

which gives us the optimal f_I:

$$f_I^M = (k + (1-\gamma)x_I^0 - A)/3. \tag{A.2}$$

If $\gamma > 1$ we see that $f_I{}^M$ is decreasing in $x_I{}^\circ$ whereas $\gamma < 1$, $f_I{}^M$ is increasing in x_I^0.

Duopoly: With incumbent and potential entrant

We saw in Section 8.3 that when the firms sell positive amounts of output, the equilibrium price equals the equilibrium output for each firm. Thus the first component of a firm's profit $p_i x_i$ just equals $(x_i)^2$ in equilibrium.

The same is true if a firm sells zero output (since then $p_i x_i = (x_i)^2 = 0$) or if a firm sells its monopoly output (as previously analysed). Thus a firm's equilibrium profits can be written as:

$$\Pi_i = [x_i(f_i, f_j)]^2 + f_i m(f_i, x_i^0), \quad i \in \{I,E\}. \tag{A.3}$$

Assuming no constraints on the x's and the f's, making use of equations (1)–(3) and substituting, the profits can be written as:

$$\Pi_i = [(A + k + 2(1+\gamma)x_i^0 - (1+\gamma)x_j^0 + f_j - 2f_i/3)]^2 + f_i(k + x_i^0 - f_i), \quad i \in \{I,E\}. \tag{A.4}$$

It can be checked that Π_i in the above equation is concave in f_i. *Setting* $\partial\Pi_i/\partial f_i = 0$ *gives us the Cournot reaction function in licence fees for firm i.* $j \in \{I,E\}$:

$$f_i = k/2 + (2(1+\gamma)x_j^0/5 + (1-8\gamma)x_i^0/10 - 2A/5 - 2f_j/5. \tag{A.5}$$

We note from equation (A.5) that the reaction functions in licence fees are downward-sloping. The reason behind this is as follows: from equation (A.3) we see that a firm's equilibrium profit has two components, – $[x_i(f_i, f_j)]^2$ which represents the revenues from direct sales and $f_i m(f_i, x_i^{\circ})$ which is the revenue from licensing. Firm i's reaction function gives us the optimal f_i for every f_j that the other firm might charge.

Assuming that both firms sell positive amounts, increasing f_i by a small amount decreases the profit from the first term (since from the Cournot game, x_i is decreasing in f_i). At the optimal f_i this must be exactly offset by the increase in revenue from licensing. Again, given that both firms sell positive amounts, the equilibrium x_i is increasing in f_j whereas firm i's revenue from licensing is unaffected by it. Thus if we now increase f_j, then a small increase in f_i from its previous optimal level increases the revenue from the licence fees by the same amount as before. However, the decrease in revenue from direct sales is larger since x_i is larger. Thus with the larger f_j profits are decreasing in i's licence fees at the previously optimal f_i. Given the concavity of the profit function in a firm's own licence fee this requires a lowering of f_i in order to reach equilibrium again. Hence the reaction function is downward sloping.

Lemma A1: *There exists* f_I^D *for the incumbent such that for all* $f_I < f_I^D$ *the optimal response of the entrant is to stay out of the market and for* $f_I > f_I^D$ *it is to enter.* f_I^D *is thus the largest licence fee for the incumbent that deters entry.*

Proof. Let f_I^D be the licence fee for which $f_E=0$ is the smallest licence fee that makes $x_E(f_E,f_I^D)=0$. When $f_E=0$, $x_E=0$, and $f_I=f_I^D$ then $\Pi_E=0$ and $\partial\Pi_E/\partial f_E=0+k>0$. Thus reducing f_E will only make Π_E negative while a bigger f_E is not an acceptable licence fee. So entry is deterred at f_I^D.

To calculate f_I^D we set x_E equal to zero and $f_E=0$ with it.

We then get

$$A+2(k-f_E)-(k+x_I^0-f_I^D)-\gamma x_I^0=0. \tag{A.6}$$

Therefore,

$$f_I^D=(1+\gamma)x_I^0-A-k. \tag{A.7}$$

Entry is deterred at any $f_I<f_I^D$ too, since then the smallest f_E that makes $x_E(f_E,f_I)=0$ must be negative, see (A.6). For $f_I>f_I^D$ the smallest f_E that makes $x_E(f_E,f_I)=0$ is positive, thus there do exist acceptable licence fees that give the entrant positive profits. Hence the entrant will enter and so there is no entry deterrence. Thus f_I^D is indeed the largest licence fee that the incumbent can charge and still deter entry.

Lemma A2: *There exists* $f_I^R>f_I^D$*, such that in the interval* $[f_I^D, f_I^R]$ *it is optimal for firm* E *to charge the largest acceptable licence fee* f_E *that makes* $x_E=0$ *and for* $f_I>f_I^R$ *the optimal response for firm* E *induces it to produce a positive amount.*

Proof. From equation (A.5) we know that the best response for the entrant to f_I^R *is* given by

$$2f_E=k+(4(1+\gamma)x_I^0/5-4A/5-4f_I^R/5. \tag{A.8}$$

For x_E to equal zero when $f_I=f_I^R$, f_E must be as given by (A.6) with f_I^R replacing f_I^D,

$$i.e.\quad 2f_E=A+k-(1+\gamma)x_I^0+f_I^R. \tag{A.9}$$

If there were no restrictions on the x's, then when firm *I* chose f_I^R and firm *E* chose its licence fee optimally, by definition of f_I^R, in the subsequent output game firm *E* would end up selling nothing. If firm *I* chose $f_I>f_I^R$ and firm *E* chose its licence fee optimally, then in the subsequent output game firm *E* would end up selling a positive amount. If, however, firm *I* chose $f_I<f_I^R$ and firm *E* chose its licence fee according to its reaction function, then in the subsequent output game firm *E* would be required to sell a negative amount. Since that is not possible *E* just sets its licence fee at the level that just makes its output in the next stage zero. Hence when there

is no entry block, in the interval $[f_I{}^D, f_I{}^R]$ firm E charges the largest acceptable licence fee f_E that makes $x_E = 0$. Firm I makes profits according to the monopolist's profit function. This is where the inequality $x_i \geq 0$ for the entrant is binding.

For $f_I > f_I{}^R$ the entrant just plays its unconstrained best response in the Stackelberg game. $f_I{}^R$, is the smallest licence fee for the incumbent above which the entrant plays its unconstrained optimal fees and output in equilibrium.

Proof of Proposition 8.2

We can prove the theorem in a few steps.

Lemma A3: *Entry is blocked if* $(1+2\gamma)\ x_I^0 \geq A + 2k$.

We find the smallest f_E that gives us $x_E(f_E, f_I^M) = 0$

From (A.6) we get:

$$A + 2(k - f_E) - (k + x_I^0 - f_I{}^M) - \gamma x_I^0 = 0.$$

Substituting the expression for $f_I{}^M$ from equation (A.2) into the above gives us:

$$f_E^* = A + 2k - (1 + 2\gamma)x_I^0/3.$$

f_E^* *is* the largest acceptable licence fee for the entrant when the incumbent charges f_I^M.

When $f_E^* \leq 0$, Π_E at f_E^* is non-positive (when $f_E^* \geq 0$, Π_E at f_E^* is non-negative).

Further, $\partial \Pi_E / \partial f_E = k - 2f_E^* > 0$.

Given the concavity of Π_E in f_E means that if $f_E^* \leq 0$, there is no acceptable licence fee for the entrant that gives it a positive profit. Hence entry is blocked if $f_E^* \leq 0$, i.e. if

$$A + 2k < (1 + 2\gamma)x_I^0.$$

$f_E^* = 0$ is the largest f_E^* for which entry is blocked since if $f_E^* > 0$ then there does exist an acceptable licence fee for the entrant that gives it positive profit.

Lemma A4: *For* $A + 2k \geq (1 + 2\gamma)\,x_I^0 \geq A + k2$, *entry is restricted with the incumbent charging* $f_I{}^M$ *and earning its monopoly profits.*

We saw earlier that in the interval $\left[f_I{}^D, f_I{}^R\right]$ firm E charges the smallest f_E that makes $x_E = 0$, firm I makes profits according to the monopolist's profit function.

We know that the incumbent makes the largest possible profits when it is a monopolist charging ${f_I}^M$.

Thus if, when there is no blocked entry, ${f_I}^M \leq {f_I}^R$ then the optimal licence fee for the incumbent is just ${f_I}^M$.

$$f_I^M \leq f_I^R \Leftrightarrow k + (1-\gamma)x_I^0 - A/3 \leq (1+\gamma)x_I^0 - A$$

which gives us the condition

$$k + 2A < 2(1+\gamma)x_I^0.$$

Putting together this condition with that for no blocked entry ${f_I}^0 > {f_I}^M$ (Lemma A3) gives us the required result:

$$A + k/2 \leq (1+2\gamma)x_I^0 \leq A + 2k.$$

Lemma A5: *For* $\mathrm{A} + \mathrm{k}/2 > (1+2\gamma)\mathrm{x}_I^0 \geq 2\mathrm{A}/5 + \mathrm{k}/5$, *entry is restricted with the incumbent charging* $\mathrm{f_I}^{\mathrm{R}}$ *and earning less than its monopoly profits.*

When $k + 2A > 2\,(1+\gamma)\,x_I^0$, then the optimal monopoly licence fee, f_I^M, is bigger than f_I^R (Lemma A4). Since the incumbent's profits in the range $[f_I^E, f_I^R]$ coincide with its monopoly profits Π_I^M, its profits must be increasing in $[f_I^D, f_I^R]$.

Call Π_I^S unconstrained Stackelberg profits for firm *I*. We get Π_I^S by substituting f_E from firm *E*'s reaction function (A.5) into firm's *I* profit function (A.4):

$$\Pi_I^S = \left[A/5 + k/2 + 4(1+\gamma)x_I^0/5 - 4f_I/5\right]^2 + f_I(k + x_I^0 - f_I)$$

with the incumbent charging f_I^R and earning less than its monopoly profits. Then, for $f_I > f_I^R$ the incumbent's profits coincide with Π_I^S, and at $f_I = f_I^R$, $\Pi_I^M = \Pi_I^S$ (since x_E in the output game equals zero when $f_I = f_I^R$ and *E* chooses its best response to f_I^R, thus making $\Pi_I^M = \Pi_I^S$). It can be checked that Π_I^S is concave in f_I.

If Π_I^S is decreasing in f_I at $f_I = f_I^R$ then given the concavity of Π_I^S, the optimal licence fee for the incumbent must be f_I^R:

$$\partial\Pi_I^S/\partial f_I = -8/5\left[A/5 + k/2 + 4(1+\gamma)x_I^0/5 - 4f_I/5\right] + k + x_I^0 - 2f_I. \quad \text{(A.10)}$$

When $f_I = f_I^R = (1+\gamma)x_I^0 - A$,

$$\partial\Pi_I^S/\partial f_I = 1/5\left[k + 2A - 5(1+2\gamma)x_I^0\right].$$

Π_I^S is decreasing in f_I at $f_I = f_I^R$ if

$$k + 2A \leq 5(1+2\gamma)x_I^0.$$

Thus for $2A/5 + k/5 \leq (1+2\gamma)x_I^0 \leq A + k/2$, the optimal f_I is f_I^R and the incumbent earns less than monopoly profits.

Lemma A6: *For* $2\text{A}/5 + \text{k}/5 > (1+2\gamma)\ x_I^0$, *entry is unrestricted and the firms just charge the unconstrained Stackelberg licence fees.*

If $k + 2A > (1+2\gamma)\ x_I^0$ then we infer from Lemmas A2 to A3 that there is no blocked entry, the incumbent's profits are increasing in the range $[f_I^D, f_I^R]$, and are also increasing at $f_I = f_I^R$. The incumbent's profits are continuous in the relevant range of $f_I \geq f_I^D$ and coincide with Π_I^S for $f_I \geq f_I^R$. Since Π_I^S is concave, the optimal licence fee for the incumbent must be given by the one that maximizes Π_I^S. Call this licence f_I^S. Setting $\partial\Pi_I^S/\partial f_I = 0$ from (A.10) we get

$$f_I^S = \left[5k - 8A - (7+32\gamma)x_I^0\right]/18. \tag{A.11}$$

Given our definition of, f_I^R, $f_I \geq f_I^R$ is optimal for $x_E > 0$.

Thus if $k + 2A > 5(1+2\gamma)x_I^0$, then the incumbent charges f_I^S, the entrant charges f_E given by its reaction function, and both firms produce positive amounts.

Proof of Corollary 8.2

From Lemmas A2 and A3 we infer that for $A + k/2 \leq (1+2\gamma)x_I^0$, the incumbent charges its optimal monopoly licence fee, f_I^M. Hence the statement of the corollary is true in this case. From Lemma A4, when $2\text{A}/5 + k/5 \leq (1+2\gamma)\,x_I \leq \text{A} + k/2$, the optimal f_I is ${f_I}^R$. But in this scenario monopoly profits are increasing in the interval $[{f_I}^D, {f_I}^R]$ as we saw in Lemma A4. Thus it must be the case that ${f_I}^M > {f_I}^R = {f_I}^*$. Finally, when $k + 2\text{A} > 5(1+2\gamma)\ x_I^0$, the optimal licence fee for the incumbent is given by f_I^S in equation (A.11).

$${f_I}^S - {f_I}^M = -\left[2A + k + 13(1+2\gamma)x_I^0\right]/18 < 0.$$

Thus when $k + 2\text{A} > 5(1+2\gamma)\ x_I^0$, ${f_I}^M > {f_I}^S = f_I^*$.

Proof of Proposition 8.3

It can be checked that in the cases when there is unrestricted entry and when there is restricted entry without monopoly output, $f_I < f_E$. When there is restricted entry with monopoly profits

$$f_I - f_E = \left[-k - 2A + (2+\gamma)x_I^0\right]/3.$$

Thus, $f_I < f_E$ if $k + 2\text{A} > (2+\gamma)\ x_I^0$,

$$f_I > f_E \quad \text{if } k + 2A < (2+\gamma)x_I^0.$$

Hence, $k + 2\text{A} = (2+\gamma)\ x_I^0$.

9 Technological competition with product differentiation

> Intense rivalry among potential innovators could result in overinvestment in innovative activity as compared to the 'social optimum'.
>
> Y. Barzel (1968)

9.1 Introduction

Asymmetries between firms are frequently observed in market competition as, for example, in the case that one firm has a more advanced technology. If there exists some difference between firms, it will certainly affect the incentives to engage in innovative activity. The asymmetries between those firms will generate different incentives to introduce new technology. Since the outcome of technological competition has an obvious effect upon the firms' market positions, firms need to take account not only of the immediate profits that it brings but also the advantage that it may confer in subsequent competition when later innovations are expected. To explore this we construct a two-period model and ask how the existence of such subsequent innovative opportunities affects the outcome of technological competition.

The analysis uncovers two different forces on determining the incentives to innovate. The competitive incentive is defined as the difference between a firm's profits if it innovates and the profits it would make if its rival innovated instead. The profit incentive is determined by calculating the increase in a firm's profits if it alone were investing in R&D. Given those forces, we explore how the incentives to innovate depend on the degree of product differentiation or niche building in product markets. If the products are less substitutable, the profit incentive dominates each firm's willingness to invest in R&D and leads to the result that the currently less-advanced firm engages in more R&D. If the products are very substitutable, then the competitive incentive determines the outcome of technological competition so that the currently more advanced firm increases its superiority in technology. Hence, the rate of product differentiation is shown to be an important determinant predicting the outcome of technological competition. The existence of a second period of competition largely reinforces the outcome from a single period race.

An alternative outcome could be expected with technological competition when firms can share both the cost and the outcome of innovative activity. The majority of models consider the case that there is only one winner in the technological competition. But alternative models develop technological competition in which each firm engaging in R&D can obtain a patent on a cost-reducing technology. Since innovative investments are often very expensive and knowledge has a tendency to leak to others, firms may have an incentive to conduct their innovative activities cooperatively.

Cooperation in R&D among competing firms can produce more rapid technological progress. Since the new technology is shared among competitors through cooperative R&D, firms can avoid the duplication of R&D. The cooperation in R&D can also be a solution to the leakage of the information. Since knowledge is inherently a public good, the research done by one firm can be used by another firm even though the latter does not have permission to use the inventive output. Cooperation in R&D is a solution to this problem because cooperative research agreements can internalize spillover externalities. However, cooperation between firms' R&D activities may not always produce a better social outcome. The potential gain from R&D will be dissipated to consumers if price competition in the product market is intense. This creates the potential collusive reduction in R&D. That is, firms could use a cooperative agreement in R&D as a vehicle to slow the pace of technological innovation.

In exploring the relationship between technological competition and product differentiation we ask how the rate of product substitutability influences technological progress. The number of firms in the market determines the intensity of price competition. However, market competition is also affected by the degree of product differentiation. When the products in the market are less substitutable, the participants in the market will experience less competition. Instead of asking how the incentive to innovate is affected by the increase in the number of firms in the market, which is a common theory in analysing the relation between R&D and product differentiation (Suzumura, 1995, Chapter 4), we consider the question of how the incentive to innovate is related to the degree of product differentiation.

When two products are close substitutes, technological competition is intensified and it drives the market to be monopolized by a firm. If two products are less substitutable, the firms have greater monopolistic power over customers. This discourages a firm from stealing its rival firm's customers through innovation and the market keeps its competitive structure unchanged. It also turns out that the effect of a spillover externality on technological progress is dependent on the closeness of the two products. As stated earlier, knowledge is inherently a public good and it is difficult to prevent its use by others. The analysis shows that the incentives to innovate are generally reduced as the rate of such leakage of knowledge increases. However, when products are very different, the firm that has less of an incentive to innovate has a tendency to raise its R&D effort as the spillover rate increases. The increase in the spillover rate helps the firm that has less of an incentive to improve its marginal contribution of technological progress and compensates for the reduction in its own competence.

Before reviewing previous lines of thought on the interaction between technological competition and market structure, in Section 9.2, we present a model of a horizontally differentiated product market in Section 9.3. Section 9.4 examines the conditions that influence the pattern of market evolution. It shows that the pattern of market evolution depends on the rate of product differentiation. In Section 9.5 we suggest how the dynamic effect of technological competition on market evolution will be different from the static one. Section 9.6 looks at research joint ventures (RJVs) in competitive markets. Section 9.7 concludes.

9.2 A review of technological competition and market structure

When firms are not identical, there is no reason to suppose that their incentives to innovate will be same. We first note the role of each individual firm's market position on the incentive to innovate. The question about the advantage or disadvantage of current market dominance on technological competition has been one of the main themes in empirical industrial economics. We ask if the firm which is already a leader has a greater incentive to perform R&D than its rival. We analyse it in the context of an oligopolistic market and explore the role of product differentiation on the development of market structure.

Although market structure may influence innovative activity, it is itself influenced by technological progress. The outcome of one patent race will change each firm's market position and affect the incentives in the next race. When a sequence of innovations is expected, the existence of such reciprocal effects requires a firm to take account not only of the immediate profits that an innovation brings, but also the strategic advantage or disadvantage it may confer in subsequent races. Hence, it is worthwhile analysing the strategic incentives to innovate in the context of a series of innovative opportunities. For this purpose, we extend the static analysis to a two-period model.

After Schumpeter (1942) emphasized the role of technology on economic growth and addressed the importance in driving the economic progress of large, oligopolistic firms, a traditional line of reasoning has been that monopoly power is a stimulus to R&D. Gilbert and Newberry (1982) (see also Dasgupta (1986)), proved that, when an incumbent monopolist is faced with potential entrants, the monopolist's incentive to remain a monopolist is greater than the entrant's incentive to become a duopolist. However, Arrow (1962) sought to establish the reverse proposition that more competitive environments would give a greater incentive. He considers the gains to the originator of an innovation, in monopoly and competition. For a drastic innovation, which leaves the inventor a monopolist, an incumbent monopolist would have less of an incentive than would a competitive firm that currently has no share in the market because the competitive firm becomes a monopoly whereas the monopolist only replaces itself. An opposite implication comes from two different forces that determine the economic profit through innovation. The firm in Gilbert and Newberry recognizes that it is in a race against its rival to be the first to innovate and so the incentive for R&D reflects the difference between the firm's profits if it innovates and the profits it would make if its rival innovated first. Because there is some dissipation of rents when the market is shared rather than monopolized, the incumbent monopolist has a greater incentive than a potential entrant. The incentive on which Arrow focuses is determined by calculating the increase in a firm's profits if it alone were investing in R&D. The incentive to innovate is independent of any strategic considerations of pre-emptive innovation. This leads to the conclusion that a competitive market structure is more conducive to innovation than a monopolistic one because the superior position from a competitive market position provides a higher return from owning the exclusive right to a new technology. The first incentive, which we owe to Beath *et al.* (1987, 1989), is called the competitive incentive, and the latter the profit incentive. Most technological competition will need to take account of those two forces on the incentive to innovate. However, both Gilbert and Newberry and Arrow show very extreme cases of those two incentives by limiting their analysis to either a monopolistic or perfectly competitive market. By introducing a differentiated products market, it appears that the results from Gilbert and Newberry and Arrow are really extreme cases rather than a mainstream result of technological competition.

Empirical studies have also investigated the relationship between technological change and market structure. They examine how a technological improvement influences market concentration. The studies have presented mixed results. In some industries, the large institutions are the major innovators. Phillips (1966, 1971) examined his theory with data on 11 industry groups and later on the aircraft industry. He found that the relative success in innovation was an important determinant of the growth of firms and of growing concentration. Mansfield (1984) on the steel and petroleum industries and Grabowski and Mueller (1978) lend support to the results of Phillips. However, there is also considerable evidence that dominant companies are slow innovators. Blair (1972), Geroski and Pomroy (1990), and Carlsson (1984) provide evidence in support of this perspective. There is no absolutely unambiguous answer. These results require theoretical models to aid in studying patterns of market evolution. Our analysis shows that the rate of product differentiation plays an important role.

We construct an oligopolistic model in which a differentiated product market is assumed. We start our analysis by drawing upon Vickers (1986). He considers a dynamic model in which two firms compete for a series of non-drastic innovations. Since his approach for computing the incentive for R&D is defined in terms of value functions, we can apply it in our analysis to predict which firm will win a patent. Our analysis first explains two patterns of market evolution. It reveals that each firm's desire for R&D is largely determined by the profit incentive when the rate of product substitutability is low, while technological competition is influenced by the competitive incentive when two products are close to perfect substitutes. The rate of product differentiation plays an important role in determining the relative magnitude of those two forces and the opposite conclusions in technological competition is derived by varying the rate of product differentiation. We also compare the static analysis to the results of a two-period model. It shows the difference which can arise between them and improves our understandings of the effect of continuous innovations on market evolution.

9.3 A simple model

We assume an economy with two firms, each producing a differentiated good. There are two firms, h and w. Their initial technologies are represented by constant marginal production costs s_{h} and s_{w}. The first period innovation provides the cost-cutting technology s_1. If allowed, the second period innovation offers the cost-cutting technology s. Here, technological progress is limited to be the cost-cutting technology, the so-called process innovation. The firm that acquires new technology s_1 or s can use it exclusively and produces its output at the cost of s_1 or s, respectively. It is assumed, without loss of generality, that $s_{\mathrm{h}} > s_{\mathrm{w}} > s_1 > s$. At the beginning of each period, they compete for new technology s_1 or s. Their competition for new technology is like a simple auction in which each firm proposes a bid which represents the maximum amount that firm will pay for the patent of the new technology and the highest bidder wins it. The winner pays – as expense for buying the new technology – the maximum that the other firm would have paid in order to have the patent rather than not to have it. The loser does not owe anything.

The product market is specified by a differentiated duopoly of the Dixit (1979) type with linearity in demands assumed as follows:

$$q_{\mathrm{h}} = a - bp_{\mathrm{h}} + cp_{\mathrm{w}}$$

$$q_{\mathrm{w}} = a - bp_{\mathrm{w}} + cp_{\mathrm{h}}$$

where q_i represents the demand for firm i and p_i its price. It is further assumed that

$$a > 0, b > c \geq 0 \tag{9.1}$$

and

$$q_h = a - bs_h + cs \geq 0. \tag{9.2}$$

The goods are independent when $c=0$, while they are substitutes when $c > 0$. When $b=c$, the goods are perfect substitutes. The assumption (9.2) implies that both firms are guaranteed to be active even after innovation.

After the race for new technology s_1 or s, the firms compete in product market where the Bertrand competition is assumed. They set prices and the non-cooperative Nash equilibrium concept is adopted as the equilibrium in the product market.

9.4 First period competition

At the beginning of the first period, there is a competition for the patent that gives the exclusive right to employ the cost level s_1 in the given period. Each firm's initial unit cost is denoted by s_h and s_w, respectively. They are myopic and do not form expectations for the next period. Each firm competes for the patent and proposes a bid for it. Then, we ask who is willing to bid more and examine how the rate of product differentiation influences the market development.

In the product market, firm i chooses its price level p_i to maximize its profit, taking p_j as given, where $i \neq j$ and $i, j = \text{h, w}$, respectively. Let Π_h (s_h, s_w) denote firm h's profit level, ignoring R&D expenses, when it has the cost level s_h and its rival has the cost level s_w. Similarly, Π_w (s_w, s_h) is defined for firm w when it has the cost level s_w and its rival has the cost level s_h. Then it is easily seen that provided that q_j is positive, the reaction function for firm i is

$$p_i(p_j) = (a + bs_i + cp_i)/2b, \quad \text{where } i, j = \text{h or w and } i \neq j. \tag{9.3}$$

It is straightforward to compute the equilibrium price from the reaction function. Since the slope of the reaction function is positive but less than one, the unique interior solution can be derived as

$$p_i^*(s_i, s_j) = \left(a(2b + c) + 2b^2 s_i + bcs_j\right)/\left(4b^2 - c^2\right), \quad \text{where } i, j = \text{h or w and } i \neq j. \tag{9.4}$$

Let g_h and g_w denote firms h's and w's incentive to win the new technology s_1, respectively. Then, firm h's incentive is given by

$$g_h(s_h, s_w, s_1) = \pi_h(s_1, s_w) - \pi_h(s_h, s_1).$$

The first term on the right-hand side (rhs) is firm h's payoff after it wins the race and the second term is its payoff when it loses the race. Similarly, firm w's incentive to win the new technology s_1 is given by

$$g_w(s_w, s_h, s_1) = \pi_w(s_1, s_h) - \pi_w(s_w, s_1).$$

From the assumption that the competition for the new technology s_1 is like a bidding game, firm h wins if and only if $g_h\,(s_h, s_w, s_1) > g_w\,(s_w, s_h, s_1)$. Let $\rho(s_h, s_w, s_1)$ denote the difference between two firm's incentives. That is,

$$\rho(s_h, s_w, s_1) \equiv g_h(s_h, s_w, s_1) - g_w(s_w, s_h, s_1).$$

Following the definition of each term on the rhs we get

$$\rho(s_h, s_w, s_1) = \pi_h(s_1, s_w) - \pi_h(s_h, s_1) - \pi_w(s_1, s_h) + \pi_w(s_w, s_1).$$

Define $\sigma\,(s_h, s_w) \equiv \Pi_h\,(s_h, s_w) + \Pi_w\,(s_w, s_h)$.
Then, the function $\rho(s_h, s_w, s_1)$ can be rewritten as

$$\rho(s_h, s_w, s_1) = \sigma(s_1, s_w) - \sigma(s_h, s_1). \tag{9.5}$$

By substituting the solution in (9.4) into each firm's profit function, the difference between the two firms' incentives can be calculated as following:

$$\rho(s_h, s_w, s_1) = b(s_w - s_h)/\left(4b^2 - c^2\right)^2 R(s_h, s_w, s_1) \tag{9.6}$$

where

$$R(s_h, s_w, s_1) \equiv (s_h + s_w)\{\left(2b^2 - c^2\right)^2 + b^2c^2\}$$
$$- 4bc\left(2b^2 - c^2\right)s_1 - 2a(b - c)(2b + c)^2.$$

From the relation in (9.6), we can find the answers to the question of how the market will evolve. If $\rho(s_h, s_w, s_1) > 0$, firm h will have a greater incentive to win the race for s_1.

Proposition 9.1: *Suppose that* $s_h > s_w > s_l$. *Then, there exists a unique* $c(s_h,$ $s_w, s_l)$ *such that*

$$\rho(s_h, s_w, s_1) > 0 \quad \text{if} \quad c \in [0, c^*(s_h, s_w, s_1)] \quad \text{and}$$
$$\rho(s_h, s_w, s_1) \leq 0 \quad \text{if} \quad c \in [c^*(s_h, s_w, s_1), b].$$

Proof. See Appendix to this chapter.

This proposition shows the effect of product differentiation on market development. It implies that the current low-cost firm will have a greater incentive to innovate than its rival firm if the rate of product differentiation is high, while the current high-cost firm will win the new technology race when the rate is low. When the products are close substitutes, the competition in the product market is so intense that the willingness to buy a new patent is determined by competitive incentive. Since a low-cost firm's profit is greater as the difference between the production costs increases, the current low-cost firm has a greater incentive to innovate. When $c = 0$, each firm is a local monopoly and its desire for R&D is determined by the profit incentive. If the high-cost firm wins the new patent, this will bring a larger cost cut and so a greater incentive to innovate. Proposition 9.1 shows that the force determining the incentive to innovate gradually moves from the profit incentive to the competitive incentive as the rate of product differentiation increases.

Interpretation of Proposition 9.1 helps us understand the debate regarding the better market structure for technological progress. The Schumpeterian hypothesis can be interpreted as a case that the two products are close substitutes, while Arrow's conclusion can be applied to the case when the two products are close to being monopolies.

From the given proposition, we can infer the effect on technological competition. We ask how the race will be dependent on the value of s_1.

Proposition 9.2: *Suppose that* $s_h > s_w > s_l$.

Then, $\partial c^*(s_h, s_w, s_l)/\partial s_l > 0$ *and* $\partial p(s_h, s_w, s_l)/\partial s_l > 0$.

Proof. See Appendix to this chapter.

The comparative statics in this proposition indicates that the critical rate of product differentiation, $c(s_h, s_w, s_1)$, decreases as the size of the innovation increases. This result comes from the fact that the current low-cost firm's incentive to innovate increases more rapidly than its rival firm's incentive as the technological improvement becomes larger. For example, suppose the new technology provides only a small cost reduction. Then, the improvement of the low-cost firm's profit may be trivial while the high-cost firm will still get a considerable reward from it. However, suppose that technological progress is quite significant. Then, the reward to the low-cost firm will not be trivial and the increase of its incentive will be relatively greater than the current high-cost firm. Therefore, as technological improvement is greater, the firm that is currently ahead in technology becomes more active in technological competition.

9.5 Inter-period competition

The asymmetry between firms' market positions generates the difference between their incentives to innovate. Through a patent race, the firm that is already ahead may become increasingly dominant or its rival may acquire the superiority. Technological competition will affect each firm's market position. The relationship between market structure and innovation is not one-directional. Therefore, it will be more desirable to analyse the relationship between market structure and innovation in a dynamic context. Here, we examine how far the results in the previous section carry over when the analysis is extended to a two-period model.

At the beginning of each period, two firms compete for the new technology. The first-period innovation provides the cost-cutting technology s_1. Its winner will produce its output at the cost of s_1. The second-period innovation offers the cost-cutting technology s. Its holder can produce its output at the cost of s. Those technologies are put up for auction. We assume symmetric demand functions for firms h and w, and make the same assumptions about the parameters as in the previous section. Again, the equilibrium concept in the product market is the non-cooperative Nash equilibrium.

To examine the magnitude of each firm's bids and the sequence of technology allocations, we need to understand how profits are determined in each period. Since two firms compete over two periods, each firm will consider not only the current profit but also the next period's profit at the time of the first round patent race. We start by calculating each firm's profit in the second period.

There are two possible outcomes in the first period; firm w won the patent of s_1 or firm h won it. Depending on these two outcomes, the second-period profit will be determined as a result of the second-period race. As we proved in the previous section, the winner of the new technology in one period competition is characterized in Proposition 1.1.

When firm h wins the first-period race, the critical rate of product differentiation is $c^*(s_w, s_1, s)$ in the second-period race, while it is replaced by $c^*(s_h, s_1, s)$ when firm w wins the new patent s_1. Since $s_w < s_h$, $c^*(s_w, s_1, s) > c^*(s_h, s_1, s)$.

Then, we consider the following three cases:

(i) $0 \leq c \leq c^*(s_h, s_1, s)$
(ii) $c^*(s_h, s_1, s) < c \leq c^*(s_w, s_1, s)$
(iii) $c^*(s_w, s_1, s) < c < b$.

Depending on the degree of product differentiation, the following three patterns are expected in the second period: the winner of s_1 loses the second-period race as in case (i), the winner in the second period is conditional on who won the first-period race as in case (ii), or the winner in the first period gets the new patent in the second period as in case (iii).

Now, each firm's incentive in the first period can be calculated by counting those results. Let g_h^0 (s_h, s_w, s_1, s) and g_w^0 (s_w, s_h, s_1, s) represent firms h's and firm w's incentive for the new technology s_1, respectively. In case (i), the incentive to innovate in the first period can be calculated as follows:

$$g_h^0(s_h, s_w, s_1, s) = \pi_h(s_1, s_w) + \pi_h(s_1, s) \\ - \pi_h(s_h, s_1) - \pi_h(s, s_1) + \pi_w(s, s_h) - \pi_w(s_1, s).$$

The first two terms on the rhs are firm h's payoff over two periods after it wins the race s_1 and the other terms are its profits over two periods minus its rival's bid for the race s when it loses the first-period race. Similarly, firm w's incentive can be computed. Then, they can be written:

$$g_{\mathrm{h}}^{0}(s_{\mathrm{h}}, s_{\mathrm{w}}, s_1, s) = \pi_{\mathrm{h}}(s_1, s_{\mathrm{w}}) + \pi_{\mathrm{w}}(s, s_{\mathrm{h}}) - \pi_{\mathrm{h}}(s_{\mathrm{h}}, s_1) - \pi_{\mathrm{h}}(s, s_1)$$

and

$$g_{\mathrm{w}}^{0}(s_{\mathrm{w}}, s_{\mathrm{h}}, s_1, s) = \pi_{\mathrm{w}}(s_1, s_{\mathrm{h}}) + \pi_{\mathrm{h}}(s, s_{\mathrm{w}}) - \pi_{\mathrm{w}}(s_{\mathrm{w}}, s_1) - \pi_{\mathrm{w}}(s, s_1).$$

Each function is computed as the difference between each firm's profit when it wins the race s_1 and the profit when it loses the first period race. If a firm wins the race s_1, it will lose the second-period race, while it can buy the patent s by paying its rival's bid when it loses the first-period race. Define

$$\phi(s_{\mathrm{h}}, s_{\mathrm{w}}, s_1, s) \equiv g_{\mathrm{h}}^{0}(s_{\mathrm{h}}, s_{\mathrm{w}}, s_1, s) - g_{\mathrm{w}}^{0}(s_{\mathrm{w}}, s_{\mathrm{h}}, s_1, s).$$

Then, $\phi(s_{\mathrm{h}}, s_{\mathrm{w}}, s_1, s)$ can be written:

$$\phi(s_{\mathrm{h}}, s_{\mathrm{w}}, s_1, s) \equiv \sigma(s_1, s_{\mathrm{w}}) - \sigma(s_1, s_{\mathrm{h}}) + \pi_{\mathrm{w}}(s, s_{\mathrm{h}}) - \pi_{\mathrm{h}}(s, s_{\mathrm{w}})$$

where $\sigma\ (s_{\mathrm{h}}, s_{\mathrm{w}}) \equiv \Pi_{\mathrm{h}}\ (s_{\mathrm{h}}, s_{\mathrm{w}}) + \Pi_{\mathrm{w}}\ (s_{\mathrm{w}}, s_{\mathrm{h}})$.

Note that the first two terms on the rhs coincide with $\rho(s_{\mathrm{h}}, s_{\mathrm{w}}, s_1)$ in (9.5). They are exactly the same as the difference between two firms' incentives for the race s_1 when they are myopic. The other two terms are the difference of payments for the new patent s.

In case (ii), firm w always wins the second-period race because $c^*\ (s_h, s_1, s) < c \leq c^*\ (s_w, s_1, s)$. *In the second period, it can produce its product at the unit cost of s by paying* $\Pi_h\ (s, s_1) - \Pi_h\ (s_h, s)$ *if it won the race* s_1 *or paying* $\Pi_h\ (s, s_w) - \Pi_h\ (s_1, s)$ *if it lost the race* s_1. *Therefore, we get:*

$$g_{\mathrm{h}}^{0}(s_{\mathrm{h}}, s_{\mathrm{w}}, s_1, s) = \pi_{\mathrm{h}}(s_1, s_{\mathrm{w}}) + \pi_{\mathrm{h}}(s_1, s) - \pi_{\mathrm{h}}(s_{\mathrm{h}}, s_1) - \pi_{\mathrm{h}}(s_{\mathrm{h}}, s)$$

and

$$\begin{aligned} g_{\mathrm{w}}^{0}(s_{\mathrm{w}}, s_{\mathrm{h}}, s_1, s) = {} & \pi_{\mathrm{w}}(s_1, s_{\mathrm{h}}) + \sigma(s_{\mathrm{h}}, s) + \pi_{\mathrm{h}}(s, s_{\mathrm{w}}) \\ & - \pi_{\mathrm{h}}(s, s_1) - \pi_{\mathrm{w}}(s_{\mathrm{w}}, s_1) - \pi_{\mathrm{w}}(s, s_1) - \pi_{\mathrm{h}}(s_1, s). \end{aligned}$$

Therefore,

$$\begin{aligned} \phi(s_{\mathrm{h}}, s_{\mathrm{w}}, s_1, s) = {} & \sigma(s_1, s_{\mathrm{w}}) - \sigma(s_1, s_{\mathrm{h}}) + \sigma(s, s_1) \\ & - \sigma(s, s_{\mathrm{h}}) + \sigma(s_1, s) - \pi_{\mathrm{h}}(s_{\mathrm{h}}, s) - \pi_{\mathrm{h}}(s, s_{\mathrm{w}}). \end{aligned}$$

The first two terms on the rhs coincide with $\rho(s_h, s_w, s_1)$ in (9.5) and the next two terms represent the difference of the incentives for the race s when the firms are myopic. The last two terms are the difference of firm w's payments for the second-period race s.

Similarly, the difference of two firms' incentives in case (iii) is calculated as follows:

$$g_h^0(s_h, s_w, s_1, s) = \pi_h(s_1, s_w) + \pi_h(s, s_w) - \pi_w(s, s_1) + \pi_w(s_w, s) - \pi_h(s_h, s_1) - \pi_h(s_h, s),$$

$$g_w^0(s_w, s_h, s_1, s) = \pi_w(s_1, s_h) + \pi_w(s, s_h) - \pi_h(s, s_1) + \pi_h(s_h, s) - \pi_w(s_w, s_1) - \pi_w(s_w, s) \quad \text{and}$$

$$\phi(s_h, s_w, s_1, s) = \sigma(s_1, s_w) - \sigma(s_1, s_h) + \sigma(s, s_w) - \sigma(s, s_h) - \pi_h(s_h, s) + \pi_w(s_w, s).$$

In each case, firm h wins the race s_1 when $\phi(s_h, s_w, s_1, s) > 0$, while firm w wins when $\phi(s_h, s_w, s_1, s) \leq 0$.

By substituting the profit function in ((A1), Appendix) into $\phi(s_h, s_w, s_1, s)$, we can figure out the condition for the pattern of market development.

Proposition 9.3: *Suppose that* $s_h > s_w > s_l > s$. *Then*

1. If $0 \leq c < c^*(s_h, s_1, s)$, $\phi(s_h, s_w, s_1, s) > 0$.
2. If $c^*(s_w, s_1, s) \leq c < b$ *and* s_h *and* s_w *satisfy*

$s_h + s_w \geq \{2a(2b + c)[3(2b^2 - c^2) - 2bc(2b^2 - c^2)(2s_l + 3s)]/3(2b^2 - c^2)^2 + 2(bc)^2$,

then $\phi(s_h, s_w, s_1, s) \leq 0$.

Proof. See Appendix to this chapter.

This proposition first shows that the existence of a subsequent race strengthens each pattern of market development when the products are close to the extreme case. Suppose that the rate of product differentiation is less than $c^*(s_h, s_1, s)$ and $\sigma(s_1, s_w) - \sigma(s_1, s_h) < 0$. Unless there exists a race in the second period, Proposition 9.1 predicts that the low-cost firm increases its dominance since $\sigma(s_1, s_w) - \sigma(s_1, s_h) < 0$. However, in the two-period competition, it is possible to observe the pattern of alternating winners in those cases because the existence of the second-period race boosts the high-cost firm's incentive. When the market products are close substitutes, we can also expect similar results, but when the rate of product differentiation is moderate, it is ambiguous.

Unlike the one-period model, the technological competition in this section deals with two new technologies. We consider how the change in the improvement of each technology will affect the race.

Proposition 9.4: *Suppose that* $s_h > s_w > s_l > s$.

(1) *If* $0 < c < c^*(s_h, s_l, s)$, *then* $\partial\phi(s_h, s_w, s_l, s)/\partial s_l > 0$ $\partial\phi(s_h, s_w, s_l, s)/\partial s < 0$.
(2) *If* $c^*(s_w, s_l, s) < c < b$, *then* $\partial\phi(s_h, s_w, s_l, s)/\partial s_l > 0$ $\partial\phi(s_h, s_w, s_l, s)/\partial s > 0$.

Proof. See Appendix to this chapter.

Proposition 9.4 shows that the effect of increasing s_l is to encourage innovation by firm h, which is the same as found in Proposition 9.2. As the improvement of the immediate innovation is smaller, the low-cost firm is less attracted to it. However, the effect of s is shown to be dependent on the pattern of market evolution. When firm w knows that it will win only s, a small improvement by s will lead firm w to be more attracted to buying s_l, which will decrease $\phi(s_h, s_w, s_l, s)$. But, when firm w knows that it will win both s_l and s, a smaller improvement on the new technologies will provide relatively less interest to firm w because firm w's expected gain will be less significant than firm h's expected gain from them.

9.6 Notes on cooperative R&D in a competitive market

Inter-firm relationships in the R&D first stage are often assumed to be either cooperative or non-cooperative, while a second-stage product market is usually characterized by a non-cooperative Cournot competition. Most of the present R&D literature concentrates on the comparison of the pure cooperative case to the pure non-cooperative first stage, given Cournot competition in the product stage. Such frameworks may not be applicable to situations where either: (a) the firms' technology is determined by a mixture of independent R&D and cooperative research joint ventures (RJV), or (b) product market competition is not Cournot. A model of a joint research venture operating competitively in a differentiated product market would allow participating firms to engage in independent R&D in addition to that performed jointly.

After Schumpeter (1942) emphasized the importance of technological progress for economic growth, much attention has centred on the relationships between product market structure and technological progress. However, the relationship between market structure and technological progress is still an open question. The ambiguity regarding this relationship also results in difficulties in analysing the benefits and costs of R&D competition. It is difficult to determine whether cooperation in R&D will result in collusion in the production stage or how the concentration ratio will affect R&D investment. Instead of analysing a complete set of interactions, economists have tackled the issues related to cooperative R&D by adopting a two-stage model and concentrating their attention on the effects of R&D competition before considering the competitive interactions of firms in the product market. They explicitly or implicitly assume that symmetric firms are effectively prohibited from collusion. Grossman and Shapiro (1986) and Katz (1986) first used this approach to analyse the effects of cooperative R&D. Grossman and Shapiro adopt the upstream and downstream perspective and describe the possible benefits and costs of a research joint venture. Katz proposes a four-stage model in which R&D activities are specified by three stages. Using this model, he discusses the general effects of cooperative R&D. Both approaches emphasize the possibility that cooperation produces greater total R&D because it can internalize the spillover externality, thereby avoiding duplication of R&D efforts and thus decreasing each firm's R&D expenditure. They also note the possibility that a RJV may have an anti-competitive effect when product market competition is intensive and suggest that the negative effect of an RJV may be prevented by allowing the participating firms to engage in independent R&D.

D'Aspremont and Jaquemin (1988) develop a constructive model and prove that cooperative R&D produces greater R&D effort in equilibrium than non-cooperative R&D when the spillover rate is relatively high and the product market is modelled as a homogeneous goods Cournot competition. In this model, both firms in the cooperative R&D venture coordinate their R&D expenditure to maximize joint profits. D'Aspremont and Jacquemin show that while one might expect cooperation to lead to a reduction of R&D expenditure by avoiding wasteful duplication, in fact cooperation actually produces greater R&D when the spillover rate is relatively high. A limitation in their model is that the firms participating in the RJV do not share their R&D outcomes but rather simply coordinate their R&D spending. This naturally leads to the question of why the firms want to coordinate their R&D expenditure. Moreover, the coordination of R&D expenditure will not necessarily avoid duplication of R&D since the participants do not share the results. Kamien *et al.* (1992) extend D'Aspremont and Jacquemin's model to the case when the R&D outcome is fully shared and still conclude that an RJV results in greater R&D output than non-cooperative R&D competition. Choi (1993) generalizes the model to the case where the R&D outcome is uncertain and finds the same outcome.

Although D'Aspremont and Jacquemin, Kamien *et al.* and Choi succeed in showing the desirability of cooperative R&D, their work does not take into account some of the negative effects of cooperative R&D which were emphasized by Grossman and Shapiro and Katz. They fail to show the anti-competitive danger of cooperative R&D and the role of independent R&D (Suzumura, 1995). Additionally, they emphasize the usefulness of working with differentiated goods markets and the possibility that product market competition may affect incentives to participate in cooperative R&D. To address these issues, Kamien *et al.* adopt a horizontally differentiated goods market. But, in their analysis, it is difficult to determine how the degree of product differentiation affects the incentives to conduct R&D. Kamien *et al.*

build on the model introduced by Dixit (1979) and apply their analysis to the case in which the R&D outcome is fully shared among the participants. By assuming Cournot competition and concavity of the objective function, they show that full cooperation produces more R&D than non-cooperation. However, their specification of the R&D production function makes it difficult to compute the effect of product differentiation on R&D outcomes in that it does not allow for an analysis of the effect of changes in the parameters on R&D activity. Moreover, their model can explain neither the anti-competitive effect of cooperative R&D nor the role of independent R&D.

A distinctive characteristic of R&D is the spillover externality effect. If R&D for the single firm is not appropriable, the incentives at the industry level to do R&D are reduced. On the other hand, restoring appropriability by licensing also incorrectly prices the good that R&D has created and may result in redundant and excessive levels of R&D. A solution to the appropriability problem can be found by cooperation in R&D. By working together, the firms under the cooperative agreement can internalize the spillover externality. Moreover, cooperation makes it possible to share the burden of R&D cost, avoid duplicated R&D efforts and save transfer costs. D'Aspremont and Jacquemin, Kamien *et al.* and Choi emphasize all of these advantages in their constructive models. In addition, by assuming Cournot competition, they emphasize that cooperative R&D produces more output than non-cooperative R&D competition if the internal spillover rate increases under cooperation. That is, they show not only the advantages of cooperative R&D but also the desirability of it. However, as Grossman and Shapiro and Katz previously pointed out, there exists the risk that the firms engaging in cooperation may use a RJV as a vehicle to jointly slow the pace of technological innovation. This paper proves that those risks can arise. By assuming Bertrand competition in the product market, we show that cooperation in R&D does not always produce more output than non-cooperative competition in R&D.

Suppose that cooperative R&D activity is solely determined by an RJV, as assumed in most of the literature. Then, if product market competition is intense, the cooperative agreement will be used to restrict the level of R&D since most of the potential gains from innovation will be dissipated to consumers. That is, when firms price compete, cooperation in R&D can be used to avoid the difficulty of a Prisoner's Dilemma, whereas firms in the non-cooperative R&D game are unable to avoid such a situation. Unlike the Cournot model, the anti-competitive effect of RJV may be shown in this situation. However, the danger of an anti-competitive effect may be reduced or may disappear altogether if the participating firms can also engage in independent R&D. Suppose that the member firms can continue their independent R&D even after agreeing to a cooperative agreement. If the RJV sets investment on R&D to zero, the member firms will return to the same non-cooperative levels of investment undertaken prior to the agreement. Total R&D output in this market will not be the same as was made when the RJV solely determined R&D expenditure. The different structure of the RJV may result in different outcomes and have different implications for R&D activity. For example, one could envisage a scheme of cooperative R&D in which the member firms in the RJV can choose their R&D effort levels independently after the joint venture's decision is made; we then examine its implications on total industry R&D activity to show how different outcomes are possible.

Extending the cooperative model to allow for independent R&D requires specification of both the parents' and the joint venture's incentive structures. We consider the case in which the RJV moves first and then the parent firms choose their independent R&D effort given the RJV's decision. Then it is assumed that each member of the RJV has access to information on the joint activity level. Moreover, the manager of the RJV is given incentives by the parent firms, who subsequently decide their own responses to the decision of the RJV. This suggests a three-stage model of RJV/product market competition. In the first stage, firms have an opportunity to join a cooperative R&D venture in which the cost and output of R&D are fully shared. At the second stage, each parent firm chooses its own R&D effort level. R&D efforts are assumed to reduce unit production costs; i.e. attention is confined to process innovation. During the third stage, firms compete in the product market with collusion in the product market prohibited.

The analysis first asks how the cooperative R&D level depends on the degree of product differentiation when independent R&D is not allowed. When two goods are close substitutes, and so product competition is intense, most of the potential gains from R&D are dissipated to consumers. It is determined that the cooperative agreement may act as a vehicle to avoid such unprofitable results. When two goods are poor substitutes this result may be reversed, and cooperation may result in greater R&D output than non-cooperation. The result would be similar in the initial case using the Cournot model except that the analysis here reveals that the degree of product differentiation has an important impact on the RJV's incentives to innovate.

Alternative schemes of cooperative R&D can change R&D incentives and its social implications. Specifically, firms may spend at least as much in the cooperative stage as they would under the non-cooperative competition when RJV allows the participants to continue independent R&D. When independent R&D is allowed, both firms know that if cooperative R&D by the RJV is not large enough, they will increase their R&D investment. But, mostly, independent R&D is less efficient than cooperative R&D because it is not shared and does not avoid the duplication on R&D investment. When cooperative R&D allows room for independent R&D, firms can save R&D by agreeing to increase their cooperative investment on R&D. Given the importance of governmental policy with respect to RJVs, understanding the implications of alternative RJV schemes seems to be worth exploring.

A scheme that includes an RJV plays an important role in investment in process innovation. In fact, this analysis shows that when independent R&D is allowed, the level of R&D in the RJV is not lower than in the purely non-cooperative equilibrium. That is, if the RJV allows the parent firms to continue their independent R&D, the danger of the collusive reduction in R&D is unlikely to be serious. Even if member firms' products are close substitutes, an antitrust authority need not worry about the anti-competitive risk of RJVs as long as the RJVs allow participants to continue independent R&D.

9.7 Conclusions

In this chapter, we analysed the effect of different time periods on R&D competition. In this context the positioning of rival firms was assumed to come from initial technologies and firms' products. Our analysis first showed the relationship between the pattern of market evolution and the degree of product differentiation through 'niche formation'. When two products are not very substitutable, the profit incentive dominates the race for the new technology, which results in alternating winners in both one period and inter-period competition. It would suggest that firms who are ahead would reinforce their advantage through branding for which they could enjoy temporary monopoly profits. However, as two products become close substitutes, the competitive incentive dominates the pattern of market evolution. In the process of examining those relations, it is also shown that current technologies and the magnitude of technological progress are important determinants.

From a different angle, as outlined in Section 9.6, when products in a market are close substitutes and price competition is intense, cooperative R&D can be used as a 'hidden tool' for avoiding unprofitable competition. However, if firms are allowed to engage in a mix of independent and cooperative R&D, then innovative output under cooperation is not less than under non-cooperative technological competition.

References

Arrow, K.J. (1962) 'Economic welfare and the allocation of resources for invention', in *The Rate and Direction of Inventive Activity: Economic and Social Factors*, Princeton: Princeton University Press.

Beath, J., Katsoulacos, Y. and Ulph, D. (1987) 'Sequential product innovation and industry evolution', *Economic Journal*, **97**, 32–43.

Beath, J., Katsoulacos, Y. and Ulph, D. (1989) 'Strategic R&D Policy', *Economic Journal*, **99**, 74–83.

Blair, J. (1972) *Economic Concentration*, New York: Harcourt, Brace and Jovanovich.

Carlsson, B. (1984) 'The development and use of machine tools in historical perspective', *Journal of Economic Behaviour and Organization*, **5**, 91–114.

Choi, J.P. (1993) 'Cooperative R&D with product market competition', *International Journal of Industrial Organization*, **11**, 553–571.

D'Aspremont, C. and Jacquemin, A. (1988) 'Cooperative and noncooperative R&D in duopoly with spillover', *American Economic Review*, **78**, 1133–1137.

Dasgupta, P. (1986) 'The theory of technological competition', in *Economic Organizations as Games*, K. Binmore and P. Dasgupta (eds), Basil Blackwell: Oxford, pp. 139–164.

Dixit, A. (1979) 'A model of duopoly suggesting a theory of entry barriers', *Bell Journal of Economics*, **10**, 20–32.

Geroski, P.A. and Pomroy, R. (1990) 'Innovation and the evolution of market structure', *Journal of Industrial Economics*, **38**, 299–314.

Grabowski, H.G. and Mueller, D.C. (1978) 'Industrial research and development, intangible capital stocks, and firm profit rates', *Bell Journal of Economics*, **9**, 328–343.

Gilbert, R. and Newberry, D. (1982) 'Pre-emptive patenting and the persistence of monopoly', *American Economic Review*, **74**, 514–526.

Grossman, G.M. and Shapiro, C. (1986) 'Research joint venture: an antitrust analysis', *Journal of Law, Economics and Organization*, **2**, 315–337.

Kamien, M.I., Muller, E. and Zang, I. (1992) 'Research joint ventures and R&D cartels', *American Economic Review*, **82**, 1293–1306.

Katz, M.L. (1986) 'An analysis of cooperative research and development', *The Rand Journal of Economics*, **17**, 527–543.

Mansfield, E. (1984) 'R&D and innovation: some empirical findings', in *R&D, Patents and Productivity*, Z. Griliches (ed.), Chicago: University of Chicago Press.

Phillips, A. (1966) 'Patents, potential competition, and technical progress', *American Economic Review*, **56**, 301–310.

Phillips, A. (1971) *Technology and Market Structure: A Study of The Aircraft Industry*, Lexington, MA: Lexington Books.

Schumpeter, J.A. (1942) *Capitalism, Socialism and Democracy*, New York: Harper & Row.

Suzumura, K. (1995) *Competition, Commitment and Welfare*, Oxford: Clarendon Press.

Vickers, J. (1986) 'The evolution of market structure when there is a sequence of innovations', *Journal of Industrial Economics*, **35**, 1–12.

Appendix

Proof of Proposition 9.1

Under the given demand functions, each firm's price is determined to satisfy

$$\text{Max}p_{\text{h}}\ \pi_{\text{h}}\ (s_{\text{h}}, s_{\text{w}}) = (p_{\text{h}} - s_{\text{h}})q_{\text{h}} \quad \text{and} \quad \text{Max}p_{\text{w}}\ \pi_{\text{w}}\ (s_{\text{w}}, s_{\text{h}}) = (p_{\text{w}} - s_{\text{w}})q_{\text{w}}.$$

Since the second-order conditions are satisfied, the reaction functions can be derived from the first-order conditions as follows:

$$p_{\text{h}}(p_{\text{w}}) = (a + bS_h + cp_{\text{w}})/2b \quad \text{and}$$
$$p_{\text{w}}(p_{\text{h}}) = (a + bS_w + cp_{\text{h}})/2b.$$

The interior solution from those two reaction functions is computed as follows:

$$p_{\mathrm{h}}(s_{\mathrm{h}}, s_{\mathrm{w}}) = (a(2b + c) + 2b^2 s_{\mathrm{h}} + bc\, s_{\mathrm{w}})/(4b^2 - c^2) \quad \text{and}$$
$$p_{\mathrm{w}}(s_w, s_h) = (a(2b + c) + 2b^2 s_w + bc\, s_{\mathrm{h}})/(4b^2 - c^2).$$

By substituting these two arguments into each firm's profit function, we get

$$\pi_{\mathrm{h}}(s_{\mathrm{h}}, s_{\mathrm{w}}) = b\{(a(2b + c) - (2b^2 - c^2)s_{\mathrm{h}} + bc\, s_{\mathrm{w}})\}/(4b^2 - c^2)^2 \quad \text{and}$$
$$\pi_{\mathrm{w}}(s_{\mathrm{w}}, s_{\mathrm{h}}) = b\{(a(2b + c) - (2b^2 - c^2)s_{\mathrm{w}} + bc\, s_{\mathrm{h}})\}/(4b^2 - c^2)^2. \qquad \text{(A.1)}$$

Therefore, the difference between two firms' incentives to innovate can be written as

$$\rho(s_{\mathrm{h}}, s_{\mathrm{w}}, s_1) = b(s_{\mathrm{w}} - s_{\mathrm{h}})/(4b^2 - c^2)^2\ R(s_{\mathrm{h}}, s_{\mathrm{w}}, s_1)$$

where

$$R(s_{\mathrm{h}}, s_{\mathrm{w}}, s_1) \equiv (s_{\mathrm{h}} + s_{\mathrm{w}})\{(2b^2 - c^2)^2 + b^2c^2\} - 4bc(2b^2 - c^2)s_1 - 2a(b - c)(2b + c)^2\}.$$

Since $s_{\mathrm{h}} > s_{\mathrm{w}}$ and $b > 0$ by the given assumption, firm w will have a greater incentive to innovate if $R(.)$ is positive and firm h will take over the advantage on costs if $R(.)$ is negative.

In order to prove the existence of $c^*(s_{\mathrm{h}}, s_{\mathrm{w}}, s_1)$, it will be enough to show that

$$R(s_{\mathrm{h}}, s_{\mathrm{w}}, s_1) < 0 \quad \text{when} \quad c = 0 \quad \text{and}$$
$$R(s_{\mathrm{h}}, s_{\mathrm{w}}, s1) > 0 \quad \text{when} \quad c = b.$$

When $c = 0$, $R(s_{\mathrm{h}}, s_{\mathrm{w}}, s_1) = 4b^3[b(s_{\mathrm{h}} + s_{\mathrm{w}}) - 2a] < 0$ since it was assumed that $a - bs_{\mathrm{h}} + cs \geq 0$. If $c = b$, $R(s_{\mathrm{h}}, s_{\mathrm{w}}, s_1) = 2b^4(s_{\mathrm{h}} + s_{\mathrm{w}} - 2s) > 0$ because it was assumed that $s_{\mathrm{h}} > s_{\mathrm{w}} > s_1$. Since the function $R(.)$ is continuous to c, by the intermediate value theorem, there exists $c^*(s_{\mathrm{h}}, s_{\mathrm{w}}, s_1)$ such that

$$R(s_{\mathrm{h}}, s_{\mathrm{w}}, s_1)|_{c=c^*} = 0.$$

The uniqueness of $c^*(s_{\mathrm{h}}, s_{\mathrm{w}}, s_1)$ can be shown by examining the second and the third derivatives of the function $R(.)$. Suppose that there exist at least two non-negative $c^*(s_{\mathrm{h}}, s_{\mathrm{w}}, s_1)$ such that $R(s_{\mathrm{h}}, s_{\mathrm{w}}, s_1)|_{c=c^*} = 0$. Let's define c_1 and c_2 to be these values and $c_1 < c_2$. Since $a - bs_{\mathrm{h}} + cs \geq 0$, $R(s_{\mathrm{h}}, s_{\mathrm{w}}, s_1)|_{c=0} = 4b^3[b(s_{\mathrm{h}} + s_{\mathrm{w}}) - 2a] < 0$.

Since $R(s_{\mathrm{h}}, s_{\mathrm{w}}, s_1)|_{c=0} < 0$ and c_1 is inf $c^*(s_{\mathrm{h}}, s_{\mathrm{w}}, s_1)$ such that $R(s_{\mathrm{h}}, s_{\mathrm{w}}, s_1)|_{c=c^*} = 0$ there must exist at least one local maximum in the interval between c_1 and c_2. For a local maximum, the first-order derivative of $R(.)$ with respect to c must be decreasing around that local maximum. From the function $R(.)$, we can compute the third derivative of it as follows:

$$\partial^3 R(s_{\mathrm{h}}, s_{\mathrm{w}}, s_1)/\partial c^3 = 6\{4c(s_{\mathrm{h}} + s_{\mathrm{w}}) + 4bs_1 + 4a\}.$$

Since all parameters are assumed to be non-negative, the third derivative of $R(.)$ is always positive for every non-negative c. The second derivative of $R(.)$ is as follows:

$$\partial^2 R(s_h, s_w, s_l)/\partial c^2 = 6\{2(c^2 - b^2)(s_h + s_w) + 4bcs_l + 2a(b + c)\}.$$

At $c = 0$, it will be $\partial^2 R(s_h, s_w, s_l) / \partial c^2|_{c=0} = 6b\ \{2a - b)(s_h + s_w)\} > 0$.

From the result that $\partial^3 R(s_h, s_w, s_l)/\partial c^3 > 0$ and $\partial^2 R(s_h, s_w, s_l)/\partial c^2|_{c=0} > 0$, it can be inferred that $\partial^2 R(s_h, s_w, s_l) / \partial c^2 > 0$ for every non-negative c.

Therefore, the first derivative of $R(.)$ will monotonically increase, which contradicts the existence of a local maximum.

Proof of Proposition 9.2

From the proof of Proposition 9.1, $c^*\ (s_h, s_w, s_l)$ satisfies $R(s_h, s_w, s_l)|_{c=c^*} = 0$. Therefore, by the implicit function theorem, we get

$$\partial c^*(s_h, s_w, s_l)/\partial s_l = -[\partial R(s_h, s_w, s_l)/\partial s_l]/[\partial R(s_h, s_w, s_l)/\partial c]|_{c=c^*}(\ldots).$$

Since $c^*\ (s_h, s_w, s_l)$ is the unique solution such that $R(s_h, s_w, s_l)|_{c=c^*} = 0$ and $\partial^2 R(s_h, s_w, s_l) / \partial c^2 > 0$, there must exist a unique c' such that $\partial R(s_h, s_w, s_l)/ \partial c|_{c=c'} = 0$ and $c' < c(s_h, s_w, s_l)$. If not, there will be another c such that $R(s_h, s_w, s_l) = 0$ which contradicts Proposition 9.2. Since $\partial^2 R(s_h, s_w, s_l)/ \partial c^2 > 0$ and $c' < c^*(s_h, s_w, s_l)$, it must be that $\partial R(s_h, s_w, s_l) / \partial c|_{c=c^*(.)} > 0$. From that $R(s_h, s_w, s_l) = 0$, we can also show that $\partial R(s_h, s_w, s_l)/\partial s_l = \partial R(s_h, s_w, s_l)/\partial s_l = 4bc(c^2 - 2b^2) < 0$.

Therefore, $\partial c^*\ (s_h, s_w, s_l) / \partial s_l > 0$.

By differentiating $\rho(s_h, s_w, s_l)$ with respect to s_l we get

$$\partial\rho(s_h, s_w, s_l)/\partial s_l = 4b^2c(2b^2 - c^2)\ (s_h - s_w)/4b^2c(2b^2 - c^2)^2.$$

Since $b > c$ and $s_h > s_w$, we get the result that $\partial\rho(s_h, s_w, s_l)/\partial s_l > 0$.

Proof of Proposition 9.3

(1) If $0 \leq c < c^*\ (s_h, s_l, s)$, then $\phi(s_h, s_w, s_l, s) = \sigma\ (s_l, s_w)\ - \sigma\ (s_h, s_l) + \Pi_w\ (s, s_h) - \Pi_h\ (s, s_w)$.

By applying the relation in (9.6), it can be written

$$\phi(s_h, s_w, s_l, s) = \{b(s_h - s_w)(2b^2 - c^2)[2a(2b + c) - (2b^2 - c^2)(s_h + s_w) + 2bc(2s_l - s)]\}/(4b^2 - c^2)^2.$$

Since $a - bs_h + cs > 0$ by assumption (9.2) and $s_h > s_w > s_l > s$,

$$2a - b(s_h + s_w) + 2cs_l > 0 \quad \text{and} \quad 2a - 2bs + c(s_h + s_w) > 0.$$

Therefore, we can derive

$$2a(2b+c)-(2b^2-c^2)(s_h+s_w)+2bc(2s_1-s)=$$
$$2b[2a-b(s_h+s_w)+2cs_1]+c[2a-2bs+c(s_h+s_w)]>0.$$

Since each term on the rhs of the function $\phi(s_h, s_w, s_1, s)$ is positive, we can conclude that $\phi(s_h, s_w, s_1, s)>0$.

(2) When $c \geq c^*(s_w, s_1, s)$, the winner in the first-period race will have a greater incentive in the second-period race s than its rival firm. Then, as in case (iii),

$$\phi(s_h, s_w, s_1, s)=\sigma(s_1, s_w)-\sigma(s_1, s_h)+\sigma(s, s_w)-\sigma(s, s_h)$$
$$-\pi_h(s_h, s)+\pi_w(s_w, s).$$

By substituting the results in (A.1) directly into the above function and rearranging terms, it is derived that $\phi(s_h, s_w, s_1, s) \leq 0$, if

$$s_h+s_w \geq \{2a(2b+c)[3(2b^2-c^2)-2bc]$$
$$+2bc(2b^2-c^2)(2s_1+3s)\}/3(2b^2-c^2)^2+2(bc)^2>0.$$

Now we need to show that there exist s_h and s_w such that the given assumptions and the related conditions are met. Suppose that $s_h=s+(a/b)-\varepsilon$, $s_h-\varepsilon=s_w$, $c=b-\varepsilon$, where ε is infinitesimal. Then the assumptions in (9.1) and (9.2) are met. Further, $\lim_{\varepsilon\to 0} \rho(s_w, s_1, s)<0$. It implies that $c \geq c^*(s_h, s_1, \text{s})$. Now we need to show only that $s_h+s_w \geq \{2a(2b+c)$ $3(2b^2-c^2)-2bc+2bc(2b^2-c^2)(2s_1+3s)\}/3(2b^2-c^2)^2+2(bc)^2$.

When ε approaches zero, the left-hand side of the above inequality is close to $4(a-bs_1+bs)/5b$, which is greater than zero because $a-bs_1+bs>a-bs_h+cs>0$. Therefore, the conclusion of the proposition (2) is established.

Proof of Proposition 9.4

(1) If $0 \leq c < c^*(s_h, s_1, s)$, from the proof of Proposition 9.3, we directly get $\partial\phi(s_h, s_w, s_1, s)/\partial s_1=4b^2c(2b^2-c^2)(s_h-s_w)/(4b^2-c^2)^2>0$ and
(2) $\partial\phi(s_h, s_w, s_1, s)/\partial s=-2b^2c(s_h-s_w)(2b^2-c^2)/(4b^2-c^2)^2<0$. If $c^*(s_w, s_1, s)<c<b$, we can similarly get

$$\partial\phi(s_h, s_w, s_1, s)/\partial s_1=4b^2c(2b^2-c^2)(s_h-s_w)/(4b^2-c^2)^2>0 \quad \text{and}$$
$$\partial\phi(s_h, s_w, s_1, s)/\partial s=6b^2c(s_h-s_w)(2b^2-c^2)/(4b^2-c^2)^2>0.$$

Therefore, the results of the proposition hold.

10 Competition in increasing returns and network industries

> Increasing returns is incompatible with perfect competition as has been known since the work of Cournot.
>
> Kenneth Arrow (1993)

10.1 Introduction to increasing returns

Most industrial sectors of highly industrialized economies are not perfectly competitive. They are usually formed by a small number of big firms with non-negligible market share. Besides being prevalent in the economy, big firms cluster around concentrated industrial structures which exhibit a skewed distribution of firm size and market share. This situation may be brought about by the intrinsic potential of dynamic technological competition to end up in (temporary) technological monopoly, so in those cases industrial competition may start out symmetric but end up asymmetric.

In this chapter we show how the competitive process proliferates in increasing returns industries where the total of all unit activities linked together yields a higher return than the sum of the individual unit activities operating separately. For this to be happening we must show that a variety of increasing returns mechanisms combine to enable the effect of an increasing returns industry.

We propose an integrated framework to provide tools and insights for explaining competition among skewed industrial structures. However, it is only a tentative step toward attempting to explain the path-dependent, indeterminate, suboptimal, locking-in nature of technological competition under increasing returns.

Because of this we partially review the literature on the dynamics of technological diffusion, substitution, and competition. The purpose of this review is to show that we cannot accurately understand industrial competition without taking into account the self-reinforcing nature of commercial success in most emerging markets. We enrich increasing returns mechanisms by incorporating a set of stronger, yet neglected, increasing returns mechanisms – reputation effects, infrastructure effects and positive network externalities – into a preliminary framework model. The resulting theoretical structure, we will argue, captures the interdependent and cumulative character of the three aspects of industrial competition: the number and size of firms, skewed industrial structures, and the nature of technological competition.

The increasing returns discussion in economics has provided important insights into the characteristics and dynamics of modern industrial economies. However, the discussion on policy applications has (mis)lead some authors and policy analysts to conclude that a completely *new economy* is emerging and that it obeys a set of rules which are totally different from those that apply to traditional sectors of the economy. While it is undeniable that the increasing return paradigms remain fairly new and revolutionary, and while there is no doubt that this paradigm is key to our understanding of new industrial sectors and their sustaining role in productivity growth, we should clarify its proper role in industrial structure and growth of the economy (see also Chapter 11). At this stage we are most concerned about the catalytic role of technological competition in increasing returns industries. Increasing returns industries are nowadays most likely to be identified with high-technology industries, in particular with information, communication and health care-related industries (Gottinger, 2003).

For those industries Shapiro and Varian (1999) have recently suggested a combination of supply-side and demand-side scale economies to explain the intrinsic aspects of technological competition. It appears, however, that this way of seeing technological competition is too simple to capture the variety and complexity of real-world businesses in those industries Thus we suggest a general framework to describe technological competition in what we are going to call the *increasing returns economy*.

Section 10.2 identifies supply-side scale economies as a major ingredient of increasing returns economies. Section 10.3 lists increasing returns properties as part of a Schumpeterian mechanism. The relationship of increasing returns and non-ergodic markets is explored in Section 10.4. Section 10.5 explores technological competitive paths subject to uncertainty of technological outcomes, whereas Section 10.6 focuses on technological competition under standard setting. Section 10.7 relates the intensity of technological competition to network externalities, and Section 10.8 provides a summary of increasing returns factors in competitive settings. Finally, Section 10.9 draws conclusions on the systemic connection of increasing returns and technological competition.

10.2 Supply-side scale economies

A first source of increasing returns assuming constant technology identifies a concentrated industry structure as a result of supply-side scale economies.

In many cases large firms are more efficient than smaller companies because of their scale: larger corporations tend to have lower unit costs. This efficiency in turn fuels further growth. However, positive feedbacks based on supply-side economies of scale usually run into natural limits. Past a certain size companies find growth difficult owing to the increasing complexity of managing a large organizational structure. From then on, negative feedback takes over. As traditional supply-side economies of scale generally become exhausted at a scale well below total market dominance, large firms, burdened with high costs, never grow to take the entire market and smaller, more nimble firms can find profitable niches. Shapiro and Varian (1999) conclude that because of this most industrial markets are oligopolies rather than monopolies.

Negative feedback generated by the difficulties of managing large organizations (scale, diseconomies) indeed interrupts the growth of the firm and the level of industrial concentration. Nevertheless, this situation may be transient, because firms may be subject to other sources of increasing returns. Large firms that go through increasing returns mechanisms other than scale economies may increase their efficiency and overcome the negative aspects of overgrown organizations. Industries in which scale diseconomies are counterbalanced by other increasing returns mechanisms, then, may begin to head toward the extreme of 'winner-takes-most' situation. The increasing returns mechanisms capable of offsetting scale diseconomies are usually related to technological progress, so in the following sections we analyse other major causes of the growth of the firm – namely, the Schumpeterian loop, cost-reducing learning, learning-by-doing, learning-by-using, and demand-side increasing returns.

10.3 Schumpeterian mechanism

The most widely accepted theory of technological change in modern economics is that of Schumpeter (1942). In the Schumpeterian world, scale economies are also present, but technology is not a constant. Here the creative role of the entrepreneur allows for the introduction of new technologies capable of displacing the established ones. Most of Schumpeter's discussion stresses the advantages of concentrated market structures involving large firms with considerable market share. In his view, it is more probable that the necessary scale economies in R&D to develop new technologies be achieved by a monopolist or by the few large firms of a concentrated industry. Besides, large firms may increase their rate of innovation by reducing the speed at which their transient rents and entrepreneurial advantage are eroded by imitators. In the absence of patent protection large firms may exploit their innovations on a large scale over relatively short periods of time – and in this way avoid rapid imitation by competitors – by deploying their productive, marketing and financial capabilities. Large firms may also expand their rate of innovation by imitating and commercializing other firms' technologies.

Schumpeter's thesis encouraged a large body of empirical literature in the field of industrial organization. Most of this literature focused on two hypotheses associated with Schumpeter's assertion: (1) innovation increases more than proportionally with firm size and (2) innovation increases with market concentration.

The most comprehensive review of the empirical evidence of the relationship between innovation and firm size and market structure is by Cohen and Levin (1989). They observe that the empirical results on the Schumpeterian relation are accurately described as fragile. They note that the lack of robust results seems to arise in part from the inappropriate attention to the dependence of these relationships on more fundamental conditions. From their overview, they draw the basic methodological lesson that the omission of important and potentially correlated variables that influence innovation can lead to misleading inferences concerning firm size and concentration.

Following Schumpeter's lead, Richard Nelson and Sidney Winter (1978, 1982) stand out for having formalized and completed many of Schumpeter's original intuitions. Whereas the connection between industrial structure and innovation has been viewed by Schumpeter as going primarily from the former to the latter, in Nelson and Winter (1982) there is a reverse causal flow, too. That is, there is clearly a circular causality suggesting a self-reinforcing mechanism between the innovations and the firm's growth. Nelson and Winter (1982) stand out not only for having recognized the endogeneous character of innovation and market structure, but also for having pointed out and modelled the mutual causality between technical change and market structure (Nelson, 1986).

Evolutionary economists (like Nelson and Winter) define innovation very broadly. It encompasses product and process innovation, opening up new markets, and acquisition of new sources of raw material. They also describe the nature of technical progress as succession of major discontinuities detached from the past and with quite transitory life span. This process of change is characteristic of certain industries, but it is not the sole kind of technological change. Technological change can also be *continuous*. That is to say, technologies improve constantly in absolute terms after their introduction. The view of technological progress as a continuing, steady accumulation of innumerable minor improvements and modifications, with only very infrequent major innovations, has two sources: (1) the accumulation of knowledge that makes possible to produce a greater volume of output from a given amount of resources, and (2) the accumulation of knowledge that allows the production of a qualitatively superior output from a given amount of resources (Chapter 4). The former source of technological progress is the result of a *cost-reducing learning process*, while the second category is the result of what is known as *learning-by-doing* and *learning-by-using*. Given that both categories of technological progress are important determinants of the number and size of firms in a given industry, we analyse them in the next sections.

Cost-reducing learning

An important aspect of technological change is costs reducing in nature. As we saw before, Porter (1980) and Henderson (1975), in the strategic field, pioneered the notion of experience curve as a source of cost reductions. In economics, Hirsch (1956) has underlined the importance of repetitive manufacturing operations as a way of reducing direct labour requirements, while Arrow (1962) has explored the consequences of learning-by-doing (measured by the cumulative gross investment, which produces a steady rate of growth in productivity) on profits, investment, and economic growth. However, the historical study on the patters of growth and competitiveness of large corporations of Alfred D. Chandler (1990) is a major and detailed contribution to our understanding of the way firms grow by diminishing costs.

Large corporations, according to Chandler, along with the few challengers that subsequently enter the industry, do not compete primarily on the basis of price. Instead they compete for market share and profits through functional and strategic effectiveness. They compete functionally by improving their products, their processes of production, their marketing, their purchasing, and their labour relations. Big corporations compete strategically by moving into growing markets more rapidly and effectively than do their competitors. Such rivalry for market share and profits make more effective the enterprise's functional and strategic capabilities, which, in turn, provide the internal dynamics for continuing growth of the enterprise. In particular, it stimulates its owners and managers to expand into distant markets in its own country and then to become multinational by moving abroad. It also encourages the firm to diversify and become multiproduct by developing and introducing products in markets other than the original ones.

Learning-by-doing

Some of the writings on industrial competition assume that firms compete mainly in cost-reducing competitive advantages, especially those achieved through scale economies, scope economies (economies of joint production and distribution), and innovation in production and organizational processes. Here technical progress is implicitly treated as the introduction of new processes that reduce costs of producing essentially unchanging products. Beyond, there is a category of learning known as 'learning-by-doing' (Rosenberg, 1982) which enhances the qualitative aspects of final products.

Western industrial societies today, Rosenberg (1982) argues, enjoy a higher level of material welfare not merely because they consume larger *per capita* amounts of the goods available. They have also made available improving forms of rapid transportation, instant communication, powerful energy sources, life-saving and pain-reducing medications, and other goods that were undreamed of one or two centuries ago. Therefore, ignoring product innovation and quality improvements in products is to overlook what has been one of the most important long-term contributions of technical progress to human welfare. Many products, such as beverages, toothpaste, soap, clothing, VCRs, TV sets can be subject to improvements. Such improvements, however, are marginal when compared with the amazing rate of development that other products and technologies can reach. Automobiles, aircraft, flight simulators, computers, and nuclear reactors are very complex technologies and, as a consequence of this, have a tremendous capacity of being enhanced. Consequently, the competitive behaviour of the firms that produce these technologies consists not only of the innovative acts they perform to improve production, organizational, and distribution processes, but also from the efforts to improve constantly their products.

Learning-by-using

With respect to a given product, Rosenberg (1982) distinguishes between the kind of learning that is internal to the production process (learning-by-doing) and that which is generated as a result of subsequent use of that product (learning-by-using). The latter category of learning begins only after a certain new product is used. In an economy where complex new technologies are common, there are essential aspects of learning that are a function not of the experience involved in producing a product but of its *use* by the final consumer.

The optimal performance of durable goods (especially complex systems of interacting components) is often achieved only after intensive and prolonged use. In the aircraft industry, for instance, the attainment of high standards of reliability is a major concern, in particular during the development stage. But it is only through extensive use of aircraft by airlines that faults are discovered and eliminated and detailed knowledge is gained about metal fatigue, weight capacity, fuel consumption of engines, fuselage durability, minimum servicing, overhaul requirements, maintenance costs, and so on.

Demand-side increasing returns

In the economy there are increasing returns mechanisms that come from the demand side of the market, not just from the supply side. For the average (risk-adverse and imperfectly informed) consumer it becomes more attractive to adopt a widespread technology or product. Minimizing the risk of purchasing a defective technology or the cost of searching for an adequate one introduces a reputation or informational feedback that may produce a disproportionately high selection of the best-selling option. Informational or reputational feedback effects occur in various situations that could be reinforced through network externalities. First, when the complexity of the technology or product in question is such that consumers try to reduce uncertainty by asking to previous purchasers their experience with these technologies (Arthur and Lane, 1993). Second, in other situations the source of uncertainty is not the complexity of the technology, but the large quantity of options the consumers face. One is bound to choose, and the best way to do so is by confining one's attention to the best-assessed items in the consumer report. Third, in a market where the quality or value of a product is defined on the basis of arbitrary and short-lived conventions, rather than strictly on the basis of lasting objective value, consumers usually tend to follow the expert's opinion. This kind of easy-to-manipulate, reputation-driven market success is typical of markets for highly symbolic products (e.g. many art markets, fashion wear and luxury items), which also will result in a disproportionately high selection of one of the options.

Finally, the most pre-eminent and common kind of reputation effects in the economy arise plainly as a result of a well-timed and very aggressive advertising campaign. This self-reinforcing mechanism – and the lasting market dominance that it causes – might be quite unrelated to relative added value, but it certainly might produce an excessive predilection for one of the options.

By moving beyond the Schumpeterian hypotheses and focusing on a more complete model of industrial competition, we have identified other fundamental determinants of technological change that affect the mutual link between firm size and market structure (Aghion and Howitt, 1998).

These determinants – which in our analysis take the form of increasing returns mechanisms – are usually studied as if they work independently from the other. But there are not many cases of industries where one single mechanism acts in isolation from the other sources of increasing returns. Therefore, the growth of the firm and the evolution of skewed industrial structure, more than the result of a single self-reinforcing mechanism, are the effect of the combination of several sources of increasing returns, which overlap and feed back upon one another.

As depicted in Figure 10.1, the unification of the resource-based loop, the Schumpeterian loop, scale economies, the different categories of learning, and demand-side increasing returns (reputation) – loops A, B, and C, respectively – constitutes a simple but useful model capable of explaining endogenously the number and growth of firms in a given industry, and in a wider context, the gap of economic performance in a given industrial sector.

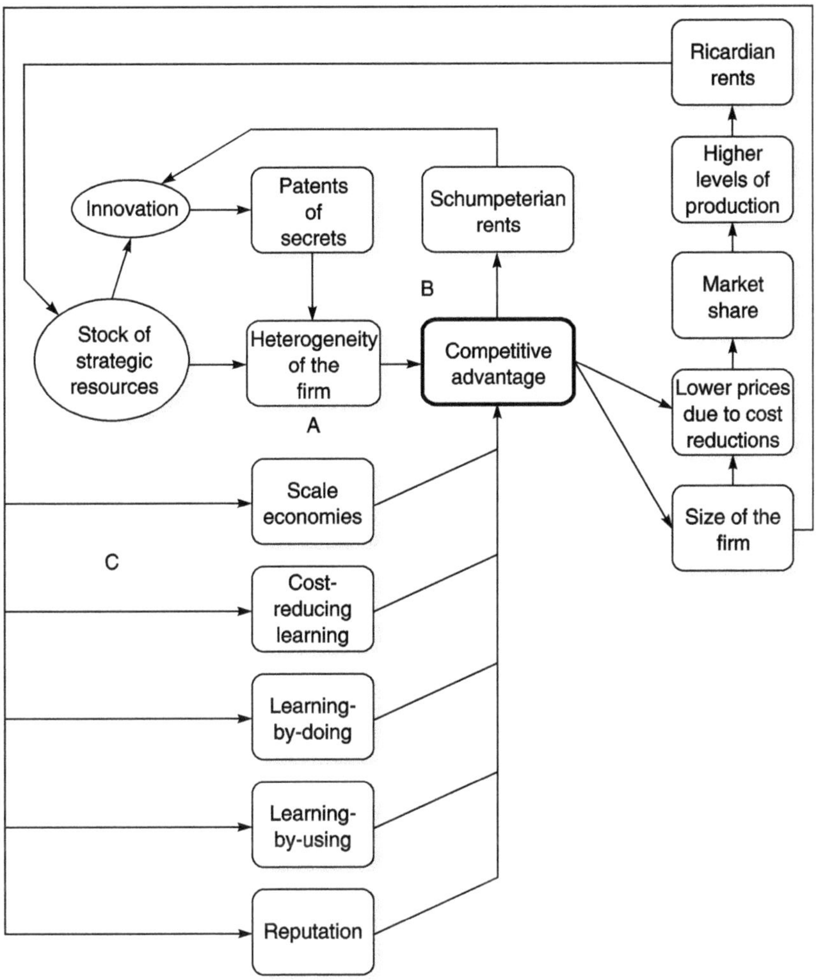

Figure 10.1 Increasing returns mechanism: a qualitative model of industrial competition.

In the model sketched in Figure 10.1 the positive relationship that runs from industrial structure to efficiency operates through the accumulation of rare resources, innovations, scale economies, reputation, and the different aspects of learning. This dynamics, over time, makes costs fall as learning accumulates, new technologies are developed and improved, and firm-specific factors are amassed and exploited due to output increases. As a result of this mutual causality, market share and production levels increase, price falls, profitability rises, and with which relatively profitable firms expand continually while unprofitable ones contract uninterruptedly.

A relevant aspect of the structural determinants of the number and size of firms in an industry suggested in this model is that, when one of them is exhausted, causing a slowdown in the growth of the firm, the other mechanisms may be activated, which may allow for a further period of continued rapid growth. When the firms of a given industry are capable of accumulating firm-specific resources, innovations, costs-reducing learning, qualitative product innovation based on learning-by-doing and learning-by-using, and reputation, these firms usually use them as strategic weapon. In doing so, they are capable not only of neutralizing but also of overwhelming the negative effects of complex, overgrown, hard-to-manage organizational structures that arise from their constant growth. The process can take a long period of time, but eventually the sources of increasing returns can drive markets toward increasingly skewed industrial structures.

For instance, in the commercial aircraft industry competition principally involves considerable development costs, continuous improvements in aircraft models, technology and product support, so this industry exhibits substantial scale economies, large scope for learning-by-doing, learning-by-using, and reputation effects. Because of this, the commercial aircraft industry has been characterized by an increasing skewed industrial structure. Recently, the structure of this industry, after the acquisition of McDonnell Douglas by Boeing, was reduced to a monopoly in the United States. In the world aircraft market Boeing only competes with European Airbus. It is obvious that the merger of the two main manufacturers of the American aircraft industry should have brought about some gain in efficiency, which counterbalanced the diseconomies owing to managing a more complex organization. Otherwise, the merger would not have taken place or would have been the result of irrational behaviour.

The structure of some industries does not seem to head toward monopoly. However, over time, their level of concentration has increased substantially. The world automobile industry, for instance, in 1970 was composed of at least 40 major competitors. In 1998, with some mergers and acquisitions, the number of main manufacturers was reduced to 17. Because of large possibilities to accumulate cost-reducing learning and the large scope for qualitative product improvements in the world automobile industry, both the number and the homogeneity of the firms competing in this industry are expected to decrease even further in the future. Here, again, benefits due to both costs-reducing learning and qualitative product innovations brought about by mergers and acquisitions are larger than any cost created by scale diseconomies. Another interesting aspect of this model is that it also offers an endogenous explanation of the number and size of firms. In contrasts with the traditional economic views – that see industrial structure (number of firms) as an exogenous variable and assume homogeneous firms – and the strategic paradigms – which are focused first and foremost in explaining heterogeneity among firms within an industry – this model recognizes that the strategic choices and acts of the firms have an effect not only on the performance and size of the firm itself, but also on the structure of market.

In summary, industrial structure is caused by a combination of various increasing returns mechanisms. Here, then, the combination of accumulation of resources, product innovation, scale economies, cost-reducing learning, learning-by-doing, learning-by-using or reputation enhances the performance of the firm and determine, to a great extent, the level of skewness of the structure of the industry where it competes.

10.4 Increasing returns and ergodic markets

Conventional economics has tended to portray most economic situations as something analogous to a large Newtonian system, with a unique and stable equilibrium solution predetermined by a given pattern of resources, preferences, and technological possibilities. Brian Arthur and his group (cf. Arthur, 1994; Arthur *et al.*, 1987), however, have shown that this conventional way of seeing economic reality overlooks important and frequent economic situations where increasing returns are conspicuous. In order to distinguish economic situations characterized by decreasing returns from those where increasing returns are dominant, Arthur *et al.* (1987) developed the theory of *non-linear Polya processes*, which describes the long-run self-organizing structures that emerge from dynamic processes where proportions are involved. The general nonlinear Polya scheme can be pictured by imaging an urn of infinite capacity to which balls of several colours are added. In the simplest case, where decreasing and constant returns prevail, the probability of a ball of a given colour to be chosen the next time is independent of the proportions of colours at the moment of the addition. In this simple sequential process, the strong law of large numbers predicts that, over time, the proportion of balls of colour i has a fixed probability $q(i)$, where $\Sigma_i q(i) = 1$. Therefore, it has a unique, predetermined outcome. Sequences of choices in these simple cases are important at the beginning of the process. However, as the process advances, different sequences of choices are averaged away by the economic forces, which are subject to constant or decreasing returns. So, no matter the sequences of choices, the system will always – with probability one – end up in the same pattern. For instance, in a coin-tossing experiment the event ‘head’ is independent of previous tosses, then the expectation of a ‘head’ in each toss is 0.5 no matter how many times the experiment is repeated. Likewise, the proportion of 6s in a dice-casting experiment will tend to 1/6. The process by which firms in an industry concentrate in different regions is like the coin-tossing or the dice-casting experiment, if the geographic preferences of each firm is not modified by the preferences of the other firms.

In more general cases – where increasing returns are present – the dynamics are completely different and the standard strong law is inapplicable. In this regime, the next ball to be added into the urn is not known, but the probability of adding one ball of specific colour depends on the present proportions of colours in the urn. In other words, the probability of an addition of the colours becomes a function of the proportions of balls of each colour at each time of choice. The case of firms deciding where to settle down illustrates this kind of non-linear Polya process. Here increasing returns can be incorporated within the model by introducing agglomeration effects. Because of agglomeration effects, additions to a specific region are not independent of previous locational choices and firms are added incrementally to regions with probability exactly equal to the proportions of firms in each region at the time. Under increasing returns, then, the process becomes *path-dependent*.

Arthur *et al.* (1987) showed that at the outset of the process proportions are not stable, but once the industry settles into a vector of proportions, locational patterns become constant at that vector with probability one. However, the constant vector is selected randomly from uniform distribution over all possible shares that sum to 1.0. This means that each time this locational process is rerun under different historical events, it will in all likelihood settle into a different pattern. Therefore, it is possible to predict that the locational pattern will tend toward a constant proportion, but it cannot be foreseen at *which* proportion it will settle down.

The interpretation of economic history is different under different regimes. Under constant and diminishing returns, the evolution of the system is *ergodic*. Ergodic structures emerge when repeated random events – that are drawn from the same distribution and are independent from previous ones – have a long-term average that approach their expected value. While other results might be possible, they have probability zero to occur. The typical example of an ergodic system is coin tossing. If a fair coin is tossed indefinitely, the proportions of heads varies considerably at the outset, but settles down to 50% with probability one. The evolution of an ergodic system, therefore, follows *a convex* probability function, which has expected motions that lead toward a unique, determinate outcome. In this regime 'historical chance' cannot influence the evolution of the systems so history is reduced to mere deliverer of the inevitable and the known.

Under increasing returns, by contrast, the process is *non-ergodic*, because small historical events become magnified by positive feedback. A non-ergodic system follows a non-convex probability function, so two or more outcomes are possible and 'historical chance' determines which of these is ultimately selected. History becomes all-important. There are some cases of non-ergodic systems in which, from the multiplicity of structures that may emerge, there are some 'corner solutions' with a single option monopolizing the choices. In this specific kind of nonergodic systems, while information on preferences, endowments and transformation possibilities allows locating and describing the various possible corner equilibria, it is usually insufficient to determine which one will be selected. In these cases, as Arthur (1994: 13) has pointed out 'there is an *indeterminacy* of outcome'.

Adoption of technologies that compete under diverse regimes can be appropriately modelled as a non-linear Polya process (cf. Arthur, 1989,

1990a, b; Arthur *et al.*, 1987). In the simplest regime, when technological competition is characterized by constant and decreasing returns, the probability of a technology of being chosen depends on its current market share. As each adoption is independent of the previous one, market share should converge to a point where they equal probability. Therefore, under constant and decreasing returns two technologies or products performing the same function will end up sharing the market according to each technology's intrinsic value and technical possibilities. Therefore, markets characterized by constant and decreasing returns can be called *ergodic markets*.

Under increasing returns to adoption, the probability of adoption depends on the numbers of adoptions holding each technology at a particular time. Markets of this kind can be called *non-ergodic* markets. Within this kind of markets there are those where increasing returns may drive the outcome toward a single dominant technology, with small events selecting the technology that takes over early on. This particular type of non-ergodic markets can be termed *tipping or indeterminate markets*. This indeterminacy relates to the 'selection problem' – how one allocation outcome is 'selected' over time by small historical events when there are several possible long-term results.

10.5 Technological competition under uncertainty and inertia

In high-technology markets the commercial success of emerging new technologies is both highly uncertain and inertial. As regards uncertainty, in addition to the problem trying to discern the true potential of a new generic technology, there is also the difficulty of foreseeing the precise direction in which the said technology will evolve. Indeed, as depicted in Figure 10.2, the emergence of a new generic technology generally opens the door not just to one specific technological path, but rather to a whole variety of possible trajectories in product design and process technology.

On the other hand, the inertial forces unleashed by commercial success are a lot more powerful than the classical models of diffusion suggests. Abernathy and Utterback (1978), Abernathy and Clark (1985) and Teece (1986), among others, have rightly underlined that in addition to the rigidities which may affect individual workers or machines and individual intermediaries or users, there are *systemic rigidities* of a much greater scope and importance. Every successful generic technology has a complex web of complementary technologies woven around its core. Once such an integrated and expensive, in terms of purchasing and using, technological system is in place, its momentum becomes enormous. Consequently, once a specific new technology becomes part of a dominant system, it will become increasingly difficult to dislodge, even by more worthy alternatives. This is depicted in Figure 10.2 where technology B (the inferior one) is foreclosed by the entrenchment of technology A

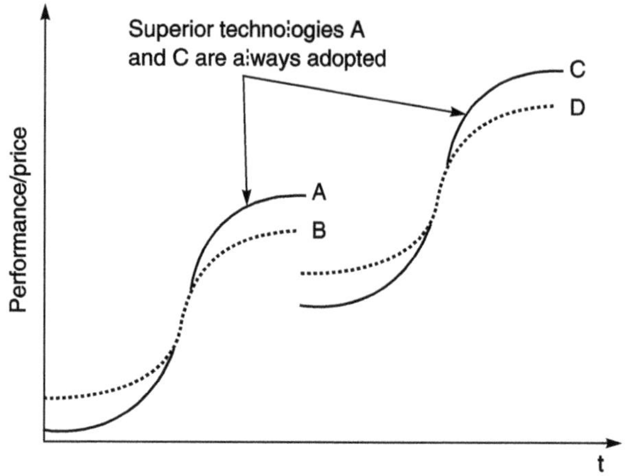

Figure 10.2 Uncertain and inertial technological trajectory: superior technologies A and C are always adopted.

(the superior one), which will only be displaced by the much superior technology C.

Uncertainty and inertia can combine to cause decisive first-mover advantages, which may grant an unassailable market dominance to an early technological trajectory. And yet, these authors do not go far enough in recognizing that in the real world, optimal technological cycles, trajectories and discontinuities, such as options A and C in Figure 10.2, are not invariable realities. They largely ignore and or minimize the self-reinforcing (and not simply inertial) nature of commercial success and the consequent unpredictability of technological evolution in general and of 'dominant designs' in particular. With this, they also overlook that a technology's success is tributary to the competitive decisions (often arbitrary and myopic) of the major players in an industry, as they are to any set of exogenous technical parameters.

Abernathy and Utterback did not believe in the research lab as an optimal selector of new technologies and they did question the optimality of selection by the market, but only to a small degree. The early articles written together by Abernathy and Utterback are thoroughly ergodic markets-oriented (cf. Abernathy and Utterback, 1978). Nowhere in them is it hinted that a *dominant design* might not be optimal or that its lasting power might not be inevitable. Later on, Abernathy and Clark (1985) made a strong case for contingency in the maturity and decline of technologies and industries. They thus rejected in no uncertain terms the deterministic view of technological life cycles. Paradoxically, however, their emphasis on historical contingency did not extend to the emergent phase of a new generic technology. They continued to suggest or imply that within a given generic technology, a specific 'dominant design' will be chosen strictly on the

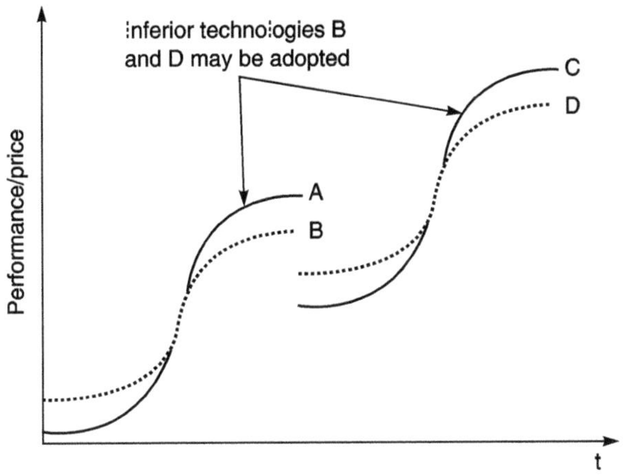

Figure 10.3 Indeterminate technological trajectory: inferior technologies B and D may be adopted.

basis of its relative merit. As for Utterback, in his more recent writings (Utterback, 1993; Utterback and Suarez, 1994), he fully acknowledges that indeterminacy characterizes both the emergence and the decline of a generic technology.

Our observations on technological competition have shown that markets, in the presence of *increasing returns to adoption*, tend to become very unstable and tipping – i.e. to discriminate sharply between winners and losers – often on the basis of minimal, perhaps almost random, market share differences among the various offers and regardless of the relative merit and potential of a new technology. From two comparable competing technologies A and B (see Figure 10.3) in a market characterized by unbounded increased returns to adoption, only one will win the race for dominance (lock-in). But, *a priori*, it is hard to determine which technology will tip the market (indeterminacy). Furthermore, it is not always sure that the market will select the superior option (sub-optimality). Then, the market is not necessarily an optimal selector of optimal technologies.

Another aspect the work on technological competition under increasing returns has shed light on is related to the implications of market instability for the management of risk. Ergodic models tacitly assume that new technologies are cheap to develop, hard to improve, stand alone, easy to appraise, easy to use, and strongly protected through patents. This implies that technologies arrive at the market fully developed before diffusion and that the process of development and the diffusion itself can be separated from each other. Ergodic models, then, introduce a very limited level of uncertainty.

In the models of indeterminate technological change, in contrast, new technologies are expensive to develop, subject to further improvements, often due to systemic nature, difficult to appraise and use and with weak patent protection. Therefore, their adoption becomes self-reinforcing not only for the *reputation effect* of market success, but also for the significant improvements that technologies accumulate during their spreading. As the *diffusion process confers value* to technologies and not only conspicuousness, a technology that initially did not deserve being chosen may end up meriting it. Consequently, according to Foray (1989), the market becomes not only a selector of adequate technologies, but a creator of dominant, superior ones. Under these circumstances, technological sponsors do not face an information problem, but an indeterminate scenario, as Arthur (1994) has indicated. Thus the most effective way to manage totally contingent and unpredictable results in unstable environments is to invest aggressively in market share as the market takes off.

10.6 Standards and increasing returns

Early models of technology diffusion explicitly recognized the general notions of adoption externalities and self-reinforcing dynamics. The epidemic model specified that the diffusion of a new technology, much as that of a potent virus, would be a self-reinforcing process since every user of the respective technology would turn into one of its ardent promoters. What these models failed to recognize, however, was that diffusion can considerably increase the value, and not simply the reputation, of the new technology. In other words, the early models were too focused on the information externalities and neglected to take into account another source of self-reinforcing in diffusion, namely, *increasing returns to adoption*.

In recent years, technological competition and the emergence of a technological monopoly over a whole market has been the privileged topic of the literature on standards (David and Greenstein, 1990). This literature has identified three processes by which technological standardization can be attained through (a) government regulation through mandated standards, (b) voluntary agreements through formal or standardization committees, and (c) market competition. The first two processes, often called *de jure* standardization, usually result in a standard with public-good characteristics. Standards selected through the market – *de facto* standards – on the other hand, are usually owned by a firm, which can therefore exclude other firms from its use. Every standardization process has its own theoretical interest, but here we will focus exclusively on *de facto* standardization, which is identified with the economic literature on technological competition under increasing returns (Besen and Farrell, 1994).

10.7 Network externalities

At the basis of what we know about technological competition is the literature on network externalities (Katz and Shapiro, 1992, 1994;

Economides, 1996), in which market size relates to increasing returns, and benefits grow with the size of competing networks.

Just as scale economies, learning, or reputation effects, positive network externalities are a self-reinforcing diffusion dynamic. Network effects, however, differ from the other self-reinforcing mechanisms in several important respects: first, while the benefits of scale economies, learning and (some) reputation effects can only be reaped at the time of purchasing the product in question, most of the benefits accruing from network externalities can be enjoyed well past the point of purchase and throughout the entire life cycle of the product. Second, network effects are considerably *forward-looking* and *less bounded* and therefore more powerful than scale and learning effects. In fact, because they cast a shadow into the future, network effects can combine with reputation effects to create extremely powerful and lasting self-reinforcing dynamics in market success. Since most of the benefits accruing from network externalities can be enjoyed throughout the full life cycle of a product, new users faced with a multiplicity of competing technical options will have to make their choices not simply as a function of what the majority of post purchasers have already chosen, but also as a function of what the majority of *interface* users are likely to choose. Interestingly, while very pessimistic user expectations about an overall emerging market can be self-dampening, optimistic expectations about the success of a specific technical format in a battle of standards could easily become *self-fulfilling*. The greater the number of people who think that a given technical option is likely to become dominant, the more new users will side with it and the more likely and pronounced the dominance will in fact become. Third, while scale economies and learning can only be a source of increasing returns to adoption and while users' learning costs (or switching costs) can exclusively be a source of inertia, both reputation effects and network externalities, in contrast, can act as both strong inertial and increasing returns to adoption.

10.8 A summary on increasing returns in industrial competition

The analysis so far has gradually recognized the central role of increasing returns mechanisms in generating and sustaining dominant firms and technologies. Clearly, while scale economies, the resource-based loop, the Schumpeterian (innovation-based) loop, reputation, and the different categories of learning help to explain some of the most basic occurrences of dominant firms and slanted industrial structures, only increasing returns to adoption – a notion intrinsically connected to Schumpeterian economics – can explain most of the instances of technological dominance which we see in contemporary high-technology markets. Thus the integration of all these increasing returns mechanisms, as Figure 10.1 shows, results in a quite complete explanation of industrial competition.

Such an explanation combines the self-reinforcing loops based upon resources and innovations (loops A and B), scale economies, learning, and reputation (loops C) with the loops based upon increasing returns to adoption. These loops are of two kinds. A set of further loops, composed of a mesh of scale economies, learning, reputation effects, infrastructure effects, and network effects, links increasing competitive advantages with increasing returns to market share. A last set of loops indicates that, if increasing returns to adoption are present and considerable, market share becomes a strategic asset well worth investing on in an aggressive manner through vigorous production capacity expansion, price reductions, infrastructure development, and alliances with manufacturers of complementary technologies, products, and services.

As a model of increasing returns mechanism in industrial dynamics, the one described here seems to be rather comprehensive. It can explain the polarized outcomes that are common in most industrial sectors and describes business competition as a dynamic and cumulative process. A final but no less interesting feature of this model of industrial competition is that it is general, in the sense that it is capable of describing the three levels of industrial competition simultaneously. It elucidates how technological adoption, the number and growth of firms, and industrial structure combine and cause each other. In other words, the general model of industrial competition gives a picture of how industrial competition is a process in which technological competition affects the size of the firm's competition in a given industry and how the growth of the firms, in turn, influences the structure of that industry. That is why technological adoption goes between conduct and performance in the chain of causality that leads from the size of the firms to industrial structure.

10.9 Conclusions: increasing return mechanisms and technological competition

The strategic importance of increasing returns to technology adoption is unquestionable. In a strictly ergodic market technological options eventually obtain the share of the market they deserve in proportion to their value and technical possibilities. In non-ergodic situations, in a tipping market, on the other hand, the winner takes all or most and the losers (no matter how worthy and how many of them there are) lose all or much. Because of this, the introduction of factors causing tipping markets determines the outcomes of technological competition. This framework not only captures the interplay of institutional arrangements, resources and network of firms and industries in industrial competition, but also delineates very concrete regularities, which can provide us with a simple but powerful tool to explain endogenously and in a dynamic way the firm's growth, industrial structure, and technological competition.

This chapter suggests a Schumpeterian model of industrial competition. In contrast to the approaches that underline a specific aspect of industrial competition and/or base their explanation on a reduced number of factors and against conventional economics that overlooks increasing returns mechanisms, this model links the increasing returns mechanisms that determine endogenously interfirm asymmetries and the kind of industrial structure which emerges during this competition process. This framework also emphasizes the fact that the emergence of dominant firms and the evolution of industrial structure are strongly intertwined with the process of technological change and diffusion.

One of the most important contributions of the work on increasing returns is its having shown that the emergence and persistence of technological monopolies is not an exogenous datum, largely determined by scientific and technical parameters, but is strongly influenced by strong market forces stemming from self-reinforcing mechanisms. In the presence of strong, global and long lasting increasing returns, the actions and omissions of the main actors in the industry in question considerably affect the final result of technological competition. To the extent to which these actors are capable of fully perceiving and exploiting strong increasing returns in emerging markets, they can ensure the entrenchment of their technology as the industrial standard by investing in those strategies that bring about market share. Once entrenched, and to the extent they are capable to exploit inertial forces, established firms can ensure the persistence of their technologies well beyond the time warranted by their relative technical value.

This chapter showed that technological monopoly should be attributed to increasing returns in general, but it is network externalities in particular that has caused some important markets to be dominated by a technological monopoly. In fact, at the level of technological competition, the only thing the other sources of increasing returns to adoption do is to exacerbate the implications of network effects, but they do not turn, isolated from network effects, a market tipping. The most obvious and direct reason for technological monopoly is that the components of a given network are compatible and constantly interconnected. The telephone and the fax are examples of networks where physical interconnection and compatibility have led to technological monopoly.

Winner-takes-all markets are associated with cases where there is (often intense) competition in innovative activity but the future market is such that competition in it is, over a reasonable timeframe, not sustainable. Thus firms compete to attain a position of dominance.

Perhaps the most famous example of a winner-takes-all market is that for operating systems for desktop PCs. It is instructive to recognize that this market benefits from massive economies of scale in production protected by intellectual property (IP) rights (very low marginal cost of supply compared with very large fixed costs of initial product development) and substantial economies of scale in consumption (due in large part to the network effects associated with the relationships between the operating system market and the related applications software markets).

Identification of such markets is important because it affects the focus of competitive concerns. Most obviously, if there are strong grounds to believe that a future market is a winner-takes-all market, it is perhaps not appropriate for a competition authority to block a merger or agreement between firms on the basis that this will create a dominant position or lessen competition in this future market. By definition, the nature of the market is such that its existence guarantees that a firm will be dominant on it, at least in the medium term. (This illustrates an important point relevant to wider issues in competition policy: it is typically better to have a situation where a firm is dominant in a relevant market than for that market not to exist at all.) Instead, any intervention must be based on the premise that the merger (agreement) lessens or distorts competition on some other, perhaps related market, or in competition in the innovative activity associated with the winner-takes-all market.

Similarly in dominance cases, if we anticipate that a market is subject to winner-takes-all properties, then it is difficult to establish a case that a firm has abused its dominant position in monopolizing this market – the market is naturally prone to monopolization. Rather, analysis of an alleged abuse of dominance associated with this market should focus on how a dominant position in a related market (perhaps an access market) could be used to distort competition in the innovative activity associated with the winner-takes-all market, or how a dominant position in the winner-takes-all market could be abused to maintain that position – in effect, used to distort competition in the innovative activity associated with the future generation of that market.

Indirect network effects may also tilt the market in favour of one of the competing technologies. In the videofilm industry, for instance, because of strong, long-lasting and global network externalities, technological monopoly emerged and the product cycle has lasted about 20 years. The strength, duration and scope of increasing returns in the videofilm markets are the direct consequence of particular technical characteristics of the competing VCR technologies. These technologies were quite similar and mature, the usage cycle of their compatible content – videofilms – is very short, and their potential substitutes have not had large enough advantages so as to replace them. In the videogame industry, in contrast, because of weak, short-lasting and local network externalities due to incomparability of videogame technologies, the long usage cycle of videogames, and the proliferation of new, more powerful new formats, different technologies have tended to share the market according to their intrinsic value.

A comparison between the videogame industry and the videofilm industry, then, allows us to show that compatibility constraints are not a sufficient condition for a virtual network technology to become a monopoly: it is necessary, besides, that usage life of content is short and core technologies are incomparable. But when the usage cycle of content is long and core technologies are comparable, any virtual network technology becomes less systemic. As the strength, duration, and scope of network externalities are reduced considerably because of the long-lasting usage cycle and technology comparability, competing virtual network technologies end up sharing the market in proportion to their value.

In this chapter we also advanced our understanding of the technical and institutional factors which are likely to affect the nature of technological competition. In doing so, we add to the dimensions of *strength and duration*, the dimension of *scope* of increasing returns.

The distinction between strength, duration and scope is useful to realize that, contrary to popular and academic literature, a market for virtual network technologies with content dimensions will not necessarily end up with a technological monopoly. With this distinction in mind and against those who think that strong indirect network externalities always act as tipping mechanisms, we also can show that strong indirect network externalities are compatible with fairly ergodic market dynamics, if the scope of such externalities happens to be rather narrow.

By taking the telecommuncations or the videofilm industry, for example, we can show that strong network externalities are *necessary* but not *sufficient* conditions to produce technological monopoly. Short usage life of content and technology incomparability are technical and necessary conditions for technological monopoly in software-intensive virtual network technologies to happen. But these technical aspects of virtual network technologies are not a sufficient condition to produce technological monopolies. In Chapter 4 we showed that network externalities require not only high levels of strength, but also to be global in scope. Under certain institutional conditions strong indirect network externalities may be rather localized, which leads to very ergodic market results. In these conditions markets are shared by the competing technologies according to their intrinsic value.

Technical and institutional factors causing different levels of strength, duration and scope of increasing returns to adoption are relevant to determine whether a market is tipping or ergodic. This has some implications. A first implication is that not all network technology is equally systemic. If there are strong network externalities but with a local scope, the systemic nature of a network becomes rather limited. In this case the systemic nature of the VCR network would be rather local compared with the actual network, which is global in nature.

Strength, duration, and scope of increasing returns are also useful to determine in a more detailed way the nature of *cooperation*. When network externalities are strong and global, content-intensive virtual network technologies become rather systemic. In these circumstances, the main sponsors of the competing technologies may produce some components of the system, but the rest of it may be out of their reach. For instance, a PC producer may be incapable of producing software or microprocessors; and microprocessor producers may not be able to produce software or hardware. VCR producers cannot produce films, in the same way that film producers cannot produce VCRs. Consequently, technological competition in markets characterized by strong and global increasing returns is more in connection with complex networks of firms than with conventional industrial array of firms producing homogeneous products. In contrast, in markets with weak and local network externalities, competition takes place mainly between firms than between networks of firms. This is so, because in this kind of markets products are not systemic.

References

Abernathy, W. J. and Utterback, J. M. (1978) 'Patterns of innovation in technology', *Technology Review*, June–July, 40–47.

Abernathy, W. J. and Clark, K. B. (1985) 'Innovation: mapping the wind of creative destruction', *Research Policy*, **14**, 3–22.

Aghion, E. and Howitt, E. (1998) *Endogeneous Growth Theory*, Cambridge, MA: MIT Press.

Arrow, K. (1962) 'The economic implications of learning by doing', *Review of Economic Studies*, **29**, 155–173.

Arthur, W. B. (1989) 'Competing technologies, increasing returns and lock-in by historical events', *The Economic Journal*, **99**, 116–131.

Arthur, W. B. (1990a) 'Positive feedbacks in the economy', *Scientific American*, February, 92–99.

Arthur, W. B. (1990b) "Silicon Valley' location cluster. when do increasing returns imply monopoly', *Mathematical Social Science*, **19**, 235–251.

Arthur, W. B. (1994) *Increasing Returns and Path Dependence in the Economy*, Ann Arbor: The University of Michigan Press.

Arthur, W. B., Ermoliev, Yu. M. and Kaniovski, Yu. M. (1987) 'Path-dependent processes and the emergence of macro-structure', *European Journal of Operational Research*, **30**, 294–303.

Arthur, W. B. and Lane, D. A. (1993) 'Information contagion', *Structural Change and Economic Dynamics*, **4**, 81–104.

Besen, S. M. and Farrell, J. (1994) 'Choosing how to compete: strategies and tactics in standardization', *Journal of Economic Perspectives*, **8**, 117–131.

Chandler, A. D. (1990) *Scale and Scope: Dynamics of Industrial Capitalism*, Cambridge, MA: The Belknap Press of Harvard University Press.

Cohen, W. M. and Levin, R. C. (1989) 'Empirical studies of innovation and market structure', in *Handbook of Industrial Organization*, R. Schmalensee and R. D. Willing (eds), Amsterdam: North-Holland.

David, P. and Greenstein, S. (1990) 'The economics of compatibility standards: an introduction to recent research', *Economics of Innovation and New Technology*, **1,** 3–41.

Economides, N. (1996) 'The economics of networks', *International Journal of Industrial Organization*, March.

Foray, D. (1993) 'Standardisation et concurrence: des relations ambivalentes', *Revue d'Economie Industrielle*, **63**, 30–45.

Gottinger, H.W. (2003) *Economies of Network Industries*, London: Routledge.

Henderson, B. D. (1975) 'The market share paradox', in *The Competitive Economy*, Y. Brozen (ed.), Morristown: General Learning Press, 286–287.

Hirsch, W. Z. (1956) 'Firm progress ratios', *Econometrica*, **24**, 136–143.

Katz, M. and Shapiro, C. (1992) 'Product introduction with network externalities', *The Journal of Industrial Economics*, **40**, 55–84.

Katz, M. and Shapiro, C. (1994) 'System competition and network effects', *Journal of Economic Perspectives*, **8**, 93–115.

Nelson, R. and Winter, S. G. (1982) *An Evolutionary Theory of Economic Change*, Cambridge, MA: Harvard University Press.

Nelson, R. (1986) 'Evolutionary modeling of economic change', in *New Developments in the Analysis of Market Structure*, J. Stiglitz and G. F. Matthewson (eds), Cambridge, MA: MIT Press.

Porter, M. E. (1980) *Competitive Strategy*, New York: Free Press.

Rosenberg, N. (1982) *Inside the Black Box: Technology and Economics*, Cambridge: Cambridge University Press.

Schumpeter, J. A. (1942) *Capitalism, Socialism, and Democracy*, New York: Harper.

Shapiro, C. and Varian, H. R. (1999) *Information Rules*, Boston: Harvard Business School.

Teece, D. (1986) 'Profiting from technological innovations', *Research Policy*, **15**(6), 285–305.

Utterback, J. M. (1994) *Mastering the Dynamics of Innovation*, Boston: Harvard Business School Press.

Utterback, J. M. and Suarez, F. F. (1993) 'Innovation, competition and industry structure', *Research Policy*, **22**, 1–21.

11 Strategy and alliance formation

> Focussing on developing and managing corporate alliances has been a fundamental part of our strategy of becoming the premier Internet media firm.
>
> We could not have accomplished our success as a single integrated firm – and neither could our partners or competitors.
>
> AOL Senior Executive (2000)

11.1 Introduction

Alfred Chandler's work, *Strategy and Structure*, illuminated the impact of the rise of the railroad and related industries during the nineteenth century on the development of the modern capitalist corporation. During the Industrial Revolution, fixed assets produced value through the creation and distribution of hard goods, often across long distances. The wide geographic, temporal and financial requirements of managing and operating a railroad compelled the invention and evolution of substantially more sophisticated and structured corporate organizations. Earlier organizational forms and business practices were incapable of effectively managing in this new environment. As such, the railroads compelled the evolution of the modern global corporation. Later, the breadth and depth of modern corporations encouraged the evolution of ever more complex functional hierarchies to coordinate disparate resources within expansive, multi-divisional firms (Chandler, 1962 and 1990).

We witness a similar phenomenon as we enter the twenty-first century. In contrast to the Industrial Age, the emerging economy primarily generates value through the creation, dissemination and application of knowledge. Since the 1980s, networking technologies have created a dynamic similar to that of the railroad in the nineteenth century, influencing options for corporate structures, relationships and competition. The most important and far-reaching impacts occur as a result of how people and firms use these tools to create value. The conflict between proprietary ownership as a necessary means for profit and the social nature of knowledge comprises the fundamental dynamic compelling the transformation of organizational forms during the present period.

In partial response to this challenge, building and maintaining corporate alliances has become an increasingly important capability for the pursuit of both operational efficiencies and competitive advantage. Alliances have been transforming competition and, as such, corporate strategy. The primary objective of this chapter is to develop and apply a new framework providing insight into the impact of alliances on strategy. A broader, but ultimately more important objective, is to begin to consider the impact of our transition to a knowledge-focused economy and the relationship of this transition to the emergence of new organizational forms.

Section 11.2 shows how managerial strategy drives competition but in turn is driven by competition. Section 11.3 shows how transaction cost economics on one side limits strategic manoeverability but on the otherside provides new opportunities to get ahead of rivals. The role of corporate governance in view of achieving superior competitive performance of the firm is discussed in Section 11.4. Section 11.5 builds the case for alliance formation to strengthen competitive performance. Section 11.6 pinpoints the dimensions of network economics in achieving leading market positions. Section 11.7 adds the time dimension as the facilitator of change management.

11.2 The dilemma of strategy: commitment, uncertainty and change

Constructing and executing strategy, proactively or passively, requires predictions, yet the future always presents uncertainty. Herein lies the dilemma of the strategist – committing resources to an uncertain future. As market and technology change accelerates, uncertainty becomes of increasing concern. In a most basic challenge to strategic planning, in the view of Porter (1980), how can firms effectively position themselves and their offerings if marketplace positions keep changing? Alternatively, executing on strategy in the real world requires firms to commit to directions they believe will provide sustainable profits. Commitment theory, as advanced by P. Ghemawat, argues that strategy requires resource investments that might be difficult or even impossible to recover from in the future if too many strategic decisions turn out to have been wrong (Ghemawat, 1991). Strategically phrased, you cannot arrive anywhere in particular if you do not commit to a direction – and you cannot know in advance if you have chosen a good destination. Developing advantageous competency and resource combinations requires substantial time and effort, so re-direction can be costly.

There exists a dilemma. How can firms balance commitments to resource and competency investments against the need to remain strategically flexible? Firms require resources and competencies that enable advantages over competitors, but resource and competency investments require long-term commitments. Moreover, the primacy of cash flow in growing and operating business complicates the matter. For small, emerging firms, cash flow presents the obvious concern that, if a firm makes financial commitments in a direction that does not prove appropriate (which is most often the case with early stage firms!), it could encounter cash flow issues threatening to its viability. The cash flow of established firms relies on existing product and services lines. Commitments to new strategic directions in response to marketplace change can threaten or 'cannibalize' cash flow from existing offerings. Particularly true of disruptive technologies, which will re-enter our discussion in a number of guises, established firms face the uncertainty not only of technology change, but also the consequent uncertainty over the market viability of their existing products (Christensen, 1997).

With this dilemma, we arrive at a broader statement of a number of strategic issues addressed by technology strategy over the past few years. Christensen's notion of disruptive technologies focuses on the challenge of firms successful in established technologies which are caught unprepared to deal with upstarts whose technologies undermine and in some cases obsolesce the dominant firms' core offerings. Hamel and Prahalad's work advises leaders to map a course to continual transformation and renewal of competencies, recognizing the condition that while marketplace change requires varied competencies over time, competencies require time to develop (Hamel and Prahalad, 1994). Ultimately, the dilemma of strategy arises from uncertainty – the fact that no one can perfectly predict the future.

Uncertainty represents a crucial but often neglected aspect of both academic and applied approaches to strategy. Porter's Five Forces paradigm seems to recommend that a careful analysis of the marketplace enables a firm to most effectively position itself and its offerings, requiring a periodic re-evaluation of market conditions. Hamel and Prahalad's 'Strategic Intent' and *Competing for the Future* (1989 and 1994) take a more dynamic approach, prescribing that firms proactively define the future and create it. A bold approach, but one that seems to de-emphasize the risks if assumptions about the future turn out to have been wrong.

Comparative analyses of the cost implications of organizational alternatives dominate the transaction cost-based strategy literature. If one cannot define alternatives prospectively, how can managers determine cost-minimizing decisions in the present? Uncertainty regarding the future presents a fundamental rationale for the need for robust corporate strategy, but much of the literature seems to avoid its implications.

Uncertainty of a different sort played an important role in the early development of transaction cost theories of the firm. In this context, vertical integration often arises due to uncertainty of supply or capabilities, or as a result of 'opportunism' that might compromise a firm's ability to meet market demand at a reasonable cost (Williamson, 1975, 1985). Ronald Coase's (1937) original insight regarding transaction costs as the rationale for the existence of firms, as well as their boundary determinant, arose as a result of his observations of the Ford Motor Company in the 1930s. Generally, Ford vertically integrated as a result of the lack of a sufficient supply of acceptable quality inputs.

Uncertainty of this kind encourages vertical integration. The well-known 'holdup problem' (Milgrom and Roberts, 1992) provides an example. It is uncertainty regarding access to and control of the resources necessary to execute. The increasing specialization, depth and breadth of the economy over the past two centuries has substantially decreased uncertainty regarding the marketplace availability of inputs or capabilities to accomplish economic objectives. If one needs something made or accomplished, one can probably (out)source it.

Certainly, the control over resources provided by vertical integration still provides a compelling rationale for the formation and boundaries of firms; however, uncertainty of another sort increasingly dominates as we move to the future. Rapid technology and marketplace change threatens firms by challenging the relevance of every product, service and business model. This represents uncertainty regarding what to execute or, by extension, what, exactly, to integrate. Which competencies and other resources will provide a competitive advantage over the long term, most effectively integrated into the firm?

While this issue presents the most significant challenge in dynamic markets, even mature industries can transform as a result of many factors, such as changing regulatory regimes, new entrants, technologies or business models, or a significant re-direction by an established player. Moreover, smaller firms typically do not have the capability or resources necessary to achieve a broadly integrated firm, even if they have the philosophical wherewithal. The greater the uncertainty regarding the future, the more firms must construct a portfolio of strategic options, without compromising the focus necessary to successfully compete. So, effective strategy requires commitments toward an uncertain future, while marketplace change threatens to undermine or even destroy a firm's value-creating capabilities.

Markets characterized by rapid technology change exhibit high levels of alliance formation compared to mature industries, which tend to exhibit consolidation and even decline (Hagedoorn, 1993). The high frequency of alliance formation in emerging industries will become evident when we look at the biotechnology and pharmaceutical industries, which evidenced a nine-fold increase in operative alliances between 1993 and 2000 (Gottinger, 2004). More mature industries encountering reorganization or transformation also experience increases in alliances, as occurred when the global chemical industry experienced severe over-capacity in the 1980s (Bower and Rhenman, 1985). Alliances and mergers allowed the industry to rationalize production in a manner that competing firms could not have accomplished individually. These alliances represented an industry-wide alternative to consolidation through mergers and acquisitions (M&As), which also occurred during this period.

So, marketplace actors apply network strategies in periods of significant change and uncertainty. This seems to contradict the notion that uncertainty engenders vertical integration. We are dealing with differing types of uncertainty, but the responses are more similar than they might appear. Vertical integration and alliances both integrate activities more closely than in the open market. To an extent, alliances represent integration by other means. None the less, alliances present significant strategic and managerial contrasts to integrated firms, not least of which is the opportunity for firms to interact with a broader, more diverse – and potentially more productive – universe of knowledge-creating opportunities than is possible within a single firm.

Let us first examine its precedents in the established organizational economics and strategy literature.

11.3 Strategy and transaction cost economics

Understanding firm behaviour and performance requires understanding of the nature of firms. As such, most academic approaches to understanding firm strategy have been grounded in the theory of the firm (TOF) literature. The classic to which we will continuously refer is the transaction cost theory of the firm, or transaction cost economics (TCE; Coase, 1937). Firms exist to decrease transaction costs for the exchange and employment of resources relative to transaction costs for similar resources in the marketplace. Over the past decade information and communication technologies (ICTs) have begun to transform the costs to manage and execute transactions between marketplace participants. Given the notion of transaction costs as a fundamental explanation for the existence and boundaries of the firm, significant changes in transaction cost dynamics in the marketplace should eventually be reflected in changes in the structure and relationships between firms. The proliferation of emerging inter-firm organizational forms over the past decade – from strategic joint ventures to deep outsourcing arrangements and virtual firms – has coincided with the expansion and increasing sophistication of ICTs. This has not been merely a correlation. Many organizing opportunities exist today for firms that were impossible or impractical prior to the past decade. Given that the Internet offers opportunities to dramatically decrease transaction costs for a wide range of transactions, a transaction cost-based theory of the firm must conceptually predict changes in the size, boundaries and structure of firms (Gottinger, 2003).

Transaction costs in the marketplace include factors such as search, selection, negotiation, fulfillment and enforcement. Within firm boundaries, costs generally include agency and control costs. The hybrid condition of any sort of inter-firm governance of resources, whether through formal contractual or informal relationships, expresses transaction costs in the form of coordination costs; that is, the costs of coordinating endeavours between autonomous firms. The evolution of ICT technologies, as well as increasingly sophisticated inter-firm management capabilities, have decreased these coordination costs, thus encouraging governance regimes between the hierarchical control of the firm, and the contractual governance of the market. The trend toward outsourcing of a broad range of activities illustrates an early and rapidly diffusing form of deep inter-firm relationships.

For example, in the early 1990s, Eastman Kodak Corporation became a pioneer in outsourcing information technology (IT) services. While many corporations outsourced various aspects of their IT operations, Kodak was one of the first large, multinational corporations to outsource the majority of its corporate IT function. After a long period of planning, Kodak outsourced to three firms: the Digital Equipment Corporation (DEC), BusinessLand and IBM. Although a Harvard Business School case on the outsourcing strategy (Applegate and Montealegre, 1995) reports overall success of the outsourcing venture, the process of building the integration and management processes between these four firms proved more challenging than Kodak's CEO originally imagined. Kodak did achieve cost savings as a result; however, the overall cost savings were lower than anticipated. Kodak found that by outsourcing, they saved costs from shifting IT personnel and infrastructure to firms whose core competencies were IT services. None the less, outsourcing introduced new costs involving developing staff and processes to manage, or rather, coordinate the inter-firm efforts. Control costs were swapped for coordination costs.

The TCE perspective biases the discussion of this example toward costs of achieving business results. Other motivations for outsourcing should include higher quality and improved corporate focus on the part of all firms in outsourcing relationships. Effective outsourcing should provide equal or higher quality service than would be possible inhouse, or at least a satisfactory level of service, at a similar or lower cost. Thus, outsourcing should represent net value creation between the contracting entities, allowing each to focus on its core competencies. While the TCE perspective approaches a successful outsourcing arrangement as a more efficient governance configuration, focusing on costs can neglect or de-emphasize potential benefits of greater value creation.

The transaction cost theory of the firm presents transaction costs reduction as the organizing principle of firms; as such, TCE should prescribe any organizational form that provides the lowest transaction costs for the economic value created. It is partly a result of the efficiency and effectiveness of the modem corporate form that a transaction cost approach has provided theoretical justification for the existence of firms. It is also partly historical. Firms, in the traditional definition, exist, so they must therefore provide transaction cost advantages over market governance, given TCE's foundation in bounded rationality (Simon, 1947; Williamson, 1985) and the path-dependent nature of organizational and institutional development (Nelson and Winter, 1982; Hannan and Freeman, 1989). Moreover, the original context within which Coase developed his transaction cost insights to understanding firm existence and boundaries likely biased the foundation of the approach. Coase extrapolated from his experience working at the Ford Motor Company in the 1930s, recognizing the company's vertical integration strategy.

He postulated that vertical integration was justified in order to decrease the transaction costs that the firm would have otherwise incurred purchasing inputs in the open market. In the 1930s the industrial production model defined by Ford strove almost exclusively for operational efficiency. TCE is fundamentally an efficiency-based approach to understanding firms. Industries and strategies defined by operational efficiencies seem best-suited to TCE. A number of researchers have asserted over the past two decades that inter-organizational relationships are often not well explained by TCE, particularly when efficiency issues do not represent the primary decision factors (Mariti and Smiley, 1983; Prahalad and Hamel, 1990; Zajac and Olsen, 1993). Decisions to pursue new firm competencies, or to form long-term R&D alliances, are unlikely to be atomizable into discrete transactions.

11.4 Corporate governance

Alliances can be a critical strategic mechanism for the long-term success of ventures; however, alliances include a broad range of types, characterized by a number of factors such as integration, co-ownership, trust and longevity. Varied forms of alliance constellations also create varied constraints for participants, from the long-term commitments into which they enter, to other potential partner opportunities which participants forego in order to build and maintain membership. For instance, membership in certain strategic blocks can preclude involvement with competing blocks, the clearest example being the automotive supplier marketplace (Nohria and Garcia Pont, 1991). In order to examine alliances effectively as strategic mechanisms, we must first identify a dimension that insightfully differentiates organizational forms.

Alliances and partnerships subsume a very broad range of potential relationships. The objective in creating a governance dimension is to consistently relate organizational forms from monolithic hierarchy (firm) to pure market relationships (purely contractual). While numerous factors present the opportunity to differentiate alliances, present research is interested in the strategic impact of hybrid organizational forms. As much of the strategic literature is based in the theory of the firm, we will develop a governance dimension grounded in this literature. The governance and contractual (Demsetz, 1993) theories of the firm prove particularly useful.

Alliances represent a variation of the governance perspective's focus on the firm or market governance decision characterized by some form of contractual relationship, whether explicit, implicit or both.

The contractual tradition, a derivative of TCE most notably represented by Jensen and Meckling (1976), Demsetz (1985, 1993) and Stinchcombe (1990), understands the firm as an agglomeration of contracts, both explicit and implicit. This perspective unifies the nature of the relationship of the firm with its employees, managers and owners, with the firm's relationship with other firms in the marketplace by characterizing them all as varied forms of contracts. The organizational sociologist Stinchcombe extends this notion to reflect inter-firm alliances as a series of explicit and implicit contracts that govern their structure in a 'pseudohierarchical' fashion; in essence, extending the notion of hierarchy beyond the confines of an individual firm (Stinchcombe, 1990). He discusses the viability of employing a dimension from 'hierarchy' to 'contract' (market) for various analytical purposes regarding the organization of information outside firm boundaries. The dimension suggested in the present research relates to this approach in that it reflects the relative degrees of coordination and control – governance – involved in inter-firm relationships.

Demsetz's contractual approach to the firm provides useful insights for developing a dimension of organization from monolithic hierarchy to pure market governance. He characterizes firms as 'a bundle of commitments to technology, personnel, and methods, all contained and constrained by an insulating layer of information that is specific to the firm, and this bundle cannot be altered or imitated easily or quickly' (Demsetz, 1993). His approach provides a strong foundation for a hierarchy/market dimension, given his interest in questioning what a firm is, in addition to the traditional TCE questions such as why firms exist, and what constitutes boundary conditions. He sees firms as a nexus of contracts, implicit and explicit. However, this could just as easily describe a constellation of corporate alliances. In order to differentiate a firm from marketplace contracts, he identifies three factors that are characteristic of firm-like coordination: specialization, continuity of association, and reliance on direction.

The notion of networks of firms can be very broadly defined. It presents integration as a dimension of corporate relationships from a monolithic hierarchy to a theoretically pure market relationship (Child and Faulkner, 1998). Gulati defines strategic alliances broadly as 'any voluntarily initiated cooperative agreement between firms that involves exchange, sharing, or co-development, and it can include contributions by partners of capital, technology, or firm-specific assets' (Gulati, 1999). The present discussion will focus on the more integrated network forms: strategic alliances and dominated networks.

So, when do network arrangements foster a competitive advantage, relative to firm or market organization? When are assets or resources worth more as part of an alliance than within an individual firm or the open market? How do networks assist in the creation, appropriation and sustainability of value?

11.5 The driving forces behind alliance formation

After all, why do firms form alliances?

The implications of network strategies vary substantially depending on the purposes for which a particular network of firms forms, as well as the purposes for which the network actually operates (which are not always identical). The creation and management of alliance constellations must be understood in light of motivations for their creation.

While transaction cost economics presents a compelling approach to understanding firm boundaries (Milgrom and Roberts, 1992, chapter 2), managers clearly engage in mergers and acquisitions (M&A), firm growth and divestiture, reorganization and other boundary-shifting initiatives for motivations quite distinct from transaction cost minimization. An extensive survey of decision-makers involved in alliances in the early 1980s found that none of them cited decreasing transaction costs as a primary rationale for their decisions (Mariti and Smiley, 1983). Marketplace competition encourages firms to minimize costs, but it also compels firms in other directions: to acquire firms to pre-empt or respond to competitors, to adjust strategic vision in the face of disruptive technologies, to innovate to create value over the long term, to name a few. Innovation by its nature requires investment in the creation of new knowledge and capabilities, whether these be capabilities new to a particular firm, or new to the marketplace. In the case of knowledge capital creation, such as in R&D partnerships, 'transaction' costs in the form of inter-firm coordination can in many cases be higher than an internally controlled effort; however, the overall value created by the combination of capabilities between firms can outweigh the increased costs of coordination. Zajac and Olsen's notion of 'transactional value' attempts to reflect the importance of value calculations in understanding inter-organizational arrangements (Zajac and Olsen, 1993).

While value creation and cost reduction reflect complementary notions for the factors that encourage firms to change their boundaries through M&A or inter-organizational arrangements, these approaches do not assist much in elucidating the complexity of motivations influencing organizational decisions. In broad terms of the field of industrial organization, cost and value creation provide appropriate generalizations. For the purposes of applied corporate strategy, cost and value paradigms are by themselves woefully limiting, particularly given an ever-uncertain future.

Economists attempt to reflect the behaviour of market actors based on preferences, available data and other decision factors, as well as provide prescriptions for the most effective economic actions as a result of the insights. The study of corporate strategy attempts to decipher why firms organize and act the way they do, how perhaps they should act regarding the development and execution of effective strategies (but often do not), and the outcomes associated with various scenarios. The fact that much of the strategy literature seeks grounding in economics reflects their complementary nature. Implicit in both of these disciplines is the notion of motivation (Milgrom and Roberts, 1992, Chapter 16). Economic behaviour, while observable in the marketplace, can only be modelled in the classical sense of optimizing functions (even given bounded rationality) by making assumptions about the motivations of market actors, known as self-interested behaviour. Strategy can only be understood by examining the motivations and purposes for which firms act. The primary motivation of any for-profit firm should be some combination of increasing share holder value and maximizing profits. This observation is too general to provide helpful insights into network strategy; none the less, understanding why firms form alliances to achieve higher profits or shareholder value can help elucidate under what general conditions firms might benefit from an alliance as opposed to governing a resource internally or sourcing it from the market.

Examining 'economic motivations' for alliances differs to some extent from understanding the broadest possible set of conditions under which alliances exist. While motivations present a form of conditions, they do not include all conditions impacting alliances. Examples of conditions not subsumed by economic motivations include the personal relationships between firm actors, chance events (that might, however, impact the economic motivations for alliances), and the alliance history of a particular firm (Gulati, 1999). In some cases, personal relationships and social networks could be econometrically analysed, by modelling factors such as trust and decreased risk, but the point remains that a number of factors exist which are not primarily economic in nature. Since corporate strategy aims to maximize the economic effectiveness of a for-profit firm, then the most pertinent factors impacting alliances should pertain to the economic and market conditions under which alliances add value.

We can examine these conditions by understanding the motivations for alliance formation. Thus why do firms form alliances, and how must firms build and execute network strategies under these varied conditions?

11.6 Three dimensions of network formation

What are the key building essentials firms might need for network formation? There are potentially limitless reasons, but all rational, optimizing motivations for forming alliances from a firm's perspective can be categorized into three types, aside from purely financial motivations.

1. *Network economics*: a firm is attempting to compete in some manner under conditions influenced by network economies. Network economies refer to situations in which network economic phenomena

significantly influence the dynamics of a particular industry or market (Gottinger, 2003).

2. *Competencies*: a firm is attempting to augment, transform or further leverage its set of internal competencies in some manner. Competencies refer to the internal decisions a firm makes regarding the competencies it develops in-house, which competencies it chooses to source through partners, and which competencies it might contract from the market.
3. *Market structure*: a firm is attempting to compete or become involved in some manner in a market or markets where the market structure compels a firm to form alliances. Most prominently, this category includes industry structure, positioning (Porter, 1980), and institutional issues that impact competition, such as government regulation and political risk (North, 1990).

Market structure refers to the environmental, institutional and competitive factors present in a particular industry, market and/or economy.

The following list presents a summary of the examples collected.

Factors favouring alliances:

- Knowledge-sharing and creation
- Distribution
- Risk-sharing
- R&D outsourcing
- Network effects
- Leveraging complementary competencies
- Manufacturing outsourcing
- Increasing focus on specific competencies
- Valuation
- News
- Entering a new product line or line of business
- Entering a new market, geographic or offering brand leverage
- Capital access investment
- Pre-emptive move
- Market spanning.

While such a list might never be complete, most factors for alliances can be dominantly and usefully categorized into one, two or three of the types presented. The taxonomy need not be exhaustive to be effective. The point is that these three categories – network economics, competencies and market structure (NCM) – provide the dimensions that most powerfully present implications for the creation and prosecution of network strategy. Competencies refer to the internal decisions a firm makes regarding the competencies it develops in-house, which competencies it chooses to source through partners, and which competencies it might leave for the market to provide. Market structure refers to the environmental, institutional and competitive factors of a particular industry, market and/or economy. Network economics refers to whether network economic phenomena significantly influence the dynamics of a particular industry or market under consideration. Network economics includes all of the issues traditionally associated with this field of economics, such as demand-side economies of scale, network effects, and positive and negative feedback loops (Shapiro and Varian, 1999). Network economics could be collapsed into the market structure dimension, but only at the expense of crucial explanatory power. The presence or absence of strong network economic effects on a market exerts such a strong influence on the formation of firm networks that it requires its own categorization.

Drivers for alliances: network economics

Industries heavily impacted by network economics are characterized by demand-side economies of scale, where the relative number of users can significantly impact the success or failure of a venture. Network economics most directly pertains to industries characterized by technologies which benefit from broad-based standards and interoperability, and/or are information-intensive. Shapiro and Varian's summary of the implications of network economics on information-intensive industries, *Information Rules*, underscores how strategy in markets influenced by network economics can differ substantially from those without such dynamics (Shapiro and Varian, 1999).

For instance, in some cases an open, give-it-away strategy can provide superior long-term prospects over a proprietary approach. This distinction presents interesting challenges to the definition and application of firm-specific assets, from the resource-based view of firm strategy. Sun Microsystem's JAVA programming environment provides a prototype. Sun has thus far invested hundreds of millions of dollars developing and promulgating JAVA with developers, succeeding in creating wide popularity and a broad installed user base. Can JAVA accurately be considered a firm-specific asset, given that Sun has pursued an open source strategy to encourage proliferation of the language? Despite the fact that the firm still maintains various intellectual property rights to the language, Sun has found some difficulty capitalizing on JAVA's success. The widespread success of JAVA within the developer community arose in large part from the open-source strategy pursued by Sun, which allowed easy access to the development environment (Chapter 7). Sun has pursued a network flexible strategy that has succeeded in driving diffusion of the technology. While widespread adoption has presented Sun with many opportunities, it has also encountered significant challenges in monetizing its success. Sun's variable success with JAVA has been significantly challenged by Microsoft's attempts to usurp the programming environment by amending its application within the Microsoft environment. Much later, in mid-2001,

Microsoft announced that JAVA would no longer be supported on Microsoft's newly introduced XP operating system, a critical challenge to JAVA, given Microsoft's dominating position.

Drivers for alliances: market structure

In many industries, market structure issues encourage firms to create alliance constellations. The monopoly position of Microsoft in the last section provides an example. Significant aggregation of market power, as in monopolies and monopsonies, can create situations in which firms that desire to enter these markets have very little choice but to ally with an existing player. If a firm decides to participate in the diamond trade, it must ally in some way with the DeBeers cartel. While DeBeers' grasp on the diamond trade has weakened somewhat of late, the firm has traditionally been the absolute dominant player, worldwide, and continues to be. There are no inherent network economics involved in the diamond industry, and in many cases, members of the DeBeers network might not require alliances for the sake of complementary competencies, were it not for the firm's effective monopoly. Any government-mandated monopoly falls into this category as well, but can also overlap with other categories, as in the case of national telecommunications monopolies (network economics), or national control of raw materials, mining and export. In the case of an oligopoly, existing players might, officially or unofficially, ally to maintain the status quo, as in the case of a cartel. The opposite can also be true. In highly disaggregated markets, firms might decide to ally in order to create economies of scale, lowering costs and developing bargaining power against suppliers. Economists use metrics such as the Herfindahl Index or a top four-firm concentration ratio to represent market concentration.

A literature exists on the implications of the level of market concentration in the industrial organization literature. The purpose here is not to develop and test many possible hypotheses related to market concentration and alliance formation, but rather to underscore that market structure plays a role in motivating the creation of networks of firms, which holds implications for network strategy.

Market structure dimensions are not limited to monopolistic or oligopolistic conditions. As evident from examples presented earlier, the market structure category includes institutional influences such as government regulation and antitrust issues that might motivate inter-organizational arrangements over acquisition or market relationships. Here we define 'institutional' in the terms developed by North in the economic field of institutional analysis (North, 1990). These institutional issues relate to the market structure category of the taxonomy given the fact that they influence the organizational decisions between firms (i.e. to integrate or ally) by impacting competitive market conditions, providing rules of the game. In this sense, it is not necessary to separate market structure and institutional issues into separate categories. On close inspection, the insights turn out on close inspection to be similar.

Drivers for alliances: competencies

The competencies dimension addresses firms allying in order to leverage competencies and knowledge between firms, create and/or learn new competencies and knowledge, or a combination of both motivations. It should be noted that alliances formed for specific knowledge-creation purposes, such as to license or share intellectual property, fit into this mould. Knowledge or intellectual property does not necessarily refer to a 'competency' *per se*; none the less, competencies represent the broader application of knowledge sets and capabilities to accomplish objectives through operational, managerial and learning processes. As such, the competencies motivation space includes any objectives for alliances characterized by knowledge creation, learning and/or transfer.

Competency-motivated firm networks take many forms. The simplest outsourcing arrangements can resemble market relationships, while some complex, long-term cooperative arrangements can resemble single firms. In fact, joint equity ventures are often structured as single firms, where the parent, cooperative firm owns equity interests and contributes capital, competencies and even people. The proliferation of outsourcing over the past few decades has occurred largely as a result of firms' interests in shedding internal resources dedicated to secondary or periphery competencies, such as data storage or call-centre management. EDS, a pioneer in information technology outsourcing, and EMC, an early leader in enterprise data storage, illustrate firms competing based on these types of relationships. In other cases, competency-based alliances reflect more than the 'hands-off' character of basic outsourcing relationships. The US automobile industry evolved over the final decades of the twentieth century to include fewer suppliers, longer-term relationships and significant supplier involvement in the design process. All of these developments greatly improved the industry's global competitiveness (Dyer, 1996; Gulati *et al*., 2000). But the industry's reorganization has not been limited to increased integration between firms.

For example, the Big Three US automakers (so-called before the creation of Daimler-Chrysler in the late 1990s) and their Tier One suppliers divested themselves of billions of dollars worth of assets during the 1980s and 1990s, typically in order to shift their mix of internal competencies. For example, the spinout of Delphi Automotive by General Motors created the largest supplier to the automotive industry worldwide, with revenues of close to $30 billion in 2000. Although the competencies spun out of GM as a result of the creation of Delphi were no longer integrated into GM, each firm required substantial long-term cooperative agreements in order for the spin-out to succeed. GM continued to demand Delphi's products and services, while

Delphi would have been stillborn without a close, continued relationship with its former parent.

Competency-based alliances usually require careful integration of operational, learning and managerial processes between firms. The breadth and depth of this integration depends on the extent of cooperation necessary between the firms to accomplish their objectives. Knowledge-creation alliances, such as cooperative research ventures, require the development of cooperative learning processes. It is partly a result of the fact that purely market relationships can hinder knowledge creation and transfer between firms that firms create closer associations for these purposes than others. Simple outsourcing or licensing arrangements might simply require clear contractual arrangements, including objectives, milestones, performance metrics, enforcement and benefit- and risk-sharing components. However, even seemingly simple outsourcing or licensing arrangements can pose more significant challenges than appear at first. In a classic business case, Eastman Kodak's pioneering decision to outsource a large portion of its corporate IT services to IBM, Business Land and the Digital Equipment Corporation proved substantially more complicated – and expensive – than had been anticipated (Applegate and Montealegre 1995). According to the case, coordination costs between Kodak and its outsourcing partners offset more of the savings as a result of outsourcing than had been anticipated.

None the less, Kodak's managers believed the arrangement to have been a success as of the writing of the case, and IT outsourcing has proliferated in the decade since.

While the integration of learning, managerial and operational processes presents challenges in competency-based alliances, these integrated processes can also provide a competitive capability for the allied firms.

11.7 The temporal dimension: representing change over time

Firm constellations change over time. Sometimes, this occurs as a natural process of growth, or a deepening of contact between firms.

At other times, participant firm objectives diverge. By definition, the NCM taxonomy underscores the influences that motivate firms to form alliances. As those influences change on a given constellation, so too will their nature and construction. The economics literature does address firm life cycles when considering long-term decisions. In these cases, economists consider 'entry and exit' of firms over the course of the term in question.

None the less, assumptions of firm perpetuity, both by researchers and practitioners, can be misleading. Rarely do firms last more than a few generations. Those that do usually undergo significant change over time. While firms like the Ford Motor Company continue to create value based on the original purpose for the company (automobile production), other venerable firms compete in completely different markets to their predecessors. Nokia's current dominance of the wireless handset and telecommunications industries bears very little relationship with its past, which has included long periods in forest products, paper and pulp, and as a conglomerate with many unrelated lines of business. Can the Nokia of 1865, the year of its founding, be considered the same firm as today? So, despite any legal continuity (and a company's legal status can change), firms change markedly over time. Given that firms are by no means perpetual, researchers should treat them as entities finite in time and seek to understand firm lifespans and growth cycles (Greiner, 1972; Jovanovic, 1982).

Similarly, hybrid forms should be treated as arrangements in flux, developed in response to, and as a result of, participant firms' motivations and changing marketplace conditions. The taxonomy provides varied insights into the life cycles of firm constellations. Consortia created to motivate standards development in an emerging industry, at the interface of network economics and competencies dimensions, would be likely to migrate further to the network economics after the standard has been developed and disseminated. Once the standards race has been won (not necessarily by the consortia), the motivation for the alliance recedes, and the consortia is likely to be disbanded unless alternative benefits have been developed between member firms. In contrast, alliances in the competencies dimension often take one of two predominant directions, assuming the competencies that motivated the alliances continue to be perceived as useful by the participants. Either firms in a competency-based alliance build integrated, mutually dependent competencies, which may encourage alliance longevity, or one or more of the participant firms decides to disengage from the alliance, perhaps having changed strategic priorities, or having acquired the competency in house. Often, firms form alliances to develop a competency in-house, with the assistance of a partner. Sometimes firms do so avowedly, while in other cases knowledge appropriation occurs in a more surreptitious manner. Firms can participate in an alliance until they have acquired necessary knowledge from their partner, and then defect from cooperation. Microsoft has become notorious for this, particularly with small firms. Andy Groves' famous 'paranoia' (Groves, 1998) has been part of the reason Intel's partnership with Microsoft has endured successfully for so long. Intel has maintained a healthy suspicion regarding its alliance with Microsoft. If one or more of the alliance members' strategic interest in particular competencies changes, competency alliances are most likely to be affected.

Market structure provides the most obvious example of how the factors in each motivation space can impact alliance evolution.

Government regulations can very quickly upset the competitive balances between constellations of firms, as has occurred in most instances of deregulation or privatization. The massive, global airline alliances (such as One World and Star Alliance), fit in the dimensions network economics and market structure.

Airline de-regulation in the US and elsewhere exerted a significant impact during the 1980s and 1990s, leading to a proliferation of airlines. Moreover, network economics heavily impacts the airline industry, characterized by communicating networks of routes, hub-and-spoke methods though anti-trust regulations continue to exist in force. As such, the proliferation of airlines during the 1980s, and the consolidation of the industry during the late 1980s and 1990s, was followed by the creation of global airline alliances. These alliances aim to leverage the benefits of broad and deep networks without running afoul of regulators. While the marketplace implications of these alliances remain in question, particularly from the consumer-value standpoint, these alliance forms exist as a result of changes in market structure initiated by deregulation and the nature of network economic influences on the industry. If regulators eliminated restrictions on airline mergers and acquisitions, the industry would likely witness a swift consolidation as a result of the significant demand-side and supply-side economies of scale which characterize the industry. This seems, however, unlikely, particularly at the international level, given the diverse interdependent regulatory regimes of the world's political authorities.

References

Applegate, L.M. and Montealegre, R. (1995) *Eastman Kodak Co.: Managing Information Systems through Strategic Alliances*, Boston: Harvard Business School Press, 1995.

Bower, J.L. and Rhenman, E.A. (1985) 'Benevolent cartels', *Harvard Business Review*, **68**, 124–132.

Chandler, A.D. (1962) *Strategy and Structure*, Cambridge, MA: MIT Press.

Chandler, A.D. (1990) *Scale and Scope: The Dynamics of Industrial Competition*, Cambridge, MA: Harvard University Press.

Child, J. and Faulkner, D. (1998) *Strategies of Co-operation: Managing Alliances, Networks and Joint Ventures*, New York: Oxford University Press.

Christensen, C. (1997) *The Innovator's Dilemma*, Boston: Harvard Business School Press.

Coase, R. (1937) 'The nature of the firm', *Economica*, **4**, 386–405.

Demsetz, H. and Lehn, K. (1985) 'The structure of corporate ownership: causes and consequences', *The Journal of Political Economy*, **93**, 1155–1177.

Demsetz, H. (1993) 'The theory of the firm revisited', in *The Nature of the Firm*, O.E. Williamson and S.G. Winter (eds), New York: Oxford University Press, pp. 159–178.

Dyer, J.H. (1996) 'Specialized supplier networks as a source of competitive advantage; evidence from the auto industry', *Strategic Management Journal*, **17**, 55–83.

Ghemavat, P. (1991) *Commitment: The Dynamics of Strategy*, New York: Free Press.

Gottinger, H.W. (2003) *Economies of Network Industries*, London: Routledge.

Gottinger, H.W. (2004) 'Biotech and pharma industries: application of network economies', *Pharmaceutical Industries*, **66**(12), 1475–1478.

Greiner, J. (1972) 'Evolution and revolution as organizations grow', *Harvard Business Review*, July–August, 37–46.

Groves, A. (1998) *Only the Paranoid Survive*, New York: Norton.

Gulati, R. (1999) 'Network location and learning. The influence of network resources and firm capabilities on alliance formation', *Strategic Management Journal*, **20**, 397–420.

Gulati, R., Nohria, N. and Zaheer, A. (2000) 'Strategic networks', *Strategic Management Journal*, **21**, 203–215.

Hagedoorn, J. (1993) 'Understanding the rationale of strategic technology partnering: interorganizational modes of cooperation and sectoral differences', *Strategic Management Journal*, **14**, 371–386.

Hamel, G. and Prahalad, C.K. (1994) *Competing for the Future*, Boston, MA: Harvard Business School Press.

Hannan, M. and Freeman, J. (1989) *Organizational Ecology*, Cambridge, MA: Harvard University Press.

Jensen, M.C. and Meckling, W.H. (1976) 'Theory of the firm, managerial behaviour, agency costs and ownership structure', *Journal of Financial Economics*, **3**, 305–360.

Jovanovic, B. (1982) 'Selection and the evolution of industry', *Econometrica*, **50**, 649–670.

Milgrom, P. and Roberts, J. (1992) *Economics, Organization and Management*, Englewood Cliffs, NJ: Prentice Hall.

Mariti, P. and Smiley, R.H. (1983) 'Co-operative agreements and the organization of Industry', *Journal of Industrial Economics*, **31**, 437–451.

Nelson, R.R. and Winter, S. (1982) *An Evolutionary Theory of Economic Change*, Cambridge, MA: Harvard University Press.

Nohria, N. and Garcia-Pont, C. (1991) 'Global strategic linkages and industry structure', *Strategic Management Journal*, **12**, 105–124

North, D.C. (1990) *Institutions, Institutional Change and Economic Performance*, Cambridge, Cambridge University Press.

Porter, M.E. (1980) *Competitive Strategy*, New York: Free Press.

Prahalad, C.K. and Hamel, G. (1990) 'The core competence of the corporation', *Harvard Business Review*, **68**, 79–91.

Shapiro, C. and Varian, H.R. (1999) *Information Rules*, Boston, MA: Harvard Business School Press.

Simon, H. (1947) *Adminstrative Behavior*, New York: Macmillan.

Stinchcombe, A.L. (1990) *Information and Organizations*, Berkeley, CA: University of California Press.

Williamson, O.E. (1975) *Markets and Hierachies. Analysis of Antitrust Implications: A Study in the Economics of Internal Organization*, New York: Free Press.

Williamson, O.E. (1985) *The Economic Institutions of Capitalism*, New York: Free Press.

Zajac, E.J. and Olsen, C.P. (1993) 'From transaction costs to transactional value analysis: implications for the study of interorganizational strategies', *Journal of Management Studies*, **30**, 131–145.

12 Aggregate technological racing: economic growth, catching up, falling behind and getting ahead

> A country's potential for rapid growth is strong not when it is backward without qualification, but rather when it is technologically backward but socially advanced.
>
> Moses Abramovitz (1986).

12.1 Introduction

Moving beyond the firm-led racing patterns revolving in a particular industry to a clustering of racing on an industry level is putting industry in different geoeconomic zones against each other as they are becoming dominant in strategic product/process technologies. Here racing patterns among industries in a relatively free-trade environment could lead to competitive advantages, more wealth creating and accumulating skill dominance in key product/process technologies in one region at the expense of others. The question is whether individual races on the firm level induce similar races on the industry level and, if so, what controlling effects may be rendered by regional or multilateral policies on regulatory, trade and investment matters.

Similar catch-up processes are taking place between leaders and followers within a group of industrialized countries (or even emerging economies) in pursuit of higher levels of productivity. Moses Abramovitz (1986) explains the central idea of the catch-up hypothesis as the trailing countries' adopting behaviour of a 'backlog of unexploited technology'. Supposing that the level of labour productivity were governed entirely by the level of technology embodied in capital stock, one may consider that the differentials in productivities among countries are caused by the 'technological age' of the stock used by a country relative to its 'chronological age'. The technological age of capital is an age of technology at the time of investment plus years elapsing from that time. Since a leading country may be supposed to be furnished with the capital stock embodying, in each vintage, technology which was 'at the very frontier' at the time of investment, the technological age of the stock is, so to speak, the same as its chronological age.

While a leader is restricted in increasing its productivity by the advance of new technology, trailing countries have the potential to make a larger leap as they are provided with the privilege of exploiting the backlog in addition to the newly developed technology.

Hence, followers being behind with a larger gap in technology will have a stronger potential for growth in productivity. The potential, however, will be reduced as the catch-up process goes on because the unexploited stock of technology becomes smaller and smaller. This hypothesis explains the diffusion process of best-practice technology and gives the same sort of S-curve change in productivity rise of catching-up countries among a group of industrialized countries as that of followers to the leader in an industry.

Although this view can explain the tendency to convergence of productivity levels of follower countries, it fails to answer the historical puzzles why a country, the United States, has preserved the standing of the technological leader for a long time since taking over leadership from Britain at around the end of the last century and why the shifts have taken place in the ranks of follower countries in their relative levels of productivity, i.e. technological gaps between them and the leader. Abramovitz poses some extensions and qualifications on this simple catch-up hypothesis in his attempt to explain these facts. Among other factors than technological backwardness, he lays stress on a country's 'social capability', i.e. years of education as a proxy of technical competence and its political, commercial, industrial, and financial institutions. The social capability of a country may become stronger or weaker as technological gaps close and thus, he states, the actual catch-up process 'does not lend itself to simple formulation'. This view has a common understanding to what Mancur Olson (1982) expresses to be 'public policies and institutions' as his explanation of the great differences in per capita income across countries, stating that any poorer countries that adopt relatively good economic policies and institutions enjoy rapid catch-up growth. The suggestion should be taken seriously when we wish to understand the technological catching-up to American leadership by Japan, in particular, during the post-war period and explore the possibility of a shift in standing between these two countries. This consideration will directly bear on the future trend of the state of the art which exerts a crucial influence on the development of the world economy.

Steering or guiding the process of racing through the pursuit of industrial policies aims to increase competitive advantage of respective industries, as having been practiced in Japan (Gottinger, 1992, 1998), in that it stimulates catch-up races but appears to be less effective in promoting frontier racing. A deeper reason lies in the phenomenon of network externalities affecting high-technology industries. That is, racing ahead of rivals in respective industries may create external economies to the effect that such economies within dominant 'increasing returns' industries tend to improve their international market position and therefore pull ahead in competitiveness *vis-à-vis* their (trading) partners (Krugman, 1997).

The point is that racing behaviour in leading high-technology industries by generating frontier positions create cluster and network externalities pipelining through other sectors of the economy and creating competitive advantages elsewhere, as supported by the 'increasing returns' debate (Arthur, 1996). In this sense we speak of positive externalities endogenizing growth of these economies and contributing to competitive advantage.

Let us briefly recall the pattern of industrial racing and the implications of the way the firms in major high-technology markets, such as telecommunications, split clearly into the two major technology races, with one set of firms clearly lagging the other technologically. The trajectories of technological evolution certainly seem to suggest that firms from one frontier cannot simply jump to another trajectory. Witness, in this regard, the gradual process necessary for the firm in the catch-up race to approach those in the frontier race. There appears to be a frontier 'lock-in' in that once a firm is part of a race, the group of rivals within that same race are the ones whose actions influence the firm's strategy the most. Advancing technological capability is a cumulative process. The ability to advance to a given level of technical capability appears to be a function of existing technical capability. Given this path dependence, the question remains: why do some firms apparently choose a path of technological evolution that is less rapid than others? Two sets of possible explanations could be inferred from our case analysis, which need not be mutually exclusive. The first explanation lingers primarily on the expensive nature of R&D in ICT industries which rely on novel discovery for their advancement.

Firms choosing the catch-up race will gain access to a particular technical level later than those choosing the frontier, but will do so at a lower cost. Take the example of Indian drug firms, focusing on the production of generic drugs (imitation) before taking up the task of investment in innovative drug research. Similarly, Chinese network technologies in rural Chinese markets against rival Western firms (Alcatel, Ericsson, Siemens), ploughing the returns into R&D investment for cutting-edge systems, thus enabling them to compete with their rivals on a global scale. How does this process on the micro level correspond to that on a macro level?

Section 12.2 reviews the seminal work on economic development and catch-up hypothesis between countries. Section 12.3 contrasts the properties of neoclassical aginst endogenous growth models. The extension of 'leapfrogging' to the aggregate level, aggregate leapfrogging, is the theme in Section 12.4. Industrial racing between nations is the theme of Section 12.5. In Section 12.6 an endogenized neoclassical growth model is established for explaining industrial racing between nations. Conclusions are drawn in Section 12.7.

12.2 Economic development: backwardness and catch-up

We can view the growth literature as consisting of three distinct theoretical explanations of growth: the first is the neoclassical growth model of the Solow type (Solow, 1956, 1957) which predicts country convergence; the second is the endogenous growth model which, in general, predicts country divergence; and the third is the correlating theory of economic development which takes a more historical perspective on the growth of countries. The neoclassical model is based on diminishing returns to factor inputs and view the long-run rate of growth as being exogenous. Endogenous (or new) growth theory considers constant or increasing returns to factor inputs and attempts to explain the forces that give rise to technological change. Development theory includes a wide variety of approaches to economic catch-up including technological change (Gerschenkron, 1962; Abramovitz, 1986), neo-institutional economics (North, 1990; Eggertsson, 1990) and theories of institutional sclerosis (Olson, 1982). These three theoretical explanations are not mutually exclusive categories and many papers belong to more than one category. For example, Abramovitz (1986) fits also within the neoclassical explanation, but we choose to place him in the development camp.

We study whether the neoclassical model can explain the empirical finding of relative income shifts. When finding the standard neoclassical model insufficient, we extend it using ideas from the development approach, thus adding some endogeneity to the resulting model. The research is not a direct extension of the early papers mentioned above. Since 1986, in particular, there has been a flood of papers which deal with the growth of nations, catch-up, and transfer of technology.

The importance of a nation's institutional framework (such as government regulation and laws) to growth is probably shared by most, if not all, people and economists. Surely the existence of positive and negative externalities, the legal environment, transaction costs, culture and environmental pressures will affect the growth of a nation (Gottinger, 1998). However, the neoclassical model neglects these factors in an effort to clearly show the implications of a competitive market structure.

In addition to factor accumulation and institutional soundness, technology also plays an important role in growth. Thus, part of our argument addresses technology diffusion, a concept dating back to Gerschenkron (1962) who proposed in *Economic Backwardness in Historical Perspective* that a backward country by the very virtue of its backwardness will tend to develop very differently from the advanced country. He states that 'industrialization always seemed the more promising the greater the backlog of technological innovations which the backward country could take over from the more advanced country' (p. 6). The important idea that catch-up might occur due to a backlog of technology will not be traced back historically. Suffice it to say that it is an idea which is far from dead, as exemplified by recent papers in the *American Economic Review: Papers and*

Proceedings, 1996 (Crafts, 1997; Romer, 1997). These ideas are clearly expressed in a paper by Abramovitz (1986), from which we derive much motivation for what follows. It is worth pointing out, however, that modern development economics emphasizes the importance of ideas, which are also emphasized in the endogenous growth models. The model we look at has some endogeneity in it, but it is mostly an extension of the neoclassical model.

Abramovitz (1986) asserts that being backward in level of productivity carries a potential for rapid growth (the catch-up hypothesis). If the level of labour productivity were given by the level of technology embodied in the capital stock, then the 'leading' country's capital stock embodies the frontier technology at the time of investment, therefore, as Abramovitz states 'the technological age of the stock is ... the same as its chronological age.'

For a follower country the technological age of its capital stock will be high relative to its chronological age. Therefore, when a leader nation invests in new capital its technology advance is limited by the advance of knowledge. For the follower, however, investments in new capital have the potential of a larger leap as the new capital could embody frontier technology. Thus, the larger the productivity gap the stronger the potential for growth in productivity, other things held constant. This is the catch-up hypothesis in its simple form. Note its similarity to the factor accumulation argument made earlier.

The argument laid out by Abramovitz does not end here though. Technological backwardness is not usually a historical accident – societal characteristics probably account for a large portion of a country's past failure to achieve a high level of productivity. These same characteristics, or social capabilities, may remain to keep a country from making the full technological leap proposed by the simple hypothesis. This implies that the simple catch-up hypothesis needs some modification. Abramovitz argues that 'a country's potential for rapid growth is strong not when it is backward without qualification, but rather when it is technologically backward but socially advanced.' Thus being technologically backward is a necessary condition for catch-up, but it is not sufficient. A follower nation must also be able to adopt and adapt the technology which is potentially available to it. These intuitively appealing ideas have been missing from the empirical growth literature and we will point out how consistent they will be with this literature.

Convergence

Baumol (1986) showed that convergence could be observed in some groupings of countries (industrialized market economies and planned economies), but not in others (less-developed economies). The groups which displayed convergence were then called 'convergence groups'. De Long (1988) criticizes Baumol for performing a regression that uses an *ex-post* sample of countries which are rich and developed at the end of the period, and argues for the use of an *ex-ante* sample. De Long then shows that such a sample does not exhibit convergence. The conclusions and results presented in the papers of Baumol and De Long may not be as relevant as their impacts on empirical growth studies. The important point for the evolution of growth theory is the fact that convergence clubs as well as divergence clubs had been identified. In particular, the world as a whole showed evidence of divergence. Two theoretical frameworks have been used to explain these results. One attempts to explain convergence (the neoclassical growth model), the other divergence (the endogenous, or new, growth model).

At first, the empirical convergence literature tried to find out what the steady state distribution of world per capita income and productivity would look like. This is the question which motivated Abramovitz (1986) and Baumol (1986). Since then the empirical growth literature has mostly attempted to explain the cross-country data; in particular, to explain convergence and rates of convergence and to interpret the findings in the context of neoclassical and/or endogenous growth theory. However, as mentioned by Romer (1986), cross-country comparisons of growth rates are complicated by the difficulty of controlling for political and social (institutional) variables that strongly influence the growth process. Thus one must attempt to control for these political and social variables.

There have been many studies using the neoclassical model and its steady-state predictions to identify the sources of growth. First of all, if all economies have the same steady state then unconditional convergence is expected; that is, economies with low initial incomes should have higher growth rates. If economies differ in their steady states, then conditional convergence should be observed; that is, after controlling for steady-state differences, initial incomes should be negatively related to growth rates. Not all convergence studies choose to consider the steady state, some take a more direct growth accounting approach. The end results are essentially the same in terms of sources of growth, but the latter approach is not able to consider the transitional dynamics.

Early catch-up literature showed much concern for technological backwardness while the endogenous growth theory highlights the importance of technology.

Endogenous growth theory emerged out of a dissatisfaction for basically two things: first, some growth theorists were not happy with the exogenously driven explanations of long-run productivity growth, and second, the data for large samples of countries did not show convergence. This led to a construction of models where the determinants of growth are endogenous. Romer (1986) and Lucas (1988) are usually mentioned as the instigators of this movement. The Romer (1986) model contains increasing returns as a result of accumulation of knowledge which has a positive external effect on the production possibilities. Lucas (1988) is based on learning-by-doing which leads to (external) increasing returns to scale in the production process. Neither model postulates a theory of technological change. However, research and development, imperfect competition, and internal returns to scale were added to this class of models in Romer (1987, 1990), Grossman and Helpman (1991), Aghion and Howitt (1992), and Jones (1995a). Technological advance is the result of R&D activity which is rewarded some *ex-post* monopoly power. This means that, as long as the economy does not run out of ideas, the growth rate can remain positive in the long run. Also, the long-run growth rate is now dependent on government policies.

Less work has been done on the magnitude of international R&D spillovers; how R&D expenditures in one country affect the growth of other nations. Nadiri and Kim (1996) provide a study of, among other things, how R&D spillovers have affected the productivity of the G7 countries. International spillovers are shown to affect the growth of total factor productivity significantly for all seven countries. The importance of own R&D to foreign R&D spillovers vary across countries, e.g. for the US, own R&D is much more important than spillovers, while for countries like Canada and Italy foreign R&D is more important than own R&D for their productivity growth.

12.3 Neoclassical versus endogenous growth models

As mentioned above, there are numerous articles that show the existence of convergence. Empirical work has been done to either verify the neoclassical model or simply to look at whether a particular variable is important for growth. This approach, if not the results, has been criticized on econometric grounds (see below).

The number of empirical papers which attempt to verify the endogenous growth model are quite few. Until recently, support for this model consisted of regressions showing the non-existence of convergence. The failure of per capita output to equalize across the developed and the developing world economies as well as the failure of growth rates of developing countries to exceed those of the industrialized West were seen as evidence that there is little observable tendency for poor countries to catch up with richer ones (see for example Pritchett, 1997). Lately there have been time-series studies which more directly attempt to find evidence of endogeneous growth.

The results from these time-series models, using a variety of criteria, are mixed. Jones (1995b, 1995c) argues that the scale effect which is predicted by the endogenous growth model is not observed in the data. Lau (1996) also finds evidence unfavourable to the endogenous growth model, while Neusser (1991) finds time-series evidence which is favourable to the exogenous growth models, at least for some countries. The results from Lau and Sin (1997) are unfavourable to both classes of growth models. On the other hand, Kocherlakota and Yi (1996, 1997) find time-series evidence slightly more favourable to the endogenous growth model using data from the United States and the United Kingdom.

After having considered the empirical results from both conditional convergence studies and tests of endogenous growth models, it is hard to conclude that either of the two models towers over the other in terms of empirical relevance. Thus we will choose to utilize the neoclassical model because of its tractability. However, given this choice of theoretical approach, we must bring up some common objections to the neoclassical growth model. Most of what follows is based on Mankiw *et al.* (1992).

First and foremost, is the neoclassical model a good theory of economic growth? In particular, can the model shed any light on growth when its steady-state growth is only due to exogenous technological progress?

As Mankiw *et al.* (1992) argue, it depends on what the purpose of growth theory is. If the goal is to explain the existence of growth, then obviously the neoclassical model is uninformative. However, if the goal is to explain why there is such variation of economic growth in different countries and in different times, then the neoclassical model's assumption of constant, exogenous technological progress is not a problem. In fact, the neoclassical model is well-equipped for shedding light on the cross-country growth experiences.

A final concern here is whether the neoclassical model is able to explain phenomena such as leapfrogging and divergence. The divergence of nations has been used as one argument for the new growth theory. However, the neoclassical model would only be inconsistent with divergence if all countries have identical steady states. If poor nations are converging to low steady states while rich nations are converging to even higher steady states, then divergence is a possibility. Also, if all or some of the countries are already in their steady states then no convergence should be observed. In fact, as pointed out by Barro (1997), even if convergence held, the dispersion of per capita output would not necessarily narrow over time. The reason is that this could depend on the weighing of the convergence force relative to the effects from shocks hitting each country. Such shocks, if independent across countries, tend to create dispersion. Leapfrogging can also be similarly explained by differences in steady states. In fact, in our extended (endogenized) neoclassical model, Section 12.6, we explore whether different steady states are sufficient to explain the observed leapfrogging using the neoclassical framework. Despite all the potential problems of the empirical convergence literature, the cross-sectional evidence leans toward the neoclassical model. For example, a recent paper by Young (1995) looks at the East Asian countries, perhaps the group least likely to fit the neoclassical model, and concludes that their success is explained by the neoclassical model.

12.4 Aggregate leapfrogging

We will argue that 'leapfrogging', i.e. shifts of relative income positions, is important and a significant characteristic of cross-country growth.

Given this, the question is whether the growth models discussed above are consistent with leapfrogging. The neoclassical growth model has attempted to explain the cross-country data in terms of convergence and rate of convergence. The Solow model predicts, in general, that countries converge to their own steady states; it assumes identical technologies in all countries and concludes that exogenous differences in saving and education are the cause of all observed disparity in levels of income and rates of growth. The neoclassical model's predictions with regards to leapfrogging are apparent: in a group of homogeneous countries (as defined by the similarity of steady states) no leapfrogging should be observed. The diminishing return to capital provides a vehicle for convergence, but there is no mechanism for shifts in relative positions. If, however, countries are approaching different steady states, then positions might change due to the transitional effects as shown by Jones (1995a). Also, if a random disturbance is added to a model which contains a convergence force, then one would expect shifts in relative income positions. As mentioned in Easterly *et al.* (1993), if there is a large dispersion of distances between countries' initial incomes and their steady states, then the transitional effect will dominate the effect of random shocks. New growth theory has appeared in reaction to the neoclassical model. These models consider non-convexities and economies of scale, and, in particular, focus on the incremental change in technology. In these models, investments into human and physical capital make either the same or an increasing contribution to output as economies become richer. Hence, the 'predictions' of the early endogenous growth models are that technical change proceed most rapidly in those countries with established advantages in technologically advanced sectors, the 'leaders'. This implies economic and technical divergence between nations and no leapfrogging should be observed. However, in general, endogenous growth models can lead to a variety of growth experiences. Different starting conditions and the fact that government policies are allowed to make a difference can lead to leapfrogging. For example, if long-run growth is a function of the amount of R&D conducted in a nation, then a follower could leapfrog by allocating funds to R&D. Lately a few endogenous growth models have in fact approached the issue of overtaking, discussing either growth miracles (Lucas, 1993) or leadership change (Brezis *et al.*, 1993). A more general treatment can be found in Goodfriend and McDermott (1994). Parente and Prescott (1994) provide a recent attempt to simultaneously account for disparity in income levels and growth miracles. Their model is based on differences in technology adoption barriers which may lead to both income differences and, if persistently reduced, to development miracles.

Modern development economics argues that a technology gap presents an opportunity for rapid growth through technology flows, but a country's ability to absorb the new technology must also be considered. A low absorption capability makes it difficult for a country to take advantage of its opportunity. Since poor developing nations typically suffer from both a large technology gap and a low absorption capacity, the predictions about rate of growth and convergence are ambiguous. However, high indicators of absorption capacity (e.g. high level of education or 'good' institutions) imply a faster rate of growth for a country which faces a given technology gap, hence the possibility of leapfrogging.

Cross-country performance in leapfrogging

Looking at the OECD sample (Summers and Heston, 1991), it is apparent that the nations' growth paths cross. They show the countries' per capita GDP relative to the US over the period 1960–1990. The US is the income leader for most of the years (Switzerland obtained the leader position a few times over the sample period). Three countries in particular shifted income positions. Japan went from being one of the poorest countries in 1960 (rank 19) to become quite wealthy (rank 8) in 1990. Japan appears to be a growth miracle. The same can be said for Norway, which advanced from rank 12 to rank 4 over the sample period. In contrast, New Zealand made a rapid descent through the relative income positions (from 3 to 17), earning the title growth disaster. However, most of the rank movements take place among the middle countries (those ranked 3 to 16 in 1960) which are close in per capita GDP levels. For these countries leapfrogging could be due to random disturbances or heterogeneous shocks. A closer examination of the rankings reveals that it is very common for two countries to switch positions, only to immediately switch back. A few examples are: Germany and the UK from 1961 to 1968, Japan and Italy between 1971 and 1980 (these two countries changed positions six times only to end up at the same place in 1980), and the US and Switzerland up until 1975. This shows that much of the rank dynamics are driven by short-term fluctuations. These rank movements are most likely due to country-specific fluctuations, such as lagged business cycles, and randomness. One way to remove this from the data is to consider a longer time period than one year for the analysis. Panel studies often consider 3–5 year time intervals to side-step the influence of business cycles.

12.5 Industrial racing between nations

The cumulative literature on industrialization has formalized the long-standing idea that development traps are the result of a failure of economic organization rather than a lack of resources or other technological constraints. The so-called 'big push' models of industrialization have shown how, in the presence of increasing returns, there can exist preferable states to advance the economic states of countries in contest with other countries. Such a view not only provides an explanation for the co-existence of industrialized and non-industrialized economies, but also a rationale for government intervention to coordinate investment in a 'big push' toward industrialization.

Moreover, unlike competing theories, these models emphasise the temporary nature of any policy. Thus, industrialization policy involves facilitating an adjustment from one equilibrium to another rather than any change in the nature of the set of equilibria *per se*.

While recent formalization makes clear the possible role for the government in coordinating economic activity, little has been said about the form such policy should take. Is there a conceptual model to analyse the question: what precise form should the 'big push' take? It is argued that while many different industrialization policies can be successful in generating escapes from development traps, the form of the policy that minimizes the costs of this transition depends on the characteristics of the economic situation at hand. Factors such as the strength of the complementarities, externalities and increasing returns, among others, all play a role in influencing the nature of a 'getting-ahead' industrialization policy. Such ideas were present in the debates in development economics in the 1940s and 1950s regarding the form of industrialization policy. The models underlying these less-formal debates inspired the recent more-formal research but the policy elements of these have not been addressed, to date, in any substantive way.

Principal among the earlier policy debates was that surrounding the efficacy and costs involved in the alternative strategies of 'balanced' versus 'unbalanced' growth. Rosenstein-Rodan (1943, 1961) and Nurkse (1952, 1953) provided the rationale for the notion that the adoption of modern technologies must proceed across a wide range of industries more or less simultaneously. It was argued that the neglect of investment in a sector (or sectors) could undermine any industrialization strategy. Reacting to this policy prescription was the 'unbalanced growth' school led by Hirschman (1958) and Streeten (1956).

They saw the balanced strategy as far too costly. The advantages of multiple development may make interesting reading for economists, but they are gloomy news indeed for the underdeveloped countries. The initial resources for simultaneous developments on many fronts are generally lacking. By targeting many sectors, it was argued that scarce resources would be spread too thin – so thin, that industrialization would be thwarted. It seemed more fruitful to target a small number of 'leading sectors' (Rostow, 1960). Then those investments would 'call forth complementary investments in the next period with a will and logic of their own: they block out a part of the road that lies ahead and virtually compel certain additional investment decisions.' (Hirschman, 1958, p. 42.) Thus, the existence of complementarity between investments and increasing returns motivated an unbalanced approach. Curiously, at the same time, '[c]omplementarity of industries provides the most important set of arguments in favour of a large-scale planned industrialization' (Rosenstein-Rodan, 1943, p. 205).

Both sides appeared to have agreed that a 'big push' was warranted, but they disagreed as to its composition. Our purpose here is to use the guidelines provided by the recent formalization of the 'big push' theory of industrialization to clarify the earlier debate of the appropriate degree of focus for industrialization policy. After all, the recent literature has stressed the roles of complementarities and increasing returns that both schools saw lying at the heart of their policy prescriptions (Murphy *et al.*, 1989).

The seminal article formalizing the 'big push' theory of industrialization is that of Murphy *et al.* (1989). In their model, firms choose between a constant returns and an increasing returns of technology based on their expectations of demand. However, these choices spill over into aggregate demand creating a strategic interaction among sectors in their technology adoption decisions. Thus, under certain conditions, there exist two equilibria: with all firms choosing the constant returns or all choosing the increasing returns technology. Clearly, in the latter equilibrium, all households are better off.

While the Murphy, Shleifer and Vishny model shows how increasing returns (and a wage effect) aggregate to strategic complementarity among sectors, it does not lend itself readily to the debate concerning the degree of balance in industrialization policy. First, the static content leaves open the question of whether the intervention should take the form of anything more than indicative planning. Second, the most commonly discussed policy instrument in the industrialization debate is the subsidization of investments. However, in the Murphy, Shleifer and Vishny example, use of this instrument biases one toward a more unbalanced policy. To see this, observe that it is the role of the government to facilitate a move to the industrializing equilibrium. This means that the government must subsidize a sufficient amount of investment to make it profitable for all sectors to adopt the modern technology. Given the binary choice set, there then exists some minimum critical mass of sectors that must be targeted to achieve a successful transition. A greater range of successful industrialization policies might be more plausible, however, if firms had the choice of a wider variety of technology to choose from. One might suppose that targeting a large number of sectors to modernize a little and targeting a small number of sectors for more radical modernization might both generate a big push. Thus, to consider the balanced approach properly, a greater technological choice space is required.

12.6 Modelling technology adoption in an endogenized growth model

Prerequisites

We endogenize a Solow-type growth model to allow for the transmission of technological knowledge across national borders. The standard neoclassical model assumes a closed economy and an exogenous constant saving rate to predict that countries converge to their own steady states determined by rates of accumulation and the depreciation rate. However, in addition to having different accumulation rates, economies also differ in levels of technology. This introduces the possibility that flows of technology may present an additional opportunity for growth. Thus, adoption of technology from abroad is one possible mechanism through which the capital stock of a nation increases, as better technology improves the productivity of the existing stock of capital. The receiving nation would therefore appear to have more capital if better capital is equivalent to more capital. The possibility of adoption of knowledge and ideas is especially clear if we take a very broad view of capital by including both human and physical capital in its definition.

Flows of technology are analogous to capital mobility and labour mobility (if each migrant carries some amount of capital) since the capital stock is in effect augmented. The extension to incorporate cross-national technology flows implies that economies are open to some extent, that is, at least ideas and technical knowledge are able to travel across national borders.

Whereas physical capital tends to flow from economies with low rates of return to those with high rates of return and labour tends to travel from low-wage to high-wage nations, technology flows from very productive economies with high levels of technology to the technological laggards. The model with technology flows will differ from a model with labour or capital mobility in that technology flows are non-exclusive; i.e. flows of technology benefit the receiving economy without hurting the source economy. In contrast, for labour and capital migration the gains in population and capital stock for the destination economy represent corresponding losses for the source economy.

Replacing the closed economy nature of the traditional Solow model by a partially open economy potentially affects a nation's steady-state and transitional dynamics. The results are similar to those derived for capital and labour mobility, which are that mobility tends to speed up an economy's convergence toward its steady state. It will also come out that technology flows might augment the level of that steady state.

We enhance the possibility of technology adoption in the Solow model of a closed economy by allowing a cross-national flow of knowledge but assume that the economy is closed with respect to foreign assets and foreign labour. Thus in this setup ideas and knowledge can flow across national borders independently of capital and labour migration.

The assumption of immobility of physical capital and labour is strong, but it serves for analytical purpose to single out some effects of technology on the growth process.

The model

The model is for the most part identical to the standard neoclassical model which assumes a Cobb–Douglas production function

$$Q_t = K_t^{\alpha}(A_t L_t)^{1-\alpha}$$

and exogenous growth for population and technological progress $L_t = L_0 e^{nt}$ and $A_t = A_0 e^{gt}$.

The only difference from the standard model appears in the equation for the evolution of capital. The capital evolution depends on an exogenous savings rate, the depreciation rate, and a technology catch-up term, $\gamma(T, T^*)$, with a benchmark term T^* of the technological leading country (region) so that

$$\dot{K}_t = sQ_t - \delta K_t + \gamma(T,T^*)_t K_t.$$

It is instructive to point out the difference to models of purely disembodied technical change. These models specify capital evolution as $\dot{K}_{it} = sQ_{it} - \delta K_{it}$ so that the stock K_t can be interpreted as new-machine equivalents implied by the stream of past investments [and δ, depreciation, is the weight that transforms each vintage investment into new-machine equivalents]. We assume, in contradistinction, that new investment might also embody differences in technical design. Thus a new 'machine' may be more efficient than an old 'machine' even if there is no difference in physical capacity. The standard capital evolution equation will then tend to understate the true productivity of the capital stock. In this setup, technology from abroad may make the existing and new capital stock more productive and therefore increase the capital stock (capital is measured in efficiency units).

Transforming the model into an 'intensive form' model so that all variables are divided by $A_t L_t$, the Cobb–Douglas production function becomes

$$y_t = f(k_t) = k_t^{\alpha}$$

and the capital evolution equation becomes

$$\dot{k}_t = sk_t^{\alpha} - (\delta + n + g)k_t + \gamma(T,T^*)_t k_t.$$

This means that the growth rate of capital intensity, k, is given by

$$\dot{k}/k = c_k = sk^{\alpha-1} - (\delta + n + g) - \gamma(T,T^*)),$$

so the effective depreciation rate $(\delta + n + g) - \gamma(T, T^*)$ includes the term $\gamma(T, T^*)$.

Thus the adoption of foreign technology acts to reduce the rate of effective depreciation. In the standard Solow model a lower rate of effective depreciation yields a higher steady state, so one expects this to be true in the present model as well.

Technology adoption function and technology gap

The new results derive from the technology adoption function, $\gamma(T, T^*)$. Assume that the adoption of technology is a function of an economy's technology gap to the leader, defined as the nation with the highest level of technology. The economy is then able to adopt some fraction of this gap every time period. The simplest definition of the technology adoption function would then be $\gamma(T,T^*)_t = \rho(T_t^* - T_t)$, where ρ denotes the technology adoption rate. The measurement of technology is difficult, as no variable captures it perfectly. Possible candidates such as number of patents or number of PhDs are elusive. As a possible proxy one can make the assumption that technology is a function of the economy's capital intensity. In particular, technology will be a logarithmic function of the intensive level of capital, $T_t = \ln(k_t)$. This implies that technology is a positive, diminishing function of capital intensity (see Figure 12.1). From these assumptions one obtains

$$\gamma(T,T*)_t = \gamma(\ln(k), \ln(k^*))_t = \rho[\ln(k_t^*), \ln(k_t)].$$

Or equivalently, $\gamma(\ln(k), \ln(k^*))_t = -\rho \ln(k_t/k_t^*)$, and the technology adoption function is decreasing in k.

The technology gap could be put in terms of $\ln(k/k^*)$, convex decreasing with larger k, k^*. Also, an important characteristic of the technology function is that it is bounded below by zero; that is, having a higher level of technology than the 'leader' will never hurt you. This point goes back to the non-exclusive character of technology so that the leader nation's rate

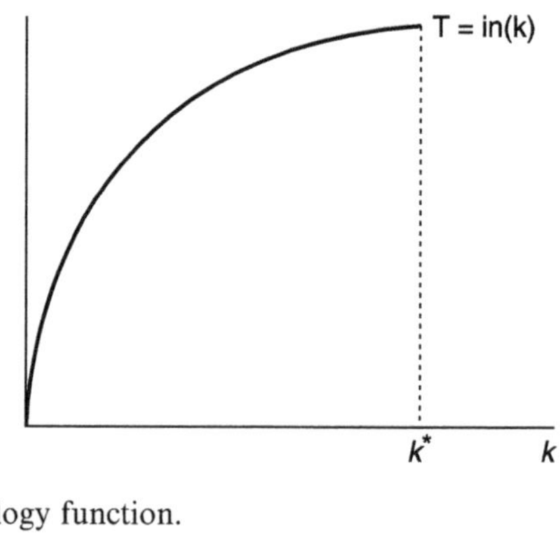

Figure 12.1 Technology function.

of growth is never hurt by the fact that the economy has the highest level of technology.

Another crucial characteristic of the adoption function is that technology flows are one-directional. That is, technology only flows from the leader to the followers. This is an assumption which would likely be violated in a multi-sectoral world, but perhaps can be justified in terms of net flows.

The steady state

Figure 12.2 is the standard Solow growth diagram augmented by the technology adoption function. The $sf(k)/k$ curve is downward-sloping as usual because of the diminishing average product of capital. The commonly horizontal line at $(\delta+n+g)$ has been replaced by the upward-sloping curve $(\delta+n+g-\gamma(T, T^*))$. The height of the effective depreciation curve is $(\delta+n+g)$ at $k > k^* = k^w$ (at benchmark) since at these capital intensity levels the technology gap facing the economy is removed (zero).

The steady state corresponds to the intersection of the $sf(k)/k$ and $(\delta+n+g)-\gamma(T, T^*)$ curves at the point k^*. In the diagram we have drawn two possible $sf(k)/k$ curves depending on the saving rate. The obvious result is that a higher saving rate $(s_2 > s_1)$ will lead to a higher steady state, but more interestingly the saving rate will also determine whether my steady state will differ from the standard Solow steady state. In the figure,

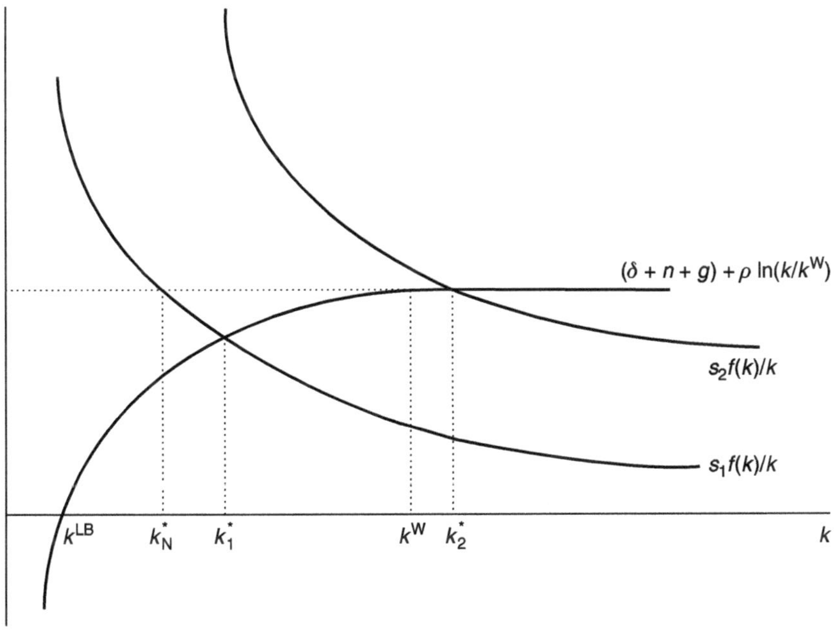

Figure 12.2 Steady state growth.

the higher saving rate corresponds to a steady state outcome which is identical to the one expected from the Solow model (as case 2). On the other hand, the lower saving rate (s_1) leads to a steady state greater than predicted by the standard neoclassical model since k^* now corresponds to a point where the follower economy is a recipient of technology in its steady state. That is, the economy will remain in steady state as a perpetual receiver of technology (as case 1). This reception of technology allows the economy to reach a steady state which is above the one predicted by its saving rate and $(\delta + n + g)$.

An implication of the model is that steady-state capital intensity is bounded below since $sf(k)/k$ is non-negative. Its lower bound is given by k^{LB} in Figure 12.2 and is equal to $k^{w}\exp\,(\delta + n + g)/\rho$.

We can also use Figure 12.2 to assess the effects of changes in the model's various parameters on the steady-state values. As seen above, the effect from an increase in the saving rate is a higher steady state. The same is true if the production function were to shift outwards from an increase in the capital-share coefficient, α. This result stems from the fact that the rate of diminishing returns from capital decreases as α increases. The other possibility is a changing effective depreciation rate. The rate of depreciation will increase from an increase in either δ, n, or g, or from a decrease in the technology adoption function resulting from either a lower ρ or a lower k^*, for any given k. The decrease in the effective depreciation rate results from the obvious (reversed) changes of parameters. An increase (decrease) of the effective depreciation rate leads to a lower (higher) steady state. One does not have to refer to this picture in order to find the steady state; this point can also be derived. The definition of the steady state is that the capital intensity does not change; that is, $k = 0$. This implies that

$$\dot{k}_t = 0 = sk_t^{\alpha} - (\delta + n + g)K_t + \gamma(T,\ T^*)_t k_t. \tag{12.1}$$

The solution of this equation for case 2 when there is no technology gap in the steady state [i.e. $\gamma(T,\ T^*) = 0$] is given by

$$k^* = [s/(\delta + n + g)]^{1/1-\alpha}.$$

The corresponding steady state output per effective unit of labour is given by

$$y^* = [s/(\delta + n + g)]^{\alpha/1-\alpha}.$$

For case 1 when the economy is a perpetual recipient of technology, the steady-state solution is given by

$$k^* = [s/(\delta + n + g) - \gamma(T,T^*)]^{1/1-\alpha}$$

or similarly,

$$k^* = [s/(\delta + n + g) + \rho \ln(k^*/k'))]^{1/1-\alpha}$$

where $k^*/k' \leq 1$. Hence, if $k^* < k'$, then the economy's steady state will be affected by the technology gap. Correspondingly, the steady state output per effective labour for case 1 is given by

$$y^* = [s/(\delta + n + g) + \rho \ln(k^*/k')]^{\alpha/1-\alpha}.$$

Transitional dynamics and rate of convergence

To assess the speed of convergence log-linearize equation (12.1) around its steady state. Since there are two kinds of steady states, there will also be two distinct convergence rates. The derivations yield that the rate of convergence for case 1 is equal to $\beta = (1-\alpha)\ (\delta + n + g)$, which is exactly the same as for the standard Solow model. However, although the rate of convergence is identical and the economy reaches its steady state in the same amount of time, the actual growth path will be very different with the present model compared to the standard Solow model. For case 2, the rate of convergence is also determined by the rate of technology adoption, ρ, and is given by $\beta = (1-\alpha)\ (\delta + n + g) + \rho\ [1 + (1-\alpha) \ln\ (k^*/k^w)]$, where the first term is the rate of convergence which is obtained from the standard Solow model, and the second term is a non-negative additional convergence factor stemming from the adoption of foreign technology. The latter term as non-negative stems from the fact that $sf(k)/k$ is always a positive number and, hence, steady-state capital has a lower bound (as described above). Thus, the 'typical' economy for which $k^* = k^w$, that is all economies have identical steady states, and when assuming that $\rho > 0$, then the above equation shows that the potential for technology adoption raises the convergence coefficient, β, above the Solow value by the amount of ρ. An interesting finding which differs from the standard model is that the rate of convergence will now depend on the steady-state position, k^*.

In the standard model the rate of convergence only depends on $(1-\alpha)$ $(\delta + n + g)$, thus the saving rate does not affect the speed of convergence and neither does the level of technology, A. These results stem from the fact that in the Cobb–Douglas case s and A produce two offsetting forces which exactly cancel each other. The two forces are: (1) given k, a higher saving rate leads to greater investment and therefore higher speed of convergence, and (2) a higher saving rate raises the steady-state capital intensity, and thereby lowers the average product of capital in the vicinity of the steady state. Again, these two forces exactly cancel in the Cobb–Douglas case. In this model, however, when $\rho > 0$, β increases with the steady-state capital intensity. Of course, this is only true as long as $k^* < k^w$ and the reason is that a higher k^* implies a higher steady-state technology adoption. A permanent improvement in the production function or a higher level of saving raise the steady state as well as increasing the rate of convergence. A final point is that the standard result that an increase in the capital share,

α, leads to lower convergence speed is possibly compromised since a higher α lowers the convergence effect from technology adoption (if $k^* < k^w$).

Convergence path

Technology adoption introduces the possibility of rapid growth in addition to being below the steady-state position. However, once the technology gap has been exploited, the economy is left with the traditional source of growth, namely the difference $sf(k)/k - (\delta + n + g)$. This is the reason why the convergence rate only depends on these factors for the case when the steady state is independent of the technology gap. However, this does not mean that an economy whose steady state is above the leader's cannot take advantage of a technology gap when such an opportunity is presented. Instead, the follower economy will be able to grow rapidly in the early stages of its catch-up due to both the diminishing return to capital effect and the adoption of foreign technology. However, once the technology gap has been bridged, the economy's capital growth is reduced to that predicted by the diminishing returns effect. Figure 12.3 shows that the rate of growth can be split up into its standard part (γ_1) and the technology adoption effect (γ_2), and that γ_2 may eventually become zero.

The fact that the convergence time will be identical to the Solow model, but that the convergence path is very different can be seen in simulations of the model. The simulations show the effect on the convergence path when an economy does or does not adopt technology when assuming identical steady states for all economies (i.e. identical saving rates). Simulations can be run for various economies which differ in initial capital stock, as well as with different adoption rates and efficiency levels of adoption (E). To contrast the present model with the standard, it would make sense in simulations to add

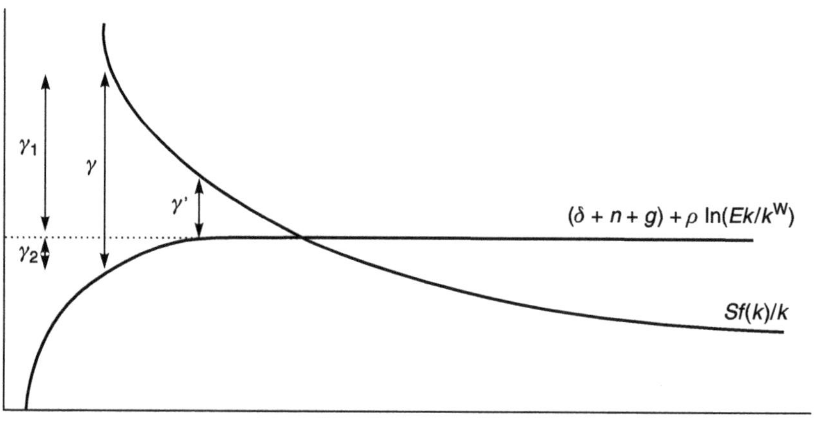

Figure 12.3 Rate of growth.

technology adoption to one of the follower countries and assume the income leader also to be the technological leader. It shows that this shifts the convergence paths significantly without changing the economies' steady states. However, although the same steady state is reached, the economy which adopts technology will have a higher level of income at any point.

A flow of disembodied technology across economies may modify the neoclassical model's steady state and rate of convergence. One also needs to consider an economy's ability to adopt and absorb this new knowledge. One reason why economies may differ in their ability to take advantage of the technology gap is through the rate of adoption, ρ. As referred to earlier, Abramovitz (1986) proposes that the abilities of countries to take advantage of the catching-up potential depends on their respective 'social capabilities', i.e. that systematic variations in social institutions and processes make some countries better or worse at catching up. Since 'follower' countries typically suffer from both a large technology gap and a low absorption capacity, the predictions about the growth effects from the technology gap is ambiguous. As Abramovitz stated: 'a country's potential for rapid growth is strong not when it is backward without qualification, but rather when it is technologically backward but socially advanced.' In other words, a technologically backward country is more likely to catch up or even get ahead if its social institutions (predominantely, education, entrepreneurship, property rights, economic freedom) lower absorption barriers. How fast the adoption rate will be, the speed of adoption for catching up, will clearly depend on its adoption capacity, its adoption learning, and its strategic selection and following through – all of those clearly shaped by institutional mechanisms. Of particular interest would be the efficiency (E) level of an economy to adopt, an important ingredient of Abramovitz's 'social capability', that is an inefficient economy may be very slow to adopt and spread available technology through institutional inertia, lack of complementary competence and inability to digest. Thus, we may identify empirically countries through having high adoption rates but low levels of efficiencies who lag significantly in their potential of catch-up.

12.7 Conclusions

This chapter has addressed several important issues related to the growth and performance of nations. We discussed and provided some tools to analyse some patterns of catching up, falling behind and getting ahead that we identify with aggregate technological racing among nations or regional economic entities.

In Section 12.6 we constructed a structural model of the Solow type based on capital accumulation rates but extended to include the impact of technology adoption. In the model adoption of technology becomes one possible mechanism through which the effective capital stock may increase. The impact of social institutions arises by allowing the potential technology gap to be modified by them. The resulting model only slightly modifies the steady state and rates of convergence predicted by a conventional neoclassical model. However, it allows for more complex growth dynamics, including non-random relative income shifts while maintaining a high degree of tractability. This model accomplishes the goal of reconciling the neoclassical model with slow technology diffusion and institutional variations as outlined in approaches to economic development.

In identifying technological racing among countries, model predictions and empirical observations indicate that follower countries, in striving for catching up, significantly benefit from the technology gap to the leader and that adoption rates vary between them. This variance could be mostly contributed to the 'social capabilities' of those countries that demonstrate various 'efficiency levels' of adoption promoted by bureaucratic efficiency (including a low level of corruption) and democratic rights (Economist, 2005) – although it is naturally difficult and needs further research to pinpoint the specific types of institutions that act 'as driving forces behind countries' inefficiencies and some of the institutional factors may show some tradeoff pattern between them, thus limiting the net impact of each of them.

A good case in point is Latin America, which carries relatively high adoption rates but overall the region fails to take advantage of its potential because of poor political and social institutions. Other interesting cases relate to Japan, Korea and the emergence of economic rivalry between China and India (Economist, 2005; North, 2005). A similar case pointing to institutional inertia relate to divergences in EU Europe (Baily and Kirkegaard, 2004).

Unlike tracing technological rivalry among corporations and industrial entities with clearly defined market conditions, the results one obtains from the aggregated data can only point toward 'more aggregate' explanations for the factors behind cross-country growth processes. Yet, a few key factors could be singled out. First of all, we can conclude that technology diffusion is an important source of growth of follower nations. Furthermore, countries seem to differ significantly in their ability to take advantage of newer and better technology. Thus, in general, any policy that allows follower nations to better adopt foreign technology should increase their growth rate, at least in the short run. Since the difference in technology adoption appears to be related to a nation's institutional efficiency, research suggests that governments are well-advised to pursue policies that increase the efficiency of markets. That is, improved technology adoption is an added motivation for the pursuit of efficient institutions. For example, consider two follower nations that face identical initial technology gaps. Their growth paths will be very different according to their ability to close this technology gap. If a nation does not incorporate foreign technology into their production it will have to rely solely on accumulation of factor inputs as a source of growth. If, on the other hand, the economy adopts better technology, then rapid growth is expected until the technology gap is removed, at which time the nation is left with factor accumulation or coming up with new ideas as the sources of growth. However, the latter nation will be richer (more productive) at all points in time. To conclude this argument, higher income can be achieved not only from increases in the savings rate (the neoclassical prediction) but also from institutional change.

References

Abramovitz, M. (1986) 'Catching up, forging ahead, and falling behind', *Journal of Economic History*, **46**, 385–406.

Aghion, P. and Howitt, P. (1992) 'A model of growth through creative destruction', *Econometrica*, **60**, 323–351.

Arthur, B. (1996) 'Increasing returns and the new world of business', *Harvard Business Review*, **72**, 100–109.

Bailey, M. and Kirkegaard, J. (2004) 'The Landlocked Continent', *The Wall Street Journal*, 6 December.

Barro, R. J. (1994) 'Democracy and growth', NBER Working Papers #4909.

Barro, R. J. (1997) *Determinants of Economic Growth: A Cross-Country Empirical Study*, Cambridge, MA: The MIT Press.

Barro, R. J. and Sala-i-Martin, X. (1992) 'Convergence', *Journal of Political Economy*, **100**(2), 223–251.

Barro, R. J. and Sala-i-Martin, X. (1995) *Economic Growth*, New York: McGraw-Hill.

Baumol, W. J. (1986) 'Productivity growth, convergence and welfare: what the long-term data show', *American Economic Review*, **76**, 1072–1085.

Brezis, E. S., Krugman, P. R. and Tsiddon, D. (1993) 'Leapfrogging in international competition: a theory of cycles in national leadership', *American Economic Review*, **83**(5), 1211–1219.

Crafts, N. F. (1995) 'Exogeneous or endogeneous growth? The industrial revolution reconsidered', *Journal of Economic History*, **55**, 745–772.

De Long, J. B. (1988) 'Productivity growth, convergence, and welfare: comment', *American Economic Review*, **78**(5), 1138–1154.

Easterly, W., Kremer, M., Pritchett, L. and Summers, L. H. (1993) 'Good policy or good luck? country growth performance and temporary shocks', National Bureau of Economic Research (NBER) Working Paper #4474.

The Economist (2005) 'The tiger in front. A survey of India and China', 5 March, 3–16.

Eggertsson, Th. (1990) *Economic Behavior and Institutions*, Cambridge: Cambridge University Press.

Gerschenkron, A. (1962) *Economic Backwardness in Historical Perspective*, Cambridge, MA: Belknap Press.

Goodfriend, M. and McDermott, J. (1994) 'A theory of convergence, divergence, and overtaking', mimeo, Federal Reserve Bank of Richmond.

Gottinger, H.-W. (1992) 'Optimal economic growth when CO2 constraints are critical', *Energy Economics*, **10**, 192–199.

Gottinger, H.-W. (1998) *Global Environmental Economics*, Boston, MA: Kluwer Academic Publishers.

Grossman, G. and Helpman, E. (1991) *Innovation and Growth in the Global Economy*, Cambridge, MA: MIT Press.

Hirschman, A. O. (1958) *The Strategy of Economic Development*, New Haven: Yale University Press.

Jones, C. I. (1995a) 'Convergence revisited', mimeo, Stanford University.

Jones, C. I. (1995b) 'R&D-based models of economic growth', *Journal of Political Economy*, **103**, 759–784.

Jones, C. I. (1995c) 'Time series tests of endogenous growth models', *Quarterly Journal of Economics*, **110**, 495–525.

Kocherlakota, N. R. and Kei-Mu Yi (1996) 'A simple time series test of endogenous vs. exogenous growth models: an application to the United States', *Review of Economics and Statistics*, **78**, 126–134.

Kocherlakota, N. R. and Kei-Mu Yi (1997) 'Is there endogenous long-run growth? Evidence from the United States and the United Kingdom', *Journal of Money, Credit, and Banking*, **29**, 2–20.

Krugman, P. (1997) 'Myths and realities of U.S. competitiveness', *Pop Internationalism*, Chapter 6, P. Krugman (ed.), Cambridge, MA: MIT Press.

Lau, S-H. P. (1996) 'Testing the Long-Run Effect of Investment on Output in the Presence of Cointegration', *Working paper*, Australian National University, Canberra.

Lau, S-H. P. and Sin, C-Y. (1997) 'Observational equivalence and a stochastic cointegration test of the neoclassical and Romer's increasing returns models', *Economic Modelling*, **14**, 39–60.

Lucas, R. E. Jr (1988) 'On the mechanics of economic development', *Journal of Monetary Economics*, **22**, 3–42.

Lucas, R. E. Jr (1993) 'Making a miracle', *Econometrica*, **61**(2), 251–272.

Mankiw, N.G., Romer, D. and Weil, D. N. (1992) 'A contribution to the empirics of economic growth', *Quarterly Journal of Economics*, **107**, 407–437.

Mauro, P. (1995) 'Corruption and growth', *Quarterly Journal of Economics*, **110**, 681–712.

Murphy, K. M., Shleifer, A. and Vishny, R.W. (1989) 'Industrialization and the big push', *Journal of Political Economy*, **97**, 1003–1026.

Nadiri, M. I. and Kim, S. (1996) 'International R&D Spillovers, Trade and Productivity in Major OECD Countries', C.V. Starr Center for Applied Economics, New York University, *Research Report* #96–35.

Neusser, K. (1991) 'Testing the long-run implications of the neoclassical growth model', *Journal of Monetary Economics*, **27**, 3–37.

North, D. C. (1990) *Institutions, Institutional Change and Economic Performance*, Cambridge: Cambridge University Press.

North, D. C. (2005) 'The Chinese menu (for development)', *The Wall Street Journal*, 7 April.

Nurkse, R. (1952) 'Some international aspects of the problem of economic development', *American Economic Review*, **42**, 571–583.

Nurkse, R. (1953) *Problems of Capital Formation in Underdeveloped Countries*, Oxford: Basil Blackwell.

Olson, M. (1982) *The Rise and Decline of Nations: Economic Growth, Stagflation, and Social Rigidities*, New Haven: Yale University Press.

Parente, S. L. and Prescott, E. C. (1994) 'Barriers to technology adoption and development', *Journal of Political Economy*, **102**(2), 298–321.

Pritchett, L. (1997) 'Divergence, big time', *Journal of Economic Perspectives*, **11**, 3–17.

Romer, P. M. (1986) 'Increasing returns and long-run growth', *Journal of Political Economy*, **94**, 1002–1037.

Romer, P. M. (1990) 'Endogenous technological change', *Journal of Political Economy*, **98**, S71–S102.

Romer, P. M. (1996) 'Why indeed in America? Theory, history and the origins of modern economic growth', *American Economic Review* 86, Papers and Proceedings, 202–206.

Rosenstein-Rodan, P. N. (1943) 'Problems of industrialisation of Eastern and Southeastern Europe', *Economic Journal*, **53**, 202–211.

Rosenstein-Rodan, P. N. (1961) 'Notes on the theory of the "big push"', in *Economic Development for Latin America*, H.S. Ellis and H.C. Wallich (eds), New York: St. Martin's.

Rostow, W. W. (1960) *The Stages of Economic Growth*, Cambridge: Cambridge University Press.

Solow, R. M. (1956) 'A contribution to the theory of economic growth', *Quarterly Journal of Economics*, **70**, 65–94.

Solow, R. M. (1957) 'Technical change and the aggregate production function', *Review of Economics and Statistics*, **39**, 312–320.

Streeten, P. (1956) 'Unbalanced growth', *Oxford Economic Papers*, 167–190.

Summers, R. and Heston, A. (1991) 'The Penn World Table (Mark 5.0): An Expanded Set of International Comparisons, 1950–1988', *Quarterly Journal of Economics* **106**(2), 327–368.

Young, A. (1995) 'The tyranny of numbers: confronting the statistical realities of the East Asian growth experience', *Quarterly Journal of Economics*, **110**, 641–680.

Index

www.ingramcontent.com/pod-product-compliance
Ingram Content Group UK Ltd.
Pitfield, Milton Keynes, MK11 3LW, UK
UKHW022013180926
475701UK00003B/338
9780415494083